CW01650609

international®
AIR POWER
REVIEW

AIRtime Publishing
United States of America • United Kingdom

international®
AIR POWER
REVIEW

Published quarterly by AIRtime Publishing Inc.
120 East Avenue, Norwalk, CT 06851
Tel (203) 838-7979 • Fax (203) 838-7344

© 2004 AIRtime Publishing Inc.
AH-1 cutaway © Mike Badrocke
Lancaster cutaway and three-views © Aerospace Publishing Ltd
Lancaster colour profiles © Osprey Publishing Ltd
Photos and other illustrations are the copyright
of their respective owners

Softbound Edition ISSN 1473-9917 / ISBN 1-880588-77-3
Hardcover Deluxe Casebound Edition ISBN 1-880588-78-1

Publisher
Mel Williams

Editor
David Donald e-mail: airpower@btinternet.com

Assistant Editor
David Willis

Sub Editor
Karen Leverington

US Desk
Tom Kaminski

Russia/CIS Desk
Piotr Butowski, Zaur Eylanbekov

Europe and Rest of World Desk
John Fricker, Jon Lake

Correspondents
Australia: Nigel Pittaway
Belgium: Dirk Lamarque
Brazil: Claudio Lucchesi
Bulgaria: Alexander Mladenov
Canada: Jeff Rankin-Lowe
France: Henri-Pierre Grolleau
Greece: Konstantinos Dimitropoulos
India: Pushpindar Singh
Israel: Shlomo Aloni
Italy: Luigino Caliaro
Japan: Yoshitomo Aoki
Netherlands: Tieme Festner
Romania: Danut Vlad
Spain: Salvador Mafé Huertas
USA: Rick Burgess, Brad Elward, Mark Farmer (North Pacific region), Peter Mersky, Bill Sweetman

Artists
Mike Badrocke, Chris Davey, Zaur Eylanbekov, Juanita Franzi, Keith Fretwell, Iain Wyllie

Controller
Linda DeAngelis

Origination by Universal Graphics, Singapore
Printed in Singapore by KHL Printing

All rights reserved. No part of this publication may be copied, reproduced, stored electronically or transmitted in any manner or in any form whatsoever without the written permission of the publishers and copyright holders.

International Air Power Review is published quarterly in two editions (Softbound and Deluxe Casebound) and is available by subscription or as single volumes.
Please see details opposite.

Acknowledgments
We wish to thank the following for their kind help with the preparation of this issue:

Major G. Dortch, BDF/AW and Margreet Scherphof

The editors welcome photographs for possible publication but can accept no responsibility for loss or damage to unsolicited material.

Contact and Ordering Information (hours: 9 am-5 pm EST, Mon-Fri)
addresses, telephone and fax numbers

International Air Power Review, P.O. Box 5074, Westport, CT 06881, USA
Tel (203) 838-7979 • Fax (203) 838-7344
Toll free within USA and Canada: 1 800 359-3003
Toll free from Australia (13-15 hours ahead) : 0011 800 7573-7573
Toll free from New Zealand (17 hours ahead): 00 800 7573-7573
Toll free from Japan (14 hours ahead): 001 800 7573-7573

International Air Power Review, Postbus 3946, 4800 DX Breda, Holland
Toll free (to our US East Coast office) from the United Kingdom, Belgium, Denmark, France, Germany, Holland, Ireland, Italy, Luxembourg, Norway, Portugal, Sweden and Switzerland (5-6 hours ahead): 00 800 7573-7573
Toll free from Finland (6 hours ahead): 990 800 7573-7573

website
www.airtimepublishing.com

e-mails
airpower@airtimepublishing.com
inquiries@airtimepublishing.com

Subscription & Back Volume Rates

One-year subscription (4 quarterly volumes), inclusive of ship. & hdlg./post. & pack.:

Softbound Edition
USA $59.95, UK £48, Europe EUR 88, Canada Cdn $99, Rest of World US $79 (surface) or US $99 (air)

Deluxe Casebound Edition
USA $79.95, UK £68, Europe EUR 120, Canada Cdn $132, Rest of World US $99 (surface) or US $119 (air)

Two-year subscription (8 quarterly volumes), inclusive of ship. & hdlg./post. & pack.:

Softbound Edition
USA $112, UK £92, Europe EUR 169, Canada Cdn $187, Rest of World US $148 (surface) or US $188 (air)

Deluxe Casebound Edition
USA $149, UK £130, Europe EUR 232, Canada Cdn $246, Rest of World US $187 (surface) or US $227 (air)

Single/back volumes by mail (each):

Softbound Edition
US $16, UK £10.95, Europe EUR 18.50, Cdn $25.50 (plus ship. & hdlg./post. & pack.)

Deluxe Casebound Edition
US $20, UK £13.50, Europe EUR 22, Cdn $31 (plus ship. & hdlg./post. & pack.)

Prices are subject to change without notice. Canadian residents please add GST. Connecticut residents please add sales tax.

Ship. & Hdlg./Post. & Pack. Rates (for back volume and non-subscription orders)

	USA	*UK*	*Europe*	*Canada*	*ROW (surface)*	*ROW (air)*
1 item	$4.50	£4	EUR 8	Cdn $7.50	US $8	US $16
2 items	$6.50	£6	EUR 11.50	Cdn $11	US $12	US $27
3 items	$8.50	£8	EUR 14.50	Cdn $14	US $16	US $36
4 items	$10	£10	EUR 17.50	Cdn $16.50	US $19	US $46
5 items	$11.50	£12	EUR 20.50	Cdn $19	US $23	US $52
6 or more	$13	£13	EUR 23.50	Cdn $21.50	US $25	US $59

Volume Twelve
Spring 2004

CONTENTS

MAJOR FEATURES PLANNED FOR VOLUME THIRTEEN
Focus Aircraft: Panavia Tornado, **Warplane Classic:** Convair B-36, **Variant Briefing:** Bristol Blenheim, **Air Power Analysis:** Russia Part 2, **Combat Colours:** No. 2 Sqn, SAAF in Korea, **Type Analysis:** Saab J 29 Tunnan, **Technical Briefing:** Mil Mi-24 upgrades, **Photo-feature:** AMI training

PROGRAMME UPDATE

Dassault Rafale

C101 was the first production single-seater, and the first aircraft to fly from the outset with Standard F2 software and systems. Its maiden 1-hour 15-minute flight was undertaken on 16 April 2003, with Frédéric Lascourreges at the controls. During the sortie the aircraft was flown supersonically.

In February 2004 the French government gave the go-ahead for the Standard F3 software development effort, awarding a Euro 659 million contract to the Dassault-led development team. Standard F3 primarily adds a nuclear strike capability to the Rafale, provided by the ASMP-A missile. However, it also allows the Rafale to reach its full conventional potential, adding anti-ship (with Exocet and ANF missiles) and reconnaissance (using the Pod Reco NG pod) capabilities. The F3 software can also support the installation of an active phased-array radar, although this has yet to be ordered for the French forces.

Giving the go-ahead to Standard F3 development has provided a welcome boost to Rafale's export hopes, which are currently focused on the Republic of Singapore Air Force's competition, in which the Rafale is pitted against the Boeing F-15 and Eurofighter Typhoon. If F3 development proceeds as planned, the aircraft should be entering service with the Armée de l'Air and Aéronavale around 2008, which fits well with Singapore's plans. The Singaporeans have been offered participation in the active radar development programme.

Production of Rafales is proceeding slowly but is expected to ramp up to 18 aircraft per year by 2006 and 24 per year by 2008. However, the full-scale production rate is currently undergoing review to stretch out costs.

Flight testing and development is working towards a 2006 initial operational capability date for the Standard F2 Rafale, which is due to enter service with EC 1/7 and EC 2/7 at Saint

B302 carries test rounds of the AASM (Armement Air-Sol Modulaire) modular bomb. As well as the 340-kg (750-lb) version seen here, a 907-kg (2,000-lb) bomb is planned.

Carrying triple ejector racks loaded with GBU-12 LGBs, this Rafale also carries the Damoclès targeting pod, incorporating a high-resolution IR sensor and laser designator.

PROJECT DEVELOPMENT

China

New trainer flies

In December 2003 it was revealed that Guizhou had flown the first prototype of its FTC-2000 advanced trainer. Possibly designated JL-9 in Chinese service, the FTC-2000 is clearly based on the Chengdu FC-1/JF-17 being developed jointly by China and Pakistan. The trainer uses a Wopen WP-13 turbojet, and features a double-delta wing. Cockpit models show the aircraft to be fitted with large multi-function displays, and the aircraft is intended to provide a modern trainer to serve the needs of an air force increasingly equipped with modern fighters such as the Su-27/30 and Chengdu J-10.

Ukraine

An-70 partnership prospects

In the continued absence of promised $48 million initial funding towards Russia's original requirement for 164 An-70 heavy-lift propfan transports, Ukraine's Antonov Design Bureau (OKB) in Kiev signed an MoU in September with China's AVIC II aerospace industrial group for possible future joint development, production and procurement. Russian air force interest now appears to have switched to the PS-90A-powered, stretched and upgraded Ilyushin Il-76MF for its heavy transport needs.

This is evidenced by its planned purchase of the two Il-76MF prototypes, and transfer of production from Tashkent, in Uzbekistan, to the Voronezh Joint Stock Aircraft Manufacturing Company in Russia. Ukraine seems unlikely to meet its requirements for 65 An-70s without an R&D and initial production partnership, despite $190 million in government funding allocations to continue development and production preparations through 2004-06.

United Kingdom

AirTanker wins FSTA

On 26 January 2004 the UK Ministry of defence announced the selection of the AirTanker consortium as its preferred bidder to fulfil its Future Strategic Tanker Aircraft requirement. AirTanker, a consortium of EADS, Cobham, Rolls-Royce and Thales, is offering the tanker version of the Airbus A330, powered by Rolls-Royce Trent 772Bs. The aircraft are to be operated under a private finance initiative, and will be based at RAF Brize Norton in Oxfordshire.

United States

Comanche axed

On 23 February 2004 the Pentagon announced the cancellation of the Boeing/Sikorsky RAH-66 Comanche helicopter. Having already spent $6.3 billion, the US Army decided to spend the production funds on non-stealthy types and upgrades of existing helicopters. A complete review of the Comanche programme and its cancellation will be featured in the next volume of *International Air Power Review*.

MOAB testing continues

The USAF's Air Armament Center conducted a successful live development test of its GBU-43/B massive ordnance air blast (MOAB) weapon at Eglin AFB, Florida, on 21 November 2003. The 9843-kg (21,700-lb) weapon was deployed from an MC-130E Combat Talon I, flown by personnel from the 46th Test Wing at Eglin, from an altitude of 6096 m (20,000 ft). Guided by an internal GPS navigation system, the weapon detonated successfully creating a plume that rose more than 3048 m (10,000 ft). Previously deployed by an MC-130H during March 2003, the latest test verified a number of modifications made to the weapon and certified it for delivery by the MC-130E.

New Sidewinder enters service

The AIM-9X variant of the Sidewinder air-to-air missile achieved initial operational capability (IOC) with the 3rd Wing's 12th and 19th Fighter Squadrons (FS) at Elmendorf AFB, Alaska on 13 November 2003.

The new missile, which is designed with an off-boresight capability, features infrared countermeasures (IRCM) resistance and jet vane control that provides it with agile turning performance. Prior to entering operational service the missile underwent more than 3,500 hours of captive flight testing, scored 18 successes in 19 guided flights during development testing, and completed 22 operational evaluation (OPEVAL) firings.

Fire Scout progress

Northrop Grumman and the Naval Air Warfare Center Aircraft Division achieved a major milestone during a 45-minute technical demonstration of the RQ-8A Fire Scout UAV at Webster Field, St. Inigoes, Maryland, on 19 December 2003. It occurred when personnel aboard a P-3C assigned to VX-20 took control of the RQ-8A after the Orion launched from NAS Patuxent River. The Orion had been

Dizier as a Jaguar replacement, and with a second Aéronavale squadron. F2 builds considerably on the limited air defence-only capability of the Standard F1 aircraft in service with Flottille 12F of the Aéronavale. F2 not only enhances air combat potential through the replacement of the Magic II air-to-air missile with the MICA IR, but also adds a significant conventional air-to-ground capability. This includes the carriage of the AASM modular bomb and SCALP EG stand-off attack missile, Damoclès targeting pod and air-to-ground modes for the RBE2 radar. Other F2 enhancements include the FSO (Front Sector Optronics) suite, Link 16/MIDS datalink and the ability to carry a buddy refuelling pod.

Elements of the Standard F2 avionics have been installed in prototype M02 and production aircraft B302, but the first aircraft to fly from the outset with the Modular Data Processing Unit (MDPU) that lies at the heart of the Standard F2 was aircraft C101, the first production single-seater, which has been flying since April 2003.

Weapons/stores clearance work continues at Cazaux and Istres, and has included testing of a triple ejector rack for bombs in the 227/340-kg (500/750-lb) class, such as the GBU-12 LGB and AASM. Damoclès has been flown on the starboard intake pylon, and SCALPs have been launched from the wing and centreline pylons. MICA IR has been test-fired several times. The previously subsonic 2000-litre (440-Imp gal) drop tanks have now been cleared for flight at speeds of up to Mach 1.6.

Sagem's AASM weapon has been carried by Rafale in tests, including carrier landings in late 2003 in which a Rafale M landed on *Charles de Gaulle* carrying six of the weapons (plus empty 1250-litre/275-Imp gal drop tanks) at an impressive weight of 15.7 tonnes (34612 lb). Live drops were scheduled to begin at Cazaux in 2004. AASM employs either a GPS/inertial guidance package, or a combined GPS/INU/optical package. It can be fitted with a range-extension kit consisting of a rocket motor and pop-out wings, which increase range to around 50 km (31 miles) for a high-altitude launch, or 15 km (9.3 miles) at low level.

Dassault is also moving ahead with Meteor integration work. Rafale can carry four of the ramjet-powered long-range air-to-air missiles, two on the aft fuselage hardpoints and two on the intermediate wing pylons, leaving two hardpoints free for drop tanks and four for MICA EM/IR weapons. Mechanical and wind tunnel tests are currently under way.

Rafales M1 and B301 demonstrate the type's tanking capability, using a buddy refuelling pod. This capability is to be used by the Aéronavale's Standard F2 aircraft. A new, larger pod is being developed which holds additional fuel, in turn increasing the Rafale's offload capacity when operating in the tanker role. In these photos B301 is carrying six GBU-12 laser-guided bombs on the triple racks, while both aircraft carry dummy rounds of the MICA IR. These weapons are both important elements in the Standard F2 capability upgrade.

equipped with an integrated tactical control system (TCS) and the tactical common data link (TCDL) and had previously received modifications associated with the Aircraft Improvement Program (AIP). The latter updated the Orion's ability to execute over-the-horizon targeting missions. During the test the TCS was used to launch, control and recover the vehicle and operate its sensors, which provided streaming electro-optical video to the Orion via the TCDL. The video data was subsequently relayed to a ground station at Patuxent River. The Fire Scout conducted its 100th test flight on 17 December 2003 and has amassed approximately 75 flight hours over the past 18 months. Northrop Grumman and its subcontractors will begin production of eight additional air vehicles during 2004, while a performance improvement programme that will develop and flight-test a four-bladed main rotor system and a weapons programme, are under consideration. The weapons programme will provide the UAV with a precision strike capability and several engineering tests associated with the programme have already been completed. These include the installation of 70-mm (2.75-in) rocket launchers equipped with advanced precision kill weapon system laser-guided rockets. Live firings are scheduled for early 2004 with unguided rockets.

EA-18G development approved

On 18 December 2003 Naval Air Systems Command received approval from the Secretary of Defense for the two-seat EA-18G airborne electronic attack (AEA) aircraft project to proceed into the system development and demonstration (SD&D) phase. SD&D, which will run from FY04 until mid-FY09, includes laboratory, ground and flight tests from component level testing through to full-up EA-18G weapons system performance flight-testing. The EA-18G team had been working under a pre-SD&D contract since September 2002. The EA-18G will carry up to five AN/ALQ-99 jamming pods as well as two AIM-120 AMRAAM and two AGM-88 HARM missiles, enabling the aircraft to defend a strike group from both surface-to-air and air-to-air threats. Under the current plan it will achieve initial operational capability (IOC) in 2009. The service subsequently awarded Boeing and Northrop Grumman a $1 billion SD&D contract on 29 December 2003.

Merlin demonstrator

AgustaWestland recently loaned an EH-101 helicopter to Lockheed Martin for use as a demonstrator for the US-101 project. The aircraft, which is the last of nine pre-production

On 18 December 2003 ShinMaywa flew the first prototype of its US-1A Kai amphibian. Although superficially similar to the US-1A, the Kai introduces fly-by-wire controls, composite wings, Rolls-Royce AE2100J turboprops and has a pressurised upper hull which allows it to operate at much higher altitudes. The new avionics suite includes a Thales Ocean Master search radar. The prototype was handed over to the Japan Defence Agency for ongoing trials on 24 March.

For AgustaWestland 12 December 2003 was a notable day, marking the first flights of EH-101s for Denmark (left, at Yeovil) and Portugal (above, at Vergiate).

EH-101s, was delivered to the contractor's facility in Owego, New York, during September 2003. Although the aircraft will be used to demonstrate a variety of missions, it will initially support the contractor's efforts to sell the aircraft as a replacement for the US Marine Corps small fleet of VH-3D Sea King helicopters that provide executive airlift for the President of the United States.

New UAV weapon tested

The US Army's new Viper Strike precision munition scored direct hits on nine intended targets during a successful flight demonstration conducted at White Sands Missile Range, New Mexico. Developed from Northrop Grumman's Brilliant Anti-Tank (BAT) submunition, the Viper Strike was delivered by an RQ-5A Hunter equipped with video and infrared (IR) sensors and laser designator. The weapons were fired at a variety of moving and stationary targets comprising T-72 tanks and armoured vehicles, a Free Rocket Over Ground (FROG) artillery rocket launcher, modified pickup and utility trucks and an automobile.

Quick Bolt missile test

The US Navy recently conducted two test firings of the Quick Bolt (QB) anti-radiation missile from an F/A-18 Hornet at NAWS China Lake test range in California. Developed under the advanced anti-radiation guided missile (AARGM) advanced concept technology demonstration (ACTD), the missile features a new multi-mode seeker combining passive anti-radiation homing, active millimetre-wave radar and global positioning system/inertial navigation system. Its seeker is being considered to upgrade the current AGM-88 HARM missile. Quick Bolt also features an embedded national tactical receiver that enables it to receive off-board targeting and a weapons impact assessment (WIA) data function which improves its ability to detect, locate and destroy enemy air defences even after they have been shutdown.

Spirit tests JASSM

The 72nd Test and Evaluation Squadron (TES) at Whiteman AFB, Missouri, is testing the AGM-158 Joint Air-to-Surface Standoff Missile (JASSM), which will be added to the B-2A's weapons suite. Guided by a global positioning system (GPS) and an inertial navigation system (INS), the low-observable subsonic missile is designed to penetrate next-generation air defence systems. The highly survivable weapon has an all-weather, launch-and-leave capability and is powered by a small engine. JASSM is launched from the bomber's standard rotary launcher assembly. It weighs 1021 kg (2,250 lb) and is 4.27 m (14 ft) long. It has also been undergoing flight-testing with the 419th Flight Test Squadron at Edwards AFB, California.

Osprey shipboard suitability

The V-22 Integrated Test Team (ITT) recently completed its second 'at-sea' period aboard USS *Bataan* (LHD 5) as part of the tiltrotor's Phase IV shipboard suitability testing. Over an 11-day period five pilots, including one from VMX-22, conducted deck-landing qualifications. The ITT also demonstrated numerous test points that were required to expand the Osprey's wind-over-deck envelope. Tests also measured the effects that hovering CH-46Es and CH-53Es had on the Osprey, when it was spotted on the flight deck behind helicopters. MV-22B No. 22 was deployed for the entire period; however Osprey No. 10 joined the project in its final two days. The Osprey will return to sea for Phase V shipboard suitability testing in April 2004.

On 7 November 2003 the V-22 ITT deployed the no. 24 MV-22B from NAS Patuxent River, Maryland, to the Canadian Forces base at Shearwater, near Halifax, Nova Scotia. The aircraft and a 67-person detachment were scheduled to remain at Shearwater through April 2004 testing the Osprey's de-ice and anti-ice systems in the cold and humid atmosphere around Halifax.

Osprey No. 25 has been delivered to Boeing's facility in Philadelphia where it is being modified to CV-22B configuration under a $61 million contract. It will be delivered to Edwards AFB, California, in November 2004.

Apache tests new rotor blades

Boeing recently began flight-testing new advanced composite rotor blades for the AH-64D Apache Longbow as part of a programme designed to enhance the helicopter's performance and reduce its overall operating costs. The blades will likely become a component of the proposed Block III upgrades for the AH-64D. The initial 30-minute test flight included a variety of manoeuvres intended to confirm the blades' structural integrity. The advanced composite rotor blades, which have been under development for the past four years, will have nearly double the service life of the standard metal blades now in service on the AH-64D and cost 25 per cent less to produce. The blades also incorporate a more efficient airfoil shape that will result in improved hover and forward flight performance. The system is also designed to support more powerful engines planned for the Block III aircraft. Other key elements planned for the Block III upgrade include the joint tactical radio system, improved target detection and acquisition equipment.

Boeing's 737-800 BBJ demonstrator lands at NAS Jacksonville, Florida, during the aircraft's tour round current P-3 Orion bases, as part of Boeing's campaign to be chosen to fulfil the US Navy's MMA (Multi-mission Maritime Aircraft) requirement. The aircraft was fitted with two representative operator consoles.

Lockheed Martin MMA

Lockheed Martin has selected Pratt & Whitney and Hamilton Sundstrand to supply the propulsion system for its entry into the MMA competition. Its advanced P-3 Orion variant will be powered by the Pratt & Whitney PW150A turboprop engine, that will provide approximately 5220 kW (7,000 shp). The engine will drive the same eight-bladed Hamilton Sundstrand NP2000 propeller that was recently approved for installation on the E-2C and will soon be tested on the C-2A. The contractor team believes that the its airframe/powerplant combination will provide the fleet with an aircraft that has a 60 per cent shorter takeoff roll, burns 27 per cent less fuel and provides 50 per cent faster thrust response, while loitering on station, over Boeing's proposed 737 variant. Both teams recently submitted their final proposals and the Navy is expected to select a winner in early 2004. The MMA is expected to enter service in 2012.

Kaman demonstrates Burro UAV

Kaman Aerospace recently demonstrated its optionally-manned, cargo-lifting Burro UAV to the US Army at Eglin AFB, Florida, Cairns AAF, Fort Rucker, Alabama, and the Redstone Arsenal in Alabama. The Burro is an experimental version of the K-Max commercial external lift helicopter. Although it retains the K-Max's ability to lift up to 2722 kg (6,000 lb) of cargo, it is also capable of delivering the cargo autonomously as an unmanned aerial platform. During the demonstration the aircraft flew autonomously from Eglin to Fort Rucker where it delivered four loads of cargo to an outlying field. The aircraft subsequently completed a 417-km (225-nm) flight between Cairns and Redstone where another cargo lift mission was completed. The demonstration was sponsored by the Advanced Systems Directorate of the US Army Aviation and Missile

Harrier GR.Mk 9 goes to sea

While HMS *Ark Royal* was at sea in February 2004 for 800 NAS's last embarkation (see p. 15), BAE Systems took the opportunity to undertake trials of the Harrier GR.Mk 9 in a carrier environment. The primary aim was to assess the performance of the Mk 9's new inertial navigation unit (IN/GPS) on a moving platform. The trials involved undertaking the various types of inertial align while on the carrier, and then flying ashore to measure the resultant drift between alignment and landing at a known, fixed position. Test pilot Simon Hargreaves performed the first carrier landing for the GR.Mk 9 in early February, after which John Lawson took over the trials flying. In addition to the IN/GPS trials, the test team also evaluated the aircraft's new mission computer and gathered data for an upcoming ground proximity warning system assessment. The carrier trials were completed on 16 February, despite some inclement weather. At the end of the month the GR.Mk 9 began weapon aiming trials and assessment of TIALD laser designator performance. The upgraded Harrier is due to enter service with the RAF and RN in 2005.

Neil Dunridge

Command's Research, Development and Engineering Center (AMRDEC).

New UAV for SOCOM

The US Department of Defense has assigned the mission design series designation CQ-10A to Mist Mobility Integrated System Technology's (MMIST) Snow Goose self-propelled, autonomous GPS-guided parachute delivery system. The CQ-10A is powered by a single ROTAX 914 engine and equipped with a high-glide ratio ram air parachute. The commercially developed 'Powered Sherpa' airborne guidance unit controls all engine functions, canopy deployment and steering lines, and the cargo module. It is capable of deploying its cargo, which would typically include leaflets, small amounts of ammunition, medical supplies and other equipment, from the air or by landing at a predetermined location. The UAV is capable of being deployed from a cargo aircraft or launched by ground personnel and can operate at a maximum altitude of 5486 m (18,000 ft). It cruises at a speed of approximately 65 km/h (35 kt). Five CQ-10As have been purchased for evaluation by the US Special Operations Command, which has a requirement for 74 UAVs.

NG joins Boeing FCS Team

Northrop Grumman has joined a Boeing-led Industry Team that is developing the US Army's Future Combat System (FCS). Boeing and Science Applications International Corporation (SAIC) will conduct systems integration, while Northrop Grumman will develop and produce the FCS programme's Class IV unmanned aerial system (UAS). It will also integrate the reconnaissance, surveillance, and target acquisition systems. The UAS will be based on the US Navy's RQ-8A Fire Scout vertical takeoff and landing tactical unmanned aerial vehicle (VTUAV). The army's RQ-8B will, however feature a four-bladed main rotor and its gross takeoff weight will be increased from 1202 kg (2,650 lb) to 1429 kg (3,150 lb).

ACS becomes joint programme

The US Navy has announced plans to join the US Army's Aerial Common Sensor (ACS) programme and will procure as many as 19 ACS-N aircraft that will replace the service's EP-3E fleet. The two prime contractors that have been short-listed by the army include Lockheed Martin and Northrop Grumman, which have respectively proposed the Embraer EMB-145 and the Gulfstream G450 as the ACS aerial platform. A winner was due for selection in March 2004.

Upgrades and Modifications

Belarus

MiG-29s updated

The air force, which operates a fleet of 43 MiG-29s, recently concluded trials of a modified example that is equipped for strike missions. Designated the MiG-29BM, the fighter was equipped with updated radar, electro-optical, fire control, and computer systems, and new cockpit displays that are intended to improve the fighter's air-to-ground capabilities. Its range has also been increased via the installation of an air-refuelling capability. The modifications will allow the fighter to carry a wider range of guided and unguided air-to-ground weapons. Belarus intends to modify all of its single-seat MiG-29s to this new configuration.

L-39 overhaul

Belarus has selected the Khan Asparuh plant in Dobrich, Bulgaria, to overhaul 36 L-39ZA trainers at a cost of $15 million. The project, which was developed jointly with Aero Vodochody, the manufacturer of the L-39, will provide the aircraft with 12 additional years of service. The trainers will allow the Belarussian Air Force and Air Defence Force to establish its own pilot training center at the 206th Attack Aviation Base at Lido.

Brazil

F-5 upgrade

On 4 December 2003 Embraer rolled out the first Northrop F-5 to be upgraded in a joint programme with Elbit Systems of Israel. The first aircraft, a two-seater, was shown wearing an 'F-5BR' logo on the fin, reflecting the overall name for the programme. However, in service the aircraft will be known as F-5EMs and F-5FMs. A total of 46 aircraft is undergoing rework, which involves a structural life extension and a complete avionics overhaul with new radar, digital NVG-compatible three-screen cockpit displays, HOTAS, helmet-mounted display, datalink and virtual training function.

The H-1 integrated test team recently completed testing of the UH-1Y's special warfare kit at MCB Quantico, Virginia. The test period, which ran for 10 days, evaluated the helicopter's ability to insert Marines using techniques other than landing. The evaluation included static line and free-fall parachute from altitudes up to 10,000 ft (3048 m), special purpose insertion extraction (SPIE) rig, rappelling, and fast-rope operations. Reconnaissance Marines assigned to Quantico supported the tests. The H-1 Upgrades ITT has completed approximately 1,500 flight test hours with three AH-1Zs and two UH-1Ys.

VX-20's E-2C Hawkeye 2000 with NP-2000 propellers is seen during carrier trials aboard USS* John F. Kennedy *(CV 67).

Pilatus PC-12 special mission aircraft c/n 518 was delivered to the Bulgarian Air Force on 19 December 2003. The new aircraft strengthens the fleet of the 16th Transport Airbase based at Sofia Airport/Vrazhdebna. The aircraft was part of an order which also encompassed six PC-9M trainers.

Russia

Upgrades delivered

On 26 December 2003 the 4th TsBPiPLS at Lipetsk received the first five Su-27SM upgraded 'Flankers' for service with the 968th IISAP. The aircraft, which have been modified with expanded air-to-surface capability, will be used to assess the new variant's operational effectiveness and to instigate new tactical procedures prior to full service entry. At least 20 more Su-27s are at the Komsomolsk-na-Amur factory (KnAAPO) for rework, and the Russian air force expects to have one regiment equipped with the type by the end of 2004.

Meanwhile, on 28 January 2004 the 344th TsBPiPLS AA at Torzhok, army aviation's combat evaluation centre, received five upgraded Mi-24PN 'Hind' attack helicopters for trials. Among the updated equipment installed in the Mi-24PN is a Zarevo fixed thermal imager in an extended nose fairing. The aircraft also has fixed undercarriage and shortened stub wings. The Mi-24PN is seen as an interim step pending introduction of the Mi-24PM (Mi-24PK-2) which features a more sophisticated equipment package centred on the GOES-342 gyro-stabilised sensor turret. Although some of the initial batch of aircraft are expected to remain at Torzhok for ongoing trials, the variant is being fielded by the 4th KA VVS i PVO (Red Banner Army of the air force and air defence), the air army which oversees operations in Chechnya.

Indonesia

Upgraded Puma

A single IPTN-built NSA 330L Puma has been upgraded by Eurocopter at its Marignane facility for service with the TNI-AU (Indonesian air force) in the VIP transport role. The core of the upgrade was the replacement of the original Turmo IVC engines with Turbomeca Makila 1A1s, as used in the Super Puma/Cougar. The aircraft also had avionics improvements fitted. Redesignated as an SA 330SM, the aircraft was due for redelivery in February 2004.

Spain

Hornet MLU

On 29 December 2003 the Ejército del Aire signed a contract with EADS-CASA covering the mid-life update of 67 Spanish F/A-18A Hornet aircraft. On 15 January 2004 the first aircraft was delivered to EADS's Getafe plant for work to begin. The first aircraft are expected to be back in service before the end of 2004.

The latest Boeing 707 Re'em (Ram) tanker was re-delivered to the Israel Defence Force/Air Force (IDF/AF) in January 2004, following conversion by the IAI Bedek Group. Re'em 290 (Boeing 707-3W6C c/n 21956) was purchased by Israel in 2000 and initially served as a transport aircraft. It is the first IDF/AF Re'em tanker equipped with a standard KC-135 boom, and previous IDF/AF Re'em tankers that are equipped with non-standard booms are to be modified accordingly. IAI Bedek is currently modifying yet another Re'em transport aircraft to tanker configuration, and the IDF/AF plans a Re'em Avionics Upgrade Programme (AUP) which, together with the required structural work, extends the tankers' service life by another 25 years.

United States

Spirit Block 30 mods complete

The B-2A programme marked a milestone on 5 September 2003 when Northrop Grumman redelivered the 21st bomber equipped with the Block 30 modifications to the USAF. The post-delivery update programme, which began in 1998, incorporated nearly 400 individual changes into each bomber. *Spirit of Arizona* was the final aircraft to be updated and returned to the USAF.

Firebees updated

The US Navy has awarded Northrop Grumman a contract to modernise its fleet of BQM-34 Firebee aerial targets with new avionics and an upgraded command and control architecture. Capable of airborne or surface launches, the BQM-34 is primarily used to simulate enemy aircraft and missiles in support of defence-readiness training. The Firebees will be updated with the same equipment currently used by the Navy's BQM-74 targets, including the autopilot installed in the current production BQM-74E. The update will provide the target with the ability to perform autonomous waypoint navigation, preprogrammed manoeuvres, and allow it to be controlled via satellite.

AMP progress

Lockheed Martin and the USAF recently began flight-testing the latest software for the C-5 Avionics Modernization Program (AMP). The AMP will replace the C-5's analog cockpit instruments and systems with digital displays and equipment, and provide it with communications and navigational systems avionics that comply with Global Air Traffic Management (GATM) requirements. A joint USAF/contractor flight crew is conducting 19 flights from Dobbins ARB using the same C-5A and C-5B aircraft that supported earlier tests. During the earlier testing, which concluded in September 2003, the two aircraft made 25 test flights accumulating 115.8 flight hours. The USAF is expected to begin its operational test and evaluation of the AMP aircraft in late 2004. Lockheed Martin has already received a $20.3 million contract covering eight AMP production kits and a second $42.8 million contract for 18 kits. Most recently it was awarded a separate $5.9 million to install the Lot I kits into eight C-5Bs. The installations will be begin in June 2004 and approximately two months will be required to equip each aircraft with the AMP equipment. Fleet-wide installation into 62 C-5As and 50 C-5Bs will be completed in 2007.

Upgraded Sea Knight

Naval Air Depot Cherry Point, North Carolina, is progressing with a $300 million engine reliability improvement programme (ERIP) that will provide the USMC's CH-46E fleet with 446 improved T58-GE-16A engines. The initial four engines were delivered to HMM-266 at MCAS New River, North Carolina, in April 2003, and the engine achieved initial operational capability (IOC) on 25 November 2003 when all of the squadron's aircraft were equipped. The -16A model features an improved engine core module comprising the compressor, combustor and high-pressure turbine, and modifications to the engine's power turbine and accessory package. It provides the CH-46E with additional power and will allow the USMC to retain the Sea Knight until the much-delayed MV-22B can be fielded.

Venezuela

Viper upgrade

Belgium's SABCA has been selected by Venezuela to update its fleet of 21 F-16A/Bs and awarded a $40 million contract. The work will begin in 2004 and be completed in 2006. It is unclear whether the work will be carried out in Belgium or Venezuela.

Super Seasprite for RAN

On 18 October 2003, the RAN's delayed SH-2G(A) Super Seasprite programme received a welcome boost during a ceremony at Nowra to mark the provisional acceptance of the first aircraft. The first two helicopters of the 11-strong order were actually taken on charge by No. 805 Sqn two days later, on 20 October, allowing air and ground crew training to commence.

Delays associated with the integration of the Litton Integrated Tactical Avionics System (ITAS) meant that, by 2001, the project was 18 months behind the original schedule, but the RAN has now decided that the first six aircraft are of a sufficiently advanced standard to allow training to begin. This has cleared the way for provisional acceptance of the first two, and the remaining four aircraft currently in Australia will be delivered during the first quarter of 2004. The 11th helicopter is at Kaman's Bloomfield, Connecticut, plant, engaged in the development and testing of the operational software version.

During the ceremony Rear Admiral Kevin Scarfe, Acting Under-Secretary Defence Materiel, noted that enormous progress had been made in recent times and, although three and a half years late, full operational capability is expected to be delivered by the end of 2004. Following trials, the first operational embarkation is expected in mid-2005. At the same ceremony, Defence Minister Senator Robert Hill noted that the project cost (in 2003 dollars) had risen to $1.016 billion but was still within budget.

The Navy is undoubtedly embarrassed by the attention that the Super Seasprite has attracted, but remains committed to the platform. When it finally enters service, the SH-2G(A) will arguably be the most capable Anti-Surface Warfare helicopter at sea, and the day that it first embarks aboard an 'Anzac'-class ship will be a red letter day for Kaman and the RAN.

Nigel Pittaway

Above: One of the first SH-2G(A)s for the Royal Australian Navy hovers near its parent vessel in Sydney Harbour. The Super Seasprites carry a hybrid serial comprising their RAN type designator (N29) with their US Navy BuAer number, in this case 161656. They also carry a tactical code, in this case 840.

The ceremony to welcome the SH-2G(A) into service was held at the RAN's principal shore base at Nowra, on the coast south of Sydney. Bringing in VIPs from Canberra for the event was a Bombardier CL-604 Challenger from No. 34 Squadron, the type having recently been leased to replace Dassault Falcon 900s on VIP transport tasks.

Procurement and Deliveries

Australia

Project Wedgetail update

Boeing recently met a significant milestone associated with Australia's Project Wedgetail airborne early warning and control programme when it successfully installed and tested the electrical power generating system (EPGS), on the first 737-AEW&C platform. The EPGS distributes and controls power to the aircraft and mission systems, including radar, navigation, communications, mission computers, displays and the flight deck. The first Wedgetail aircraft will support the airworthiness flight test programme beginning in spring 2004, while the second aircraft, which is also undergoing modifications, will be used to test the airborne mission system beginning later in 2004.

The first aircraft (A30-001) was displayed at Boeing Field, Seattle, on 20 March, complete with RAAF titles and No. 2 Squadron's markings on the fin. In Boeing's customer-specific designation system, the Wedgetail is a 737-7ES.

Brazil

Super Tucano delivered

Embraer recently delivered the first ALX Super Tucano multi-purpose military turboprop to the Força Aérea Brasiliana (FAB) at its Gavião Peixoto facility in São Paulo. The ALX is equipped with advanced avionics that will allow the aircraft to fly day/night border surveillance missions as part of Brazil's Amazon Surveillance Program, operating with R-99A/B AEW and surveillance platforms. It is capable of carrying 1500 kg (3307 lb) of ordnance on five underwing and fuselage hard points.

China

New-batch 'Flankers'

According to reports, deliveries began in early 2004 from KnAAPO of the first batch of 12 Su-30MK2 aircraft (from an order for 24). The MK2 differs from the earlier batch of 76 Su-30MKKs by incorporating the ability to fire the Kh-31A active radar-guided anti-ship missile, leading to informed speculation that the aircraft will be used to equip a regiment of the PLANAF (Chinese naval aviation). Further Su-30 purchases are expected to be of the Su-30MK3 version, which increases yet further the aircraft's weapons repertoire through the addition of Kh-29L laser-guided missiles and KAB-1500L laser-guided bombs, guided either by the aircraft's existing 52Sh optronic system (Kh-29L only) or by an externally mounted designator pod.

Denmark

Challengers delivered

In late 2003 the first of three Bombardier CL-604 Challengers was handed over to the Royal Danish Air Force following fitting out to MMA (Multi-Mission Aircraft) configuration by Field Aviation of Toronto. Operated by Eskadrille 721 at Vaerløse, the Challengers are equipped with a Telephonics AN/APS-143B(V)3 search radar in an underfuselage canoe fairing, and a FLIR Systems Star Safire electro-optical turret under the rear fuselage. The latter is retractable and is covered by doors when not in use. As well as maritime surveillance, ice reconnaissance, fishery patrol and SAR cover, the Challengers are also available for transport and medevac duties. In addition to home-based operations, they will operate in the Faroe Islands and Greenland.

Ecuador

Maritime patrol supplement

Spanish government credits totalling some $24 million are being made available for the supply to Ecuadorean Naval Aviation (ANE) of a CASA/Airtech CN-235MP Persuader maritime patrol aircraft. This would enter ANE service in 2005, following earlier deliveries of single CN-235Ms in 1988-89 to the Ecuadorean navy and army aviation.

India

Hawk deal finalised

After 19 years of negotiations, India finally signed an order for 66 BAE Systems Hawk Mk 132s on 26 March 2004. This followed formal selection of the Hawk to fulfil India's Advanced Jet Trainer requirement in September 2003. The first 24 aircraft will be built at BAE Systems' Brough factory (and flown from Warton), while the remaining 42 will be assembled at HAL's Bangalore plant using kits supplied from the UK. The first aircraft are due for delivery in 2007, and the last in 2013. The contract also covers the training of 75 Indian Air Force pilots at RAF Valley.

Spotted at Goodwood airfield in West Sussex, England, this SA 319B Alouette III was undergoing refurbishment before delivery to the Sierra Leone Air Force, serialled 401.

Central America update

The small and rarely seen Fuerza Aérea Nicaragüense (Nicaraguan air force) is currently upgrading one of its Mi-17s (serial number 334) to serve as a VIP transport (above). The Escuela de Aviación operates a few light aircraft for basic training, including this recently repainted Piper PA-28-235 Cherokee (below).

Two rarely seen aircraft of the Fuerza Aérea Guatemalteca (Guatemalan air force) are the two Bell 412EPs of the presidential flight (above). They operate from Base Aérea La Aurora (the military side of Guatemala City's international airport) and do not wear any serials. The main transport helicopter fleet, which comprises the Bell UH-1Hs and Bell 212s operated by 1 Escuadrón, is being repainted in a new light grey scheme with blue trim on the fin. This Bell 212 (below) also carries 'Rescate' (rescue) titles.

Air Arm Review

Australia

Shadow 200 proposed

BAE Systems Australia has teamed with AAI Corporation to compete for the Commonwealth of Australia JP 129 Surveillance Programme. 'Team Shadow' will propose AAI's RQ-7A Shadow 200 tactical UAV as its solution to the requirement. The JP 129 requirement will enhance Australia's land-based aerial surveillance capabilities and will also support selected maritime operations.

Austria

F-5 lease

To cover a gap in air defence capability between the retirement of the Saab J 35Ö Draken at the end of 2005, and the delivery of the first Eurofighter Typhoon, the Österreichische Luftstreitkräfte (Austrian air force) announced on 9 March 2004 that it was to lease 12 Northrop F-5Es from the Swiss Air Force. The deal also covers armament and logistic support at the Zeltweg and Graz-Thalerhof bases. The first four aircraft are due to arrive in July 2004 to allow conversion to begin. Austria had earlier looked at leasing Typhoons pending delivery of its own aircraft, but this was discarded as being too costly.

Belgium

Helicopter force restructures

On 1 March 2004 the Belgian army's light helicopter group was renamed as the Helicopter Wing, and will report to the Operational Command Air. This follows the reallocation of the School of Light Aviation to Belgian Air Component control in 2003. Under current plans the Helicopter Wing will restructure with three squadrons plus a conversion unit operating Agusta A 109BAs, with a liaison flight continuing to fly the Alouette II until the type's planned retirement in 2007. Retirement is also planned of the five remaining BN-2 Islanders by the end of 2005.

Canada

Goose Bay future in doubt

Canada's Department of National Defence is worried about the future of the NATO low-level flying programme, which is conducted at CFB Goose Bay, Labrador. The Royal Netherlands Air Force has already pulled out of the programme and Germany recently announced that it would end its participation in 2006. The German departure would significantly increase the costs to the UK and Italy as the remaining participants. The DND, which invests $27 million annually, is reportedly looking at ways to reduce the costs to its participants.

Maritime helicopter progress

The Department of National Defense formally issued a Request for Proposals for its Maritime Helicopter Project on 17 December 2003. The project includes 28 helicopters, a simulation and training suite, integrated logistic support, ship modifications, and a 20-year in-service support contract. The DND expects to receive proposals from Sikorsky and AgustaWestland for the Sea King replacement and will likely select a winner between the former's S-92 and the latter's EH-101 in mid-2004. The DND ruled that it would not consider a bid from Lockheed Martin and NH Industries, which had been expected to respond with a proposal for the NH90. The aircraft had apparently failed to meet a number of DND requirements during a series of pre-bidding trials.

Israel

More Longbows possible

The Israel Defence Force/Air Force is reportedly looking to purchase six additional AH-64D helicopters at a cost of $220 million. The air arm has already ordered nine examples of the Apache Longbow and plans to exercise an option that would provide three further gunships. Deliveries from the existing order are expected to begin in 2005. The IDF/AF also intends to upgrade its existing AH-64As to the later configuration.

Libya

Russian aircraft transfers

While maintaining large inventories of mainly Russian military aircraft and equipment, the LARAF has a poor serviceability record through spares and technical services support deficiencies, mainly derived from UN arms embargoes on Libya. Alternative sources are being sought to improve LARAF aircraft availability, although some of its excess military equipment is now being transferred to other African nations. In addition to pairs of Antonov An-26s, plus Mil Mi-8 and Mi-24 helicopters passed on to the Namibian Defence Forces, the air force of Zimbabwe has reportedly

The Aeronautica Militare Italiana took formal delivery of its first Eurofighter Typhoon on 29 March 2004. The aircraft – IT002/MM55093 – was handed over to the 4º Stormo at Grosseto, where it will be initially assigned to 9º Gruppo..

Above: El Salvador's transport capacity has been enhanced through the addition of this Piper PA-23-Aztec F, although it sustained minor wingtip damage.

Above: This ex-French Magister has been painted to resemble the 'Torogoz' – El Salvador's national bird, and is used for display purposes.

Below: The FAES's long-serving Hughes TH-300C rotary-wing trainers, originally in olive drab colours, have now received this more appropriate scheme.

Left: The Fuerza Aérea de El Salvador (also known as the Fuerza Aérea Salvadoreña) has a small fleet of UH-1M 'Cazador' gunships, which have recently been repainted in this three-tone scheme. The most recent helicopter acquisitions are four Bell 412EPs – when they have been fully integrated into service a batch of UH-1Hs will be upgraded to Huey II standard.

received small donations of the same aircraft types.

NATO

Baltic air defence

On 29 March 2004 seven new nations joined the NATO alliance (Bulgaria, Estonia, Latvia, Lithuania, Romania, Slovakia and Slovenia). A few days earlier, NATO announced that it would establish a detachment for air defence of the Baltic states at the Lithuanian air base at Zokniai. The detachment comprises four F-16s drawn on a rotational basis from the Belgian, Danish and Dutch air forces, with support provided by other NATO nations, including a Norwegian-manned air traffic service. Deployments to Lithuania began on the day the country joined NATO, the lead rotation being Belgian-manned. While work at Zokniai is completed, the detachment is operating from Vilnius airport. An alternate base is planned for Estonia.

TCA designated

Although the aircraft have already seen many years of service, three Boeing 707-300C series transports operated by the NATO Airborne Early Warning & Control Force (NAEW&CF) were recently assigned the mission design series (MDS) designation CT-49A. Referred to as trainer cargo aircraft (TCA), the 707s can be configured to carry passengers, cargo or a combination of the two. They are used both as pilot training platforms and as support aircraft operating between the main NATO Air Base at Geilenkirchen, Germany, and deployment locations in Greece, Italy, Norway and Turkey.

New Zealand

Helicopter requirements

On 12 March 2004 the New Zealand Ministry of Defence issued a letter to several helicopter manufacturers inviting them to respond to a requirement to replace the RNZAF's 14 operational Bell UH-1H Hueys, the first of which was delivered in 1966. It is thought the replacement deal covers 12 aircraft. At the same time, the RNZAF is looking for around nine training helicopters to replace its ancient Bell 47 fleet. Front-runners for this requirement are the recently retired Republic of Singapore Air Force AS 550 Fennecs.

Peru

Widespread upgrades planned

After years of stagnation through funding problems, the mainly Russian-equipped Peruvian air force (FAP) has been authorised to negotiate large-scale upgrade and modernisation programmes for its variegated combat and support fleets. Priority is apparently being given to 16 Antonov An-32 tactical transports, 40 tri-service Mil Mi-8T/Mi-17s, plus 16 army Mi-24V and three Mi-26 helicopters. Upgrades are also being negotiated for the FAP's 16 MiG-29S/UBs and 30 or so Sukhoi Su-20/22s, as well as its 10/2 Dassault Mirage 2000P/DP fighter-bombers, and 10 Cessna A-37Bs. Resumed US military aid to Peru has included recent deliveries of two sensor-equipped Cessna OT-47B Citations IIs for narcotic interdiction roles. The Mirage 2000Ps of Escuadrón 412 at La Joya recently underwent limited upgrades from provision of Thales ATLIS II targeting pods for operation with Rafael Opher laser-guided bombs, and MBDA Magic 2 Mk 2 AAMs.

Saudi Arabia

AF seeks fighter update

The Royal Saudi Air Force is reportedly interested in purchasing new fighters and an upgrade for its Tornado IDS and ADF fleet and has entered into negotiations with BAE Systems. The deal is believed to be worth as much as $4.5 billion and includes a mix of up to 100 new and upgraded aircraft. As part of the Tornado programme each aircraft would undergo an inspection to determine whether an upgrade is viable. Those aircraft deemed to be unsuitable would be replaced by newer aircraft. The new aircraft would likely include the Eurofighter Typhoon. BAE had previously proposed delivering 50 examples to the RSAF at a cost of $2.25 billion.

Eagle deployment approved

Following a US agreement that lifts restrictions on their deployment, Saudi Arabia has been granted permission to permanently base F-15S strike fighters at Tabuk AB. The restriction had allowed the US to stop the Saudis from deploying the aircraft within their own borders. The F-15S had been temporarily deployed to Tabuk as a result of the coalition invasion of Iraq in early 2003. The base is located approximately 150 km (80 miles) from Israel.

Ukraine

Bombers destroyed

The Ukrainian Government recently destroyed the last of 12 Tupolev Tu-22M bombers that had been

On 1 November 2003 the 118th Fighter Squadron, Connecticut ANG, celebrated the 80th anniversary since being extended federal recognition, and painted up one of its A-10s in a special scheme. The original 118th Aero Squadron had been formed in August 1917 and had seen service in France in a support function. The Guard unit which was formed in 1923 was initially the 118th Observation Squadron, flying Curtiss Jennies. Based at Bradley Field, Windsor Locks, the 118th has flown the A-10 since 1979.

In January 2004 this Falcon 50 from the Burundi government made a rare visit to Europe, bringing the president to Rotterdam for talks with Dutch ministers. The government operates this single aircraft to cater for its VIP transport needs. The small army air wing has a handful of light aircraft and helicopters, of which the SF.260 and Gazelle are armed.

assigned to the 13th Heavy Bomber Division at Poltava Air Base. By July 2004 19 additional Tu-22Ms based at Ozernoye air base in Mykolaiv will be destroyed, along with 225 air-to-surface missiles. These bombers had been assigned to the 15th Heavy Bomber Regiment. It is expected that 20 Tu-22Ms remaining at Nezhin Air Base will follow, although the Ukraine has not yet approved plans for their destruction. Although Ukraine inherited the world's third-largest nuclear arsenal, following the collapse of the Soviet Union in 1991, it renounced nuclear weapons. It subsequently transferred its 1,300 nuclear warheads to Russia for destruction, and eliminated dozens of strategic bombers. The last of its intercontinental ballistic missile (ICBM) silos were destroyed in October 2002.

United Kingdom

Sea Kings to move

Having operated from Lossiemouth since 1974, No. 202 Squadron's 'D' Flight is scheduled to move to nearby RAF Kinloss in May 2005 after construction work on a new facility has been completed. The flight maintains a search and rescue detachment using Sea King HAR.Mk 3s. Along with the flight, the Search and Rescue Co-ordination Centre is also due to move to Kinloss.

Jetstreams retire

After 30 years of service in the multi-engined training role, the BAe (Scottish Aviation) Jetstream T.Mk 1 was retired from RAF service on 19 March 2004. The aircraft were initially delivered to 5 FTS at Oakington, but were then reassigned to 3 FTS at Leeming. The METS (Multi Engine Training School) transferred to 6 FTS at Finningley in 1979, becoming No. 45 (Reserve) Squadron in 1992. It moved to Cranwell in 1995. Replacing the Jetstreams are seven Beech King Air 200s, operated by Serco under a five-year PFI contract.

United States

Osprey delivered for test

Marine tiltrotor test squadron VMX-22 received the first of 16 MV-22Bs recently when Osprey no. 41 (BuNo 165941) was accepted at Bell Helicopter's facility in Amarillo, Texas. The squadron will use four of its aircraft in support of the Osprey's operational evaluation (OPEVAL), which is scheduled to begin in late 2004 and run through mid-2005.

Having been deemed surplus following the arrival of more F-16s, the Rockwell OV-10C Bronco was retired from Royal Thai Air Force service on 31 March 2004. The first batch of 16 new-build aircraft was delivered from June 1971, and served with 411 Squadron at Chiang Mai in the northern part of Thailand. A second 16-aircraft batch had been delivered by 1973. On 8 September 2003 the Government of Thailand donated four OV-10Cs to the Philippine Air Force in a formal ceremony at the 250th Presidential Airlift Wing hangar at Villamor Air Base in Pasay City. They are to be used for counter insurgency and in the fight against terrorism. A second batch of Thai OV-10Cs is scheduled to be delivered in September 2004.

Deployment for new squadron

The Air Force Reserve Command's 939th Air Refueling Wing recently commenced a 45-day deployment to Incirlik AB, Turkey, from its home station at Portland IAP, Oregon. The deployment marked the first for the wing since completing the transition between the combat rescue and air refuelling missions in April 2003. Although normally operated by the 64th Air Refueling Squadron, while deployed the wing's four KC-135Rs are attached to the 90th Expeditionary Air Refueling Squadron at Incirlik.

New Airlift Squadron

The USAF reactivated the 48th Airlift Squadron as part of the 314th Airlift Wing at Little Rock AFB, Arkansas, on 5 December 2003. The unit becomes the first active component squadron to operate the C-130J and will serve as an operational training unit. Squadron personnel, including 14 pilots and 10 loadmasters, began training in June 2003 using aircraft borrowed from the Air Force Reserve Command's 403rd Wing. The squadron's first stretched CC-130J will be delivered in early 2004.

Starlifters to Stratotankers

The Air Force Reserve Command's 459th Airlift Wing (AW) at Andrews AFB, Maryland, was redesignated the 459th Air Refueling Wing (ARW) on 1 October 2003. The unit has transitioned to the KC-135R, having received serial 59-1469 as its first aircraft. It had previously flown the C-141C.

Starlifter depot closes

The Warner Robins Air Logistics Center (WR-ALC) rolled out the final C-141 to undergo programmed depot maintenance (PDM) on 17 October 2003. Each of the Starlifters had undergone PDM every five years, however, the USAF intends to retire the last of these aircraft in 2006. As a result there is no longer a requirement to perform PDM and C-141C serial 65-0248 became the final aircraft to be overhauled. It was returned to the 452nd Air Mobility Wing at March Joint Air Reserve Base, California. Over the past 30 years WR-ALC had conducted more than 1,800 C-141 PDMs.

Army command post in Iraq

A pair of Sikorsky UH-60Ls equipped with Raytheon's Army Airborne Command and Control System (A2C2S) recently underwent operational testing with the 4th Infantry Division's Aviation Brigade in Iraq. While deployed the flying command post logged more than 2,000 hours of operating time in support of combat and peacekeeping missions. The aircraft were deployed shortly after passing flight safety and airworthiness tests. The A2C2S is a kit that includes antennas, structural supports, attachment points and equipment associated with the AN/ASC-38, including workstations. Although removable, the kit can only be installed in aircraft that have been equipped with an installation kit. The modified aircraft, which include a third example that is assigned to the 1st Cavalry Division at Fort Hood, Texas, are known unofficially as EUH-60Ls. Raytheon is under contract to deliver 34 operational systems for installation in UH-60Ls.

Citation delivery

The Headquarters & Headquarters Squadron at MCAS Futenma, Okinawa, recently took delivery of a second Cessna UC-35D Citation Ultra.

Block 60 Vipers for Arizona

The Arizona Air National Guard's 162nd Fighter Wing, at Tucson IAP is scheduled to receive the first of 15 Block 60 F-16s in 2004. The aircraft are owned by, and will support training of, the United Arab Emirates AF personnel. The fighters have been referred to unofficially as the F-16E/F.

Tomcat twilight

The Commander Naval Air Force Atlantic Fleet (COMNAVAIRLANT) has announced that the 'Checkmates' of VF-211 will transition from the F-14A to the F-14B when the squadron returns from its current deployment aboard the USS *Enterprise* (CVN 65). The decision advances the planned retirement for the earliest Tomcat model and means that the current cruise will be the last for the F-14A. The recent transition of VF-154 to the Super Hornet leaves only VF-211 and VF-101, which serves as the fleet readiness squadron, operating the F-14A. Advancing its retirement will allow the Navy to shut down the maintenance programme for the Pratt & Whitney TF30 engines that power the F-14A. In related news Northrop Grumman will take over the maintenance of the remaining Tomcats assigned to VF-101.

Patrol Wing reorganised

Commander Patrol and Reconnaissance Force Atlantic was redesignated Commander Patrol and Reconnaissance Group Atlantic on 1 December 2003. Located at Naval Station Norfolk, Virginia, COMPATRECGRULANT is responsible for Patrol and Reconnaissance Wing Five at NAS Brunswick, Maine, and Patrol and Reconnaissance Wing Eleven at NAS Jacksonville, Florida.

Starlifters retired

The 452nd Air Mobility Wing at March JARB, California, retired the first of its C-141Cs when serial 66-0136 departed for AMARC on 2 October 2003. Seven additional aircraft departed the base during October 2003 as the wing prepared to transition to the C-17A. The first Globemaster III however is not scheduled to arrive until late 2005.

Galaxies retired

The first of 13 C-5A Galaxies destined for retirement from the USAF inventory arrived at the Aerospace Maintenance and Regeneration Center (AMARC) at Davis Monthan AFB, Arizona, on 4 November 2003. Serial 70-0458 had last been assigned to the

Sea Harrier – disbandments and specials

On 31 March 2004 the Royal Navy's 800 Naval Air Squadron was disbanded at its Yeovilton base. The squadron is retiring its Sea Harrier F/A.2s as the UK's Joint Force Harrier moves to an all-Harrier GR.Mk 7/9 fleet. 800 NAS is due to reform at RAF Cottesmore on the Harrier GR.Mk 7/9 on 1 April 2006. Before its disbandment, 800 NAS embarked aboard HMS *Ark Royal* for two weeks in February for Exercise Final Endeavour, marking the squadron's last carrier operations (right). It also painted up a Sea Harrier in a special scheme as 'Red 22' (below right). 899 NAS, the Sea Harrier OCU, is due to disband in 2005, and it too has painted up a special scheme, based on previous 'Admiral's Barge' colours (below). 801 NAS, the last deployable 'Shar' squadron, is due to disband in 2006 before conversion to the GR.Mk 7/9. The Harrier is due to serve until replacement in both RAF and RN squadrons by the Lockheed Martin F-35 JSF. In the late spring of 2004 'Ark' is due to enter a two-year refit at Portsmouth.

433rd Airlift Wing (AW) at Kelly Field/Lackland AFB, Texas, but was delivered to AMARC via Travis AFB, California, by the 60th AW. AMARC will take delivery of 12 additional C-5As. Serial 69-0004 has been retired to Warner Robins Air Logistics Center at Robins AFB, Georgia, where it will support non-destructive and destructive testing that will determine the feasibility of updating selected C-5As for further service.

Air Guard Globemaster III

Boeing handed over the first C-17As to the Mississippi Air National Guard's 172nd Airlift Wing at its Long Beach, California, facility on 17 December 2003. Serial 02-1112, which wears the name *Spirit of the Mississippi Minutemen*, was delivered to the unit's home base at Jackson International Airport the following day. The 183rd Airlift Squadron will receive seven additional Globemaster IIIs during the first half of 2004.

Eastern bases possible

The USAF may soon announce plans to station up to four squadrons in Bulgaria. The service had previously stationed tanker aircraft at the Sarafovo airport, near Burgas approximately 402 km (250 miles) east of Sofia in support of Operations Enduring Freedom and Iraqi Freedom. Graf Ignatievo AB and Bezmer AB in central Bulgaria are reportedly being considered, along with the former Ravnets air base, which is located approximately 20 km (12.5 miles) west of Burgas on the Black Sea.

Aggressor expansion

In the light of increased demand on its services, the 64th Aggressor Squadron (previously the Adversary Tactics Division of the 414th CTS) at Nellis AFB, Nevada, has increased its complement from seven to 11 F-16s. As well as its regular commitments to providing 'Red' air during Red/Green/Maple Flag exercises and adversary duties in support of Air Warfare Center evaluation work, the squadron is increasingly tasked with support of F/A-22 Raptor evaluation. Further growth of the squadron is planned.

Shadow deployed to Korea

The RQ-7A Shadow 200 tactical unmanned aerial vehicle (TUAV) was recently deployed to the Republic of Korea. The TUAV system was fielded by the 102nd Military Intelligence Battalion (MIB), which is based at Camp Mobile near Dongducheon. The battalion is a component of the 2nd Infantry Division (Mechanized).

Patrol fleet reductions

As part of an overall reduction in its maritime patrol fleet the navy will retire nearly a third of its P-3C Orions over 12-24 months. As a result, the fleet will be reduced from 227 to approximately 150 aircraft. The reduction is partially due to corrosion problems with the Orion.

Air Mobility Warfare Center

The USAF Mobility Weapons School was established recently adjacent to McGuire AFB at Fort Dix, New Jersey. It is a component of the Air Mobility Warfare Center (AMWC), which is also located at Fort Dix and had previously been known as the USAF Combat Aerial Delivery School (CADS). The school's three weapons squadrons (WPS) train weapons instructors for the C-17, C-130 and KC-135 airlifters. The 5½-month long weapons instructor courses provide graduate-level training to AMC's top instructor pilots and navigators. None of the squadrons is assigned its own aircraft, which are borrowed from operational units as required. AMC is currently reviewing the need to establish weapons courses for other mobility aircraft.

USAF Mobility Weapons School – Fort Dix, New Jersey

Squadron	Location	Aircraft
29th WPS	Little Rock AFB, AR	(C-130E/H)
57th WPS	McGuire AFB, NJ	(C-17A)
509th WPS	Fairchild AFB, WA	(KC-135R)

The 34th Combat Training Squadron (CTS) at Little Rock AFB, which had been assigned to the CADS, was transferred to the newly established USAF Mobility Operations School. The MOS is also a component of the AMWC at Fort Dix.

In related news, the 33rd Flight Test Squadron (FLTS) at Fort Dix was redesignated as the Air Mobility Command Test and Evaluation Squadron (AMC TES) on 1 October 2003. The unit, which is tasked with testing air mobility unique weapons systems, subsystems and equipment, reports directly to AMC at Scott AFB.

As part of Japan's first post-war international military deployment, two C-130Hs are operating in Iraq as part of the reconstruction/stabilisation force. The aircraft, one of which is seen here departing Komaki for Kuwait on 26 January 2004, were given an enhanced defensive aids suite, including missile-warning detectors and flare dispensers. The light grey scheme is also new. This aircraft wears the badge of Dai 401 Hiko-tai.

Defending Belize

Belize Defence Force Air Wing

Above: This BN-2A is the most recent BDF/AW acquisition, arriving in 2000 as a replacement for the crashed BN-2B. The Defender force is used for transport and reconnaissance duties, the latter centring on anti-drug work and economic patrols.

Right: The single Cessna 182J is used widely for small transport tasks. Belize has over 40 airstrips which can be used by the Cessna and BN-2s.

Sandwiched between Mexico and Guatemala, the small nation of Belize operates one of the smallest air wings in the world. Formerly the Crown Colony of British Honduras, Belize achieved internal self-government in 1964, and changed to its present name in 1973. On 21 September 1981 full independence was achieved (with Belmopan as the capital), although Belize remains closely tied to the UK. From the mid-19th century until 1991 Guatemala claimed Belizean territory (either in total or a part of the south of the country), mainly as a means of ensuring access to, and expansion of, its small Caribbean coastline.

In January 1978 the Belize Defence Force was formed as a local infantry force. On 4 August 1981 the Air Wing was established to provide support to the BDF. Brigadier Marcenard of the British Army was tasked to conduct a feasibility study to identify the necessary equipment, which resulted in two Britten-Norman BN-2B Defenders arriving in August 1984, together with support staff from the UK. The choice of the Defender was an obvious one as a number of civilian Islanders were flying with local air taxi operators.

The BDF Air Wing was established at Belize City's Philip S.W. Goldson International Airport, with the motto 'Arborem Super Vigemus' ('above the trees we flourish'). In 1991 the first Belizean commander, Captain Gillet, took over from a British commander, and in 1997 Belize assumed full responsibility for the Air Wing.

Expansion of the fleet began in 1995 with the delivery of a single Slingsby T67 Firefly for training, and a year later a Cessna 182J was added. In October 1998 one of the Defenders was lost in a swampy area near Indian Church: a replacement came in 2000 in the form of a BN-2A. The current commanding officer is Major Dortch, whose wish is to upgrade the BDF/AW from piston-powered aircraft to turbines, such as the Cessna 208 Caravan (now in widespread local civilian service).

Organisationally, the Air Wing is a constituent of the BDF, alongside 1 and 2 Battalions, Logistics Wing and the Maritime Wing. The Air Wing has a strength of five pilots, seven engineers and five support personnel. Pilots fly around 300-400 hours annually. The turnover of pilots means that one has to be trained every two years or so. Basic training is conducted in the US, UK or Canada, although more advanced training is conducted in Belize using the T67. Multi-engined training was previously performed at Oxford in the UK, but is now undertaken locally. Engineers receive instruction in the US or UK.

The small Air Wing has a variety of tasks, including troop transport, VIP transport, casevac/medevac, search and rescue, and patrol of the economic exclusion zone in support of the nation's important fishing and tourism industries. Anti-narcotics operations form an important part of the Air Wing's activities. Drugs are cultivated in the remote areas of the country, and the BDF/AW is heavily involved in spotting fields. The nation is also on the cocaine route from South America, and the BDF is involved in anti-smuggling operations. Light air support can be provided by the BN-2B, which can carry two 7.62-mm machine-gun pods under its wings. BDF/AW aircraft work closely with US DEA forces in anti-narcotics operations.

British presence

Throughout the long-running quarrel with Guatemala the British provided forces in Belize to counter any possible invasion. Shows-of-force were undertaken by Canberras and even Valiants in the 1950s and 1960s, and in November 1975 RAF Harriers were rushed to the country to stave off an impending invasion by Guatemala. Although this danger passed, another in July 1977 saw the Harriers return, and this time they stayed, established as 1417 Flight. 1563 Flight was formed with Pumas. Army Air Corps helicopters were deployed with 25 Flight, which flew Scouts until 1979 and Gazelles thereafter.

In 1991 Guatemala gave up its claim to the Toledo district of Belize and tensions in the region subsided. In 1993 the Harrier GR.Mk 3s of 1417 Flight were withdrawn, although one (ZD669/D) remained at Belize City, preserved outside the main terminal at the international airport. The Pumas returned home in 1994.

Today, permanent British air assets in Belize are restricted to three Bell 212s operated by the Army's 25 Flight. These replaced Gazelles and Lynxes in 2003, and are tasked with supporting the British Army Training Support Unit Belize (BATSUB, based at Price Barracks in Ladyville), which conducts jungle warfare training. The Bells can also be called upon by the Belizean government, and also support the BDF/AW in casevac duties, especially at night.

Dick Lobuis; additional material by David Donald

Left: The single T67M260 Firefly allows the BDF/AW to conduct training locally. The shed in the background houses the patrol boats of the BDF's Maritime Wing.

Below: Two of the three Bell 212s (ZJ965/B and ZJ969) from the British Army's 25 Flight prepare to lift off. The Bells are supplied to the Army under contract by Bristow Helicopters.

BDF/AW aircraft

Serial	Type	c/n	
BDF-01	BN-2B-21	2137	crashed 10/98
BDF-02	BN-2B-21	2136	
BDF-03	Cessna 182J	56975	
BDF-04	T67M260	2235	
BDF-05	BN-2A-21	277	

The legendary IV

Dassault Mirage IVP

In 2004 the Dassault Mirage IV celebrates its 40th anniversary in service with France's Armée de l'Air. Although retirement looms in 2005, this legendary and classic combat aircraft still plays a key role in French foreign affairs and has specific capabilities that will be lacking in its successor. Forty is quite a respectable age, but this did not prevent the French government from sending two of these strategic reconnaissance aircraft to Iraq in February 2003 in support of the United Nations.

The Mirage IV is one of the most attractive types ever designed by the famed French aircraft manufacturer Dassault. On the ground it is impressively big, and leaning backwards on its main undercarriage its sharp nose points slightly upwards, giving it a distinctive, somewhat arrogant, air. Once on the runway, when the pilot applies full afterburner, the beast kicks in. Anyone who has witnessed the Mirage F1, its smaller tactical sister, light its afterburner knows the potential of the SNECMA Atar 9 engine to trade fuel for noise. Imagine, then, the result in the Mirage IV, which has two such powerplants.

Bomber

In addition to a lot of noise, these powerful engines produce more than enough thrust to fulfil the aircraft's original delicate mission, which was the delivery of a nuclear bomb behind the Iron Curtain. Early in the nuclear age this could be accomplished only one way: at high speed from high altitude. The IV entered service with the French Forces Aériennes Stratégiques (Strategic Air Forces) in 1964. In total, the Armée de l'Air received 62 Mirage IVAs spread among nine strategic bomber squadrons.

In the mid-1980s, 12 of the many survivors went through a modification programme that enabled them to employ the ASMP (Air-Sol Moyenne Portée), a missile with a nuclear

Above: What entered service as the Mirage IV began life as the Mirage IVC, one of a number of proposals based on the original Mirage design that included the smaller, single-engined IVA and IVB. The IV, as built, was essentially an extrapolation of the successful delta fighter configuration with larger wings and two engines.

Below: A rare sight – two ERS 1/91 Mirage IVPs fly along France's Atlantic seaboard. The active fleet typically numbers four aircraft, although three others are either in flyable storage or undergoing scheduled maintenance. The Mirages are expected to serve until the end of June 2005.

Right: Today the Mirage IVP is very rarely seen in clean configuration, without even the CT52 reconnaissance pod. Operational flights in high-threat areas are flown supersonically, which precludes the use of underwing tanks unless they are jettisoned.

Above: A Mirage IVP lands at Mont-de-Marsan at the end of a training sortie. The aircraft wear the arms of Gascogne on the fin, comprising a shield with two lions and two wheatsheaves.

Left: An important aircraft in ERS 1/91's inventory is the Alpha Jet, which not only provides extra flight hours for Mirage IV crews, but also allows them to practise some of the evasive manoeuvres that might be required in wartime, but which would overstress the Mirage IV unnecessarily.

charge. Other modifications included a new radar, improved electronic countermeasures systems, and new navigation equipment. Consequently, the aircraft was designated Mirage IVP (P for Pénétration). The remainder were destined for retirement as the Mirage 2000N entered service.

In 1996 the nuclear mission came to an end, but that did not result in the end of the IV. The 12 newest aircraft had received special wiring that allowed the use of the CT52 reconnaissance pod, which became available in 1972. With this system, two bomber squadrons were given the important secondary mission of long-range reconnaissance. This mission remained the sole job of Escadron de Bombardement 1/91 'Gascogne' at Base Aérienne Mont-de-Marsan, later renamed Escadron de Reconnaissance Stratégique 1/91 'Gascogne' and commonly known as the ERS. Ever since, the air force has employed the IVP over many (potential) military and political hot spots.

Lieutenant Colonel Didier Priolet is the commanding officer of ERS 1/91. He spent eight years flying the Mirage 5 and F1CT at BA Colmar, in the process logging more than 60 combat missions over the Balkans and Africa. In total, he has 2,000 hours on fighter aircraft. He arrived at ERS 1/91 in the summer of 2001 as the deputy squadron commander. In September 2003 he assumed command, and will remain in charge until retirement of the IVP at the end of June 2005, when the squadron is projected to have six flyable aircraft.

This image was taken from a high-flying Mirage IVP and shows Sarajevo Airport in Bosnia. The sensor used was the Wild RC8F camera, which produces images on large square negatives. These are used primarily for mapping/survey purposes.

The Mirage IVP fleet has been very active over former Yugoslavia. This image, recorded over Bosnia in 1994 from an altitude of 7,800 ft (2377 m), shows war-damaged houses, most of which have been burnt out and are roofless.

Although operationally the Mirages are mainly used at high altitude, low-level photography skills are maintained in training. This oblique view from 2,500 ft (762 m) shows the Phare de la Vierge on the Breton coast to the north of Brest.

Left and below: Mirage IVPs were deployed to the Gulf twice to assist the United Nations effort to hunt down Iraq's weapons of mass destruction, the first time in 1998 and again in 2003. Operating from Al Kharj in Saudi Arabia, the Mirages flew high-altitude missions to monitor suspected WMD facilities at a time when ground-based inspectors were denied any access. These two images of aircraft 'BZ' were taken over Iraq during the course of the 14 operational missions the deployment flew in support of UNMOVIC between 26 February and 18 March 2003.

Below: The ERS 1/91 detachment arrived at Al Kharj on 21 February after a non-stop flight from Istres in southern France, having been refuelled en route by the C-135FR tankers of ERV 1/93 'Bretagne'. An important job for the first day after arrival was the application of large, temporary 'UN' titles on the fin.

According to Priolet, it is amazing that, in spite of its age, maintenance of the Mirage IV is not a nightmare. "Roughly, we need only around 15 hours for level one maintenance per flying hour, as a tactical Mirage squadron. We do not have specific maintenance problems."

Only seven of the 12 Mirage IVPs are airworthy, and the squadron has five in its inventory. One of those is usually at Base Aérienne Bordeaux-Mérignac for level two maintenance, and two are kept in flyable storage at Chateaudun, the storage base of the air force.

Supersonic

The reconnaissance systems are all old-fashioned, wet film cameras. The canoe-shaped CT52 pod that is semi-recessed under the centreline is configurable for low- or high-level reconnaissance. The BA (Basse Altitude) low-level configuration comprises four 114 x 114-mm (4.5 x 4.5-in) Omera 35 cameras in the forward part of the pod. Three cameras have a focal length of 150 mm (6 in): one camera is positioned in the nose, and two obliquely. The fourth camera is positioned vertically and has a 75-mm (3-in) focal length.

The HA (Haute Altitude) high-level camera set is placed aft of the low-level set. The three Omera 36 114 x 114-mm cameras are set behind each other and have 600-mm (23.6-in) focal length lenses. The fourth HA camera is the Wild RC8F; with a huge 240 x 240-mm (9.5 x 9.5-in) sized negative and a lens of 152 mm (6 in), this camera is used for large-area, high-resolution cartography and gives the IVP a unique capability. For low-level night operations, the ERS 1/91 reconnaissance systems specialists can replace the HA cameras with an SAT Super Cyclope infra-red line scanner.

The reconnaissance pod has no digital systems but the squadron has its own advanced scanner and transmission equipment. "Immediately after landing, we develop the films and the interpreters conduct a first interpretation," Lieutenant Colonel Priolet explains. "We digitise relevant imagery and transmit it through our Station Légère d'Interprétation to the CAOC, JCS or, for instance, the UN during our operations for UNMOVIC. Not too long ago it took us days to get the material to New York, now it is only a couple of hours."

"We are able to take photographs from 500 ft [152 m] up to 50,000 ft [15240 m]," describes Priolet. "In combat operations we usually fly very high because we are much less vulnerable there." If the security situation permits, the IVP flies a subsonic profile, but if there is a reasonable threat it goes supersonic. "The Mirage IV has two important defensive assets, of which speed is one. We are allowed to go up to Mach 2.2. Theoretically we can go faster, because when we reach this Mach number we are still accelerating. However, certain parts of the aircraft become too hot as a result of the friction, despite an outside air temperature of -55° C [-67° F]. We cannot touch certain metal parts inside the cockpit because we will burn our fingers."

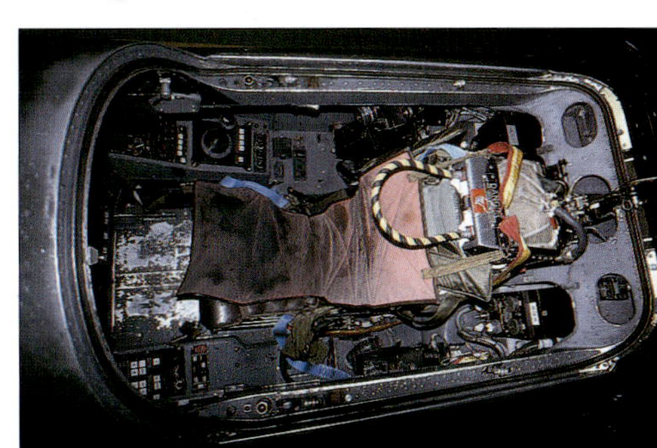

The WSO's aft cockpit has two very small windows for outside vision, and a Slaved Optical System periscope which is particularly useful when employing the Wild cartographic camera. The main displays in the WSO's cockpit remain off-limits to cameras.

Left: The CT52 pod hinges open to provide access to the cameras and film magazines.

Right: Ground crew prepare a Mirage IVP before it taxis for a training mission. This view highlights the CT52 reconnaissance pod on the centreline, and also the size of the underwing drop tanks, which add 2500 litres (550 Imp gal) each to the fuel capacity. In its original IVA nuclear bomber form the Mirage carried a single AN22 nuclear bomb semi-recessed under the central fuselage: in the IVP this gave way to a single ASMP missile on a centreline pylon.

Below: Mont-de-Marsan has played host to the Mirage IV since 1964 and the early days of the 'Force de Frappe'. As well as the few ERS 1/91 aircraft, Mont-de-Marsan today houses a light transport squadron (ETE 42) and the CEAM – the Armée de l'Air's principal test and evaluation organisation.

Priolet stresses the importance of the aircraft's ability to fly supersonic for a long period. "Only the IV can do that. During operations over Kosovo we started from Base Aérienne Solenzara in Corsica, refuelled over the Adriatic Sea, climbed to 50,000 ft [15240 m] and accelerated to Mach 1.8 prior to entering the war zone. After a supersonic flight of half an hour we returned to the tanker, before heading back to Solenzara or making another run over the Balkans."

Internal fuel capacity is 14000 litres (3,080 Imp gal), which increases to 19000 litres (4,180 Imp gal) with two external tanks. The tanks must be jettisoned in the event of supersonic flight. Despite the aircraft's ability to refuel in flight, endurance is limited to 15 hours of flight because engine oil must be replenished. Generally, however, unless operational necessity demands otherwise, crews do not fly more than eight hours because the cockpit is not very comfortable.

The second defensive system concerns electronic warfare. The French are renowned for the powerful and effective ECM systems in their aircraft. "The integrated external and internal electronic warfare suite is very good and receives upgrades continuously. Soon we will fit a video system in the cockpit that takes pictures of the electronic warfare display. Consequently, we can more easily interpret EW data after the flight and learn from our mistakes or shortcomings," says Priolet.

The ECM systems incorporate the Barax Nouvelle Génération jammer pod under the port wing, an internal chaff/flare system in addition to the Boz chaff/flare dispenser under the starboard wing, an internal jammer system to jam SA-6 surface-to-air missiles, and internal Serval electronic support measures equipment that receive incoming threat signals. According to Lieutenant Colonel Priolet, the Mirage IV has a huge chaff/flare capacity.

Icon

Since 1996 the Mirage IV has participated in almost every military operation in which France has been involved. Pilots and WSOs of other aircraft types know that assignment to ERS 1/91 offers a chance to be in the thick of action. This is one reason why there is no shortage of voluntary crews for an assignment to the very last Mirage IV squadron.

Another reason is that the Mirage IV is an icon within the French Air Force: it has been the face of nuclear power in France for decades. Of course, it is old, but for 'real airmen' it is a challenge to fly such an aircraft.

ERS 1/91 has six crews. They each fly 200 hours a year on the IV and another 50 on the squadron's sole Alpha Jet trainer. With the 'A-Jet' they train for specific profiles which they are not allowed to perform with the IV. Lieutenant Colonel Priolet: "The IV is limited to only 4 *g* when it flies with external fuel tanks,

A Mirage IVP refuels from a C-135FR during its ferry flight to Saudi Arabia in February 2003. The Mirage IV has supported numerous UN/NATO operations, having seen service over Bosnia, Iraq and Kosovo in the 1990s. It also flew recce missions over Afghanistan during Operation Enduring Freedom.

and 5 g without. We can pull 8 g but that would damage the aircraft. Such a manoeuvre would not be ruled out in war time, so we train for it with the 'A-Jet'."

The ERS's commander decided to join the air force the day he saw a IV flying. "It is a distinctive, big and beautiful aircraft," he says, almost lyrically. "I recall my first flight vividly. Like all new Mirage IVP pilots, I had to trust to my study of the manuals before taking off for the first time. We do not have simulators, never had, and we do not have any planes fitted with dual controls. I can tell you that my first flight was very exciting."

Iraqi mission

At the time of writing the ERS had no real world contingency commitments, but such a condition can change abruptly – as in early 2003, when tensions regarding Iraq heightened. Accordingly, in February 2003 the squadron was back in the Gulf region. "We had the very same task as in 1998. We flew for the weapons inspectors of UNMOVIC and also had the status of inspectors. We transmitted our photo material directly to the UN headquarters in New York, where we had a former Mirage IV pilot as liaison. In total, we logged 14 missions from Al Kharj in Saudi Arabia between 21 February and the end of March," recounted Priolet.

The first mission was flown on 26 February and the last on 18 March, which was the day US President George W. Bush gave the 48-hour ultimatum to Iraq. The Mirages were over Iraq for up to three hours during each mission.

ERS 1/91 still has an *ad hoc* mission over Kosovo, which it flies from Mont-de-Marsan. The squadron is fully prepared to deploy quickly and trains for rapid deployments abroad two to four times a year. A typical ERS 1/91 deployment comprises two aircraft, three crews and 30 specialists for maintenance, communications, and photo-interpreters.

Lieutenant Colonel Priolet maintains that his ERS generally flies only strategic reconnaissance missions during operations. "We usually get our tasking from our Joint Chiefs of Staff. However, it is also possible we will receive taskings from a Combined Air Operations Centre. During our operations over Afghanistan in support of Enduring Freedom, for example, we got our tasking from the allied headquarters in Saudi Arabia." Between October 2001 and February 2002, two Mirage IVPs, operating from an undisclosed air base in the United Arab Emirates, logged 77 missions.

Ground crew prepare to remove the film magazines from a Mirage IVP just returned from the first mission over Iraq of the 2003 deployment, on 26 February.

Successor

When Mirage IV operations finally end, in 2005, the type's successor will not be ready. The Armée de l'Air is currently developing the Pod de Reco Nouvelle Génération (New Generation Reconnaissance Pod), which should enter service in 2007 with the Mirage 2000N as an interim strategic reconnaissance asset.

Important advantages of the new reconnaissance system are its digital cameras for day and night photography, and its ability to transmit pictures in real or near-real time to a ground station. In addition, the cameras can move from side to side, looking obliquely, compared to those of the IVP which look vertically. The crew can steer the cameras, but they can also be pointed automatically towards a pre-programmed target.

The Mirage 2000N, though, has some distinct disadvantages. It has a lower maximum speed and its fuel capacity is much smaller, meaning that it cannot undertake lengthy fuel-guzzling reconnaissance runs and its range is significantly shorter at supersonic speed. According to Priolet, "The 2000N will have to do its reconnaissance work in pairs, just as its tactical brothers have done for years. This allows the wingman to escort the leader, always looking outside the cockpit to detect an eventual missile departure, and be the mission back-up. We do not need that because we maintain much higher speeds at higher altitudes, so we are much less vulnerable."

Nonetheless, the 2000N will be a capable reconnaissance system despite its disadvantages. It is the Rafale, however, that will be the ultimate strategic reconnaissance platform for the French Air Force when it becomes available in 2009. This is only a few years after the 2000N becomes strategic reconnaissance-capable, but there is no alternative. As related above, the IV is in very good mechanical condition, but it simply cannot linger on forever. Says Lieutenant Colonel Priolet, "Dassault conducted a thorough technical examination early in the last decade and determined the aircraft could fly until 2005. Should we wish to keep it operational after that year, they would have to do a new study for a life extension programme. This would be far too expensive for only seven jets, so the air force decided early not to do it."

Gert Kromhout

'AX' taxis through a sandstorm at Al Kharj during the early 2003 deployment for operations over Iraq. The two-letter codes are derived from the 'last two' of the aircraft's individual callsigns (in the range F-THAA to F-THCI), which date from Force de Frappe days.

Swiss air force

Photographed by Marnix Sap/MIAS

In 2004 the Schweizer Luftwaffe celebrates its 90th anniversary, the same year that the Army XXI programme comes into effect, reducing significantly the size of the Swiss air force. Responding to the current nature of the European military balance, Switzerland has decided to pull large numbers of aircraft out of service, resulting in the disbandment of several units (Fliegerstaffel 1, 10 and 13 were disbanded at the end of 2003) and the closure of several war bases (Interlaken, Buochs and Turtmann).

In line with this, a significant restructuring programme has been completed, seeing the integration of Fliegerbrigade 31 and Flugplatzbrigade 32, which is now known as Lehrverband Flieger 31 (Air Training Command).

From 1 January 2004, the Swiss Air Force will have three F/A-18 Hornet squadrons and three of F-5 Tigers at its disposal. They operate from three war-bases, Payerne, Sion and Meiringen, with the latter being the last with a cavern hangar structure. Training takes place at the airfields of Emmen, Locarno and Dübendorf, although the latter is planned to cease Tiger operations by 1 January 2005 and Hornet operations around 1 January 2007.

The Überwachungsgeschwader 85 (UeG or Surveillance Wing), headquartered at Dübendorf, will oversee training and operations of the Kommando Piloten Schule Luftwaffe 85 (Pilots School – HQ at Sion), Kommando Lufttransport 85 (Air Transport – HQ at Alpnach) and Kommando Luftverteidigung 85 (Air Defence – HQ at Payerne) under which the different Flieger Staffel fall according to role.

Despite the recent round of cuts, Switzerland maintains a credible defence based on the F/A-18 and F-5 (left). The outcome of F-5 replacement studies is tied to the limits of the future defence budget, which is currently funding a significant Hornet upgrade programme. As well as air defence, support of the Swiss Army and civil population is an important role, at the heart of which lies the Super Puma/Cougar (above). Local products include the Pilatus PC-7 trainer (below) and PC-9 target tug (far left).

Boeing F/A-18C/D Hornet

On 17 June 1992 the Federal Council approved SFr 3,495 million for the purchase of 26 F/A-18C and eight F/A-18D Hornets, together with armament, spares and support. Powered by F404-GE-402 turbofans, these aircraft could only be procured after a national referendum was held on 6 June 1993, which clearly showed the Swiss public to be in favour of modernising its Air Force and approving the new fighter programme to equip three squadrons.

The first F/A-18D (serial number J-5231) and the first F/A-18C (serial number J-5001) were built by McDonnell Douglas in St Louis and were used for preliminary trials and training, but the remainder of the order (seven F/A-18Ds and 25 F/A-18Cs) was assembled in Switzerland by Schweizer Flugzeuge und System AG (Swiss Aircraft and Systems Company, formerly known as F+W, now RUAG Aerospace) at Emmen. The referendum delay allowed Switzerland to specify the AN/APG-73 radar for its Hornets, and on 25 January 1996 the first F/A-18 Hornet was delivered. The last F/A-18C (J-5026) of the order was delivered on 2 December 1999.

The first squadron to be operational with the Hornet was Fliegerstaffel 17 and was commissioned in September 1997. The second squadron to convert was Fliegerstaffel 18, which changed its F-5E/Fs for Hornets in 1998. The third and final squadron to re-equip was Fliegerstaffel 11, which traded in its Tigers in November 1999. All three squadrons are part of the Überwachungsgeschwader (UeG), or Surveillance Wing, and unlike the Swiss Tigers, the aircraft are exclusively flown by professional pilots ('Profis'). On 22 October 2003, the Meiringen cavern shelter complex for the permanent protection of the F/A-18 fleet based there during exercises or wartime was officially handed over.

Switzerland aims to have the Hornets in service for 5,000 flying hours or 30 years, and in order to achieve this service life, design modifications were carried out including strengthening the airframe by constructing some of the frames out of titanium. Other modifications include the adaptation of the electronic equipment to equipment already in service in Switzerland, such as the radio and IFF, together with another modification enabling existing stocks of Sidewinder missiles to be deployed. A low-drag pylon has been specially developed to jointly accommodate one AIM-120B AMRAAM or one AIM-9P-5 Sidewinder guided missile.

Budgeted for in Rüstungsprogramm (procurement programme) 2003, the Swiss F/A-18 fleet will undergo an intensive upgrade (Kampfwertsteigerungsprogram) worth SFr 407 million, whereby the US Navy, Boeing and RUAG Aerospace are responsible for the execution. The package includes the incorporation of the Raytheon AIM-9X Sidewinder, electro-optical helmet-mounted visor, MIDS datalink system (linked into the Florako air defence guidance system) and related update of the software to incorporate the new systems. The updated aircraft are expected to be delivered between 2008 and 2009.

The possibility of equipping some Hornets with a sophisticated reconnaissance package interlinking with the SAR capability (Synthetic Aperture Radar) of the AN/APG-73 radar is currently under review, but no funds are reserved in the 2004 budget.

013
J-5013
061
J-3061

Northrop F-5E/F Tiger II

Switzerland took delivery of 98 F-5Es and 12 F-5Fs following two separate deals with the USA. The first contract (Peace Alps I), worth SFr 1.17 billion, was signed in 1976, covering the supply of 66 Northrop F-5Es and six F-5Fs. The first 13 F-5Es and all F-5Fs were supplied by Northrop, while the remaining single-seaters were locally assembled by FFA at Emmen. A follow-on batch was ordered in 1981 under the name Peace Alps II, covering the delivery of 32 F-5Es and six F-5Fs.

Now veterans, the Swiss Tigers are in the process of being phased out, and part of the Army XXI programme was the dissolution of Flieger Staffel 1 and 13 at the end of 2003, reducing the fleet from 85 to 54 aircraft in three squadrons. The 12 F-5F twin-seaters are now used primarily in the advanced training and jet conversion role, following the Hawk T.Mk 66's premature retirement (the Hawk's last operational flight took place on 13 December 2002). Ausbildungsstaffel 16 continues to use the F-5F aircraft for ECM training, employing the Vista 5 (Erijammer A-100) ECM pod. The Swiss Air Force acquired some eight ACMI range instrumentation pods for the F-5E for training at the ranges in Switzerland.

Interest has been shown by the Philippines to buy the surplus aircraft, but a more realistic customer could be the Força Aérea Brasileira, which has shown interest in buying 15 F-5Es and having RUAG convert them into two-seaters, boosting its small fleet of instructional aircraft. A first real customer is the Naval Air Systems Command Headquarters, Patuxent River, Maryland, which is procuring up to 32 low-houred F-5Es in a fixed-price contract estimated at $18,557,950. The base award of $570,950 is for the initial requirement of one aircraft, whereby the contract consists of a basic award and five options. These F-5Es will provide a one-for-one replacement of current US Navy Reserve F-5Es. If all options are exercised, the work will be completed in September 2007. The first F-5E (J-3026) was airlifted onboard a US Navy C-130 to the USA on 1 May 2003.

The 'Patrouille Suisse' has been the official jet aerobatic team of the Swiss Air Force since its formation in 1964, initially operating the Hawker Hunter, but after 30 years of shows, the 'Patrouille Suisse' flew its last Hunter display in 1994. Replacement came in the shape of the faster and more manoeuvrable F-5E. All the members of the 'Patrouille Suisse' are professional military pilots from the Surveillance Wing and practise this additional task together during the respective air show season only. A display lasts approximately 18 minutes and is always closely monitored from the ground by the commanding officer. In 2004 the 'Patrouille Suisse' celebrates its 40th anniversary.

Pilatus PC-9 (above)

Since 1991 11 Pratt & Whitney Canada PT6A-62-powered PC-9s have been in use exclusively with Zielflug Staffel 12 for target-towing duties in support of the five anti-aircraft regiments and single mobile anti-aircraft missile regiment in the country. For this role they received a high-conspicuity colour scheme. The PC-9s are also used for electronic countermeasures training with ground- and air units, employing the Ericsson Erijammer A-100 (Vista-5) ECM pod, a role previously undertaken by Hunter Mk 68 two-seaters of Flieger Staffel 24. Both unit badges are portrayed on the fuselage of the PC-9.

The PC-9's normal operational base at Samedan in Engadin will be closed for regular military use, and the aircraft will be shared between Payerne and Dübendorf.

Pilatus PC-7 Turbo-Trainer (left and below)

Some 41 Turbo-Trainers were originally procured for use by the Piloten Rekruten Schule (Pil RS) at Magadino, the Flying Training Organisation (FTO) of the Schweizer Luftwaffe which was administered by the Bundesamt für Ausbildung der Luftwaffe (BAALW, or Federal Office of Air Force Training).

Today military and flying training is conducted separately. The first year comprises basic military education, while in the second year the flying course starts and cadets receive their ICAO Joint Aviation Requirement (JAR) – Air Transport Pilot (ATP)-training over a 16-month period. This training is outsourced to the civil Swissair Aviation School (SRAS).

After this civil training, military flying continues on the PC-7 at the Pil RS. The course traditionally starts around the end of August and over a period of 70 weeks 125 hours are flown on the PC-7 and F-5F Tiger (or Alouette III for helicopter pilots). The Hawk is no longer part of the course. Annually some 1,000 19-year olds present themselves for the FVS course to become a military pilot. Between 35 and 42 men and women start their course at the Pil RS. Only a quarter will eventually complete the course successfully. As the fourth and final step of the three-year training the newly trained pilots are further trained as instructors.

Some 18 PC-7s have been selected to receive a cockpit upgrade, to conform their instrumentation to JAR standard. In 1987 a nine-aircraft formation display team made its show debut. In 1989 the 'PC-7 Team' was officially formed.

Eurocopter AS 532UL Cougar

As a result of the good experience gained with the Super Puma, in 1998 a project called Transporthelikopter TH98 was launched, covering the procurement of an additional 12 transport helicopters worth SFr 320 million. The order was for the newer AS 532UL Cougar. This 'glass' cockpit Cougar is equipped with an Electronic Flight Instrument System (EFIS) instead of a conventional horizon and HSI. The EFIS Primary Flight Display (PFD) provides information on air and ground speed, and climb and descent speeds, as well as barometric and radar altitude. Revolutionary in the Cougar cockpit is the Moving Digital Map System. A big LCD screen shows maps of varying scale, which moves according to the direction of the helicopter. In another mode the relief of the countryside can be shown, with colours changing to indicate collision danger relative to the flying altitude of the helicopter and the mountains surrounding it. Cables and other dangerous points were also added to the digital maps, which are periodically updated. The moving map is controlled by GPS/INS with metric precision. Also in the Swiss Cougars is a Flight Management System (FMS) supplied by Canadian Marconi. Besides the direct input of navigation points and routes, the complete flight plan can be made on the Sirinam mission planning system and transferred via a data link into the helicopter. The first Swiss Cougar (T-333) made its first flight at Marignane on 15 May 2000, but remained in France for further testing until arrival in Switzerland in 2002.

This and Cougar T-331 were produced in France, the remaining 10 helicopters being assembled by RUAG Aerospace at Emmen, and all 12 were officially handed over on 27 September 2002 at their base at Alpnach. T-331 (above) was delivered on 2 May 2001 and is operated purely in the VIP role, for which it received luxury leather seating and cabin sound protection, in addition to a dedicated air-conditioning system for which a climate pod is scabbed on to the port forward fuselage. The six helicopters serialled T-332 to T-337 are equipped with fixed polyvalent air intakes. These so-called EAPs improve the employment of the Cougar in heavy snowfall and icy environments, but also allow the helicopters to operate in a desert environment where they work as sand filters. Eventually, all the Cougars will be equipped with the ISSYS self-protection system (left and below). Initially exclusively operated by professional crews from Lufttransport Staffel 6, the Cougars now serve with all helicopter units.

Eurocopter AS 332M Super Puma

Part of the Air Force Rüstungsprogramm 1986 was the procurement of three Eurocopter AS 332M1 Super Pumas, locally known as TH86 (Transport Helicopter 86, the latter referring to the budget year). The first was delivered on 14 August 1987. Swiss experience with the Super Puma in the difficult mountainous terrain was highly satisfactory, encouraging the air force to enlarge the fleet in Rüstungsprogramm 1989 by purchasing 12 additional Super Pumas (now called TH89) to boost transport helicopter capacity. The three earlier Super Pumas were modified to TH89 standard.

Both militia ('miliz') and professional ('profi') air force crew operate the fleet of Super Pumas.

Sud-Aviation Alouette III

The Swiss Air Force took its first nine Sud-Aviation SA 316B Alouette IIIs into service in 1964. Coincidentally, that same year the second batch of 20 Alouette IIs was delivered, the helicopter type it would later entirely replace.

Two additional batches followed: in 1966 a total of 15 was delivered directly from the factory in Marignane, France, and between 1972 to 1974 a further 60 were locally produced under licence by F+W at Emmen.

Being the most numerous helicopter in service, the fleet of 84 Alouette III helicopters was for a long time the backbone of all Air Transport Squadrons, at its peak equipping Lufttransport Staffel (LT St) 1, 2, 3, 4, 5, 6 and 8.

Army XXI specifies a significant reduction in the Alouette III force, downsizing from 60 to a fleet of 35 by 1 January 2004, coincidentally exactly 10 years after the last Alouette IIs were withdrawn from use. The remaining mixed Lufttransport Staffels continue to operate out of Dübendorf, Payerne and Alpnach. A small detachment remains at Locarno. Current planning calls for the replacement with two squadrons of possibly armed helicopters.

Raytheon (Beech) King Air 350C

A single Beechcraft King Air 350C has been used for aerial survey and transport since May 1993. The aircraft wore a civil registration (HB-GII), but by early 2004 had been re-serialled as T-721. Wearing Swiss Air Force titles, it is jointly operated by the air force's Ressort Luftaufklärung (RLA) and the Flugdienst und Koordinationsstelle für Luftaufnahmen (KSL) of the Federal Office of Topography.

The aircraft can be equipped with two Leica RC-30 aerial camera systems or two Leica PAV-30 gyro-stabilised camera mounts with lenses of 15 cm, 21 cm and 30 cm focal length, plus a Trimble 4000ssi GPS receiver. Camera control is by the GIM system, which provides an interface between navigation and camera systems.

Pilatus PC-6 Turbo-Porter

The Swiss Air Force procured six PC-6-H2M Turbo-Porters in 1976, to join the 12 piston-engined Porters which had been operated since 1967. The latter underwent modifications to H2M standard (PC-6 retrofit) by the installation of PT6 turboprops, a programme completed in 1981. In 1997, the remaining fleet of Turbo-Porters underwent cockpit modifications. Today the Porter serves principally as a tactical transport for personnel and equipment, and is further utilised in scouting operations and for the dropping of paratroopers of Fallschirmaufklärer Kompanie 17 (left). In addition they can be transformed into fire-fighters. In the event of war the Porters would be incorporated into a utility aircraft squadron, Lufttransportstaffel 7. Some 16 remain in service and they are receiving a new camouflage scheme as they undergo depot-level maintenance.

Dornier Do 27

The Dornier Do 27 was the first aircraft to enter production in Germany after World War II and some seven Dornier Do 27H2s were delivered to the Swiss Air Force in 1958 (two) and 1959 (five), coming directly from the manufacturing plant at Oberpfaffenhofen, West Germany.

Of the seven aircraft procured, one crashed in 1969 and four were sold in 1981, leaving only two to remain in active service. These continue to provide sterling service and in 2004 both aircraft are the only fully military aircraft belonging to the Fachdienst Luftaufklärung, itself part of the Überwachungsgeschwader (UeG). V-601, at over 45 years of age, is the oldest aircraft currently in the Air Force inventory.

Originally planned for retirement in 1980, the Dorniers are still in service due to their unrivalled suitability as camera platforms. The high-wing layout provides 150° coverage beneath the aircraft, while the engines have been modified with an underslung exhaust pipe to avoid obstructing photography.

Learjet 35A and Falcon 50

Two Learjets were bought second-hand from Swiss Air Ambulance in 1988. One remains in service (right) and is normally based at Dübendorf as part of the Ressort Lufttransport. It is operated alongside a single Falcon 50 (below) on VIP flights, which are mostly operated out of Bern on government duties.

de Havilland Canada DHC-6 Twin Otter (above)

Alongside the King Air, a camera-equipped DHC-6-300 Twin Otter is operated exclusively by the Flugdienst und Koordinationsstelle für Luftaufnahmen (KSL) of the Federal Office of Topography, with the crew drawn from the RLA. In March 2003 the civil registration (HB-LID) was changed for a military serial as military control facilitates aircraft and equipment certification. The aircraft can be fitted with the same camera fit as the 350C, albeit with a GPS receiver Leica SR-9500. Camera control is provided by the ASCOT (Aerial Survey Control Tool).

British Aerospace Hawk T.Mk 66

As part of the Army XXI restructuring, the Hawks were retired prematurely from the advanced training role, the last mission being flown on 13 December 2002. Student pilots now progress straight from the PC-7 to the F-5F. Twenty Hawks were delivered to Emmen on 8 November 1989 and one was lost in a crash. The fleet logged 19,570 hours in total, with the low-time aircraft (U-1258) flying just 687 hours. The high-time aircraft is U-1251 with 1,376 hours. The remaining 19 aircraft are stored pending sale.

Ranger ADS-95 (left)

Operated by Drohnenabteilung 7, Ranger ADS-95 drones are used for reconnaissance, surveillance, target acquisition, target designation, electronic intelligence and radio relay. Replacing the earlier ADS-90 prototype system, four systems – each equipped with seven drones – were produced by a consortium of Oerlikon-Contraves, SF Schweizerische Unternehmung für Flugzeuge und Systeme (now RUAG), Israel Aircraft Industries and Tadiran in 1995, a project worth SFr 350 million. Powered by a 38-hp (28-kW) engine, the skid-equipped drone is catapulted in full auto-pilot mode by its hydro-pneumatic launcher, after which it can reach an altitude topping 18,000 ft (5486 m) with a mission range of up to 93 miles (150 km). Standard payload is the Multimission Optronic Stabilised Payload (MOSP) Mk III with auto-track capability providing real time, day and night information. Various other configurations with TV, FLIR, Combi TV/FLIR, and TV or FLIR with laser designator are also possible.

Dassault Mirage IIIRS

On 4 August 1965, the first of 18 Mirage IIIRS arrived in Switzerland, direct from Dassault. Another 17 aircraft were assembled at Emmen. In 1994, under the ISMA (Improved Swiss Mirage Aircraft) programme, the IIIRS fleet received locally developed canards, modern RWR, chaff and flare systems, new ejection seats, improved avionics and infra-red camera pods.

After 38 years of service, the era of the 'Amir' (Aufklärer-Mirage) and remaining trainers came to an end on 17 December 2003 during a formal farewell ceremony and flypast at Dübendorf. The 'Amir' fleet accumulated 55,000 flight hours. In its last months, the last European squadron to operate the Mirage III – Fliegerstaffel 10 – painted two aircraft in special schemes designed by Hansueli Oberholzer (who also created the 'Patrouille Suisse' scheme). These were 'Black' (illustrated) and 'White'. Flieger Staffel 10 was disbanded, and in January 2004 its parent unit – Ressort Luftaufklärung (RLA) – was renamed Fachdienst Luftaufklärung. With the Mirage's retirement, the Schweizer Luftwaffe will be without any fast-jet reconnaissance capability for the foreseeable future.

Mirage III RS
R-2110

Russian Federation

Part 1: Introduction, overall force structure and VVS direct-reporting units

Crippled by more than a decade of neglect and underfunding, the Russian air forces now stand at the beginning of a small revival. Through the large-scale withdrawal of older types the once vast fleet has been pared down to more realistic numbers, while the Russian Ministry of Defence is engaged in programmes of updating its remaining aircraft and of cautious procurement.

In December 1991 the Union of Soviet Socialist Republics collapsed, transforming 15 Soviet republics into 15 independent states. The largest of them – and still the largest country in the world – is the Russian Federation with an area of 17075200 km² (6,593,250 sq miles), representing 76 per cent of the territory of the former USSR. The country extends more than 4000 km (2,485 miles) north to south and almost 10000 km (6,214 miles) west to east. Russia's borders are the longest in the world: the land borders stretch 20322 km (12,628 miles) and the coastline for 38240 km (23,762 miles). Russia has 145 million inhabitants and its capital city is Moscow.

The Russian Federation came into being on 25 December 1991, when the Supreme Council of the Russian Soviet Federative Socialist Republic changed the name of the newly independent republic to the Russian Federation. Russia declared itself to be the successor to the USSR and took over all Soviet assets (as well as its debts). It also assumed the place of the USSR as a permanent member of the UN Security Council.

Formation of the Russian air forces

The formerly powerful armed forces of the USSR were divided between the new states. In spite of some efforts to retain for at least some time a common structure under the name of the Commonwealth of Independent States (CIS), each state began to organise its own armed forces. The first state to do so was Ukraine: on 24 August 1991, even before the formal disintegration of the USSR, the Ukrainian parliament declared that it was taking under Ukrainian control all military units stationed on its territory; the Ukrainian Ministry of Defence was established on the same day. Only four months later, in December 1991, the leaders of the CIS agreed that each republic was entitled to its own armed forces.

The Russian armed forces were not formally created until May 1992. Delaying this process was in the best interests of Russia which, under the pretext of promoting common armed forces, was moving troops to locations that would be advantageous to itself. In Soviet times, only 40 per cent of all the USSR's combat aircraft had been stationed on Russian territory. Moreover, the aircraft stationed in Russia were usually older, whereas the latest-model strike aircraft were stationed mainly in border regions. This arrangement, detrimental to an independent Russia, was later altered in both number and quality of aircraft. Russia, as the heir of the USSR, took over and brought back to its territory all air units from eastern Europe, including the very strong 16th Air Army from East Germany, the 4th Air Army from Poland, and units from Hungary and Czechoslovakia. Within the ex-Soviet states, Russia acquired all the aircraft from the Baltic States (Lithuania, Latvia and Estonia), most of the aircraft from the Caucasian republics (Georgia, Azerbaijan), and some units from Belarus (Su-24M and Tu-22M bombers). Kazakhstan retained tactical aviation assets but passed 40 Tu-95MS strategic bombers to Russia.

Sometimes, the aircraft were withdrawn to Russia in accordance with international agreements (for instance, with Belarus or Kazakhstan); other times, the aircraft took off and never came back. Regiments that followed the latter route from Azerbaijan included the 60th OShAP attack regiment from Sitalchay with Su-25s, the 82nd IAP fighter regiment from Nasosnyi with MiG-25PDSs, and the 976th BAP bomber regiment from Kurdamir with Su-24s. Withdrawal of the 882nd RAP reconnaissance regiment with MiG-25RB and Su-24MR aircraft, stationed in Dallyar in the western part of Azerbaijan, began on 9 June 1992. Only eight MiG-25s had managed to take off before a group of several hundred Azerbaijan soldiers burst onto the airfield. They blocked the airstrip and cut the aircraft's tyres, making departure impossible. In this way, Azerbaijan captured 16 Su-24s and MiG-25s, as well as one transport Il-76 that had arrived on the preceding day to take home the families of Russian military personnel.

This method of aircraft appropriation failed only in Ukraine, which took over many air units, including such valuable ones as the 184th TBAP regiment with Tu-160 strategic bombers (10 years later, Russia bought eight of them), the 182nd TBAP regiment with Tu-95MS strategic bombers (three of them were later bought by Russia), the 403rd tanker regiment with Il-78 air tankers, and a whole division of Il-76 transport aircraft. At the time of the USSR's disintegration, 114 Tu-22M medium bombers were stationed on Ukrainian territory (some were later given to Russia as part of the Black Sea Fleet), as were several hundred tactical combat aircraft. Only one relocation from Ukraine was accompanied by scandal: on 14 February 1992, pilots of six Su-24Ms of the 7th BAP regiment in Starokonstantinov flew their aircraft to Russia in protest against their incorporation into the Ukrainian armed forces. The pilots took the regimental colours with them.

The movement of so many aircraft to Russia caused many logistic and social problems. Many regiments were disbanded and their aircraft and crews dispersed among other units. New bases often did not have sufficient living quarters for the personnel. These extensive, enforced relocations of the air forces disturbed their structure: aircraft were moved not according to strategic needs, but to places that could accommodate them. Only after a few years, when the number of aircraft and units had been considerably reduced, could they be deployed in more rational ways.

As the result of these operations, in 1993 Russia controlled about two-thirds of the aviation assets of the former USSR. In total, Russia took over 70 per cent of Soviet military power, including such key elements as the monopoly over strategic nuclear weapons (small nuclear forces remaining in Belarus, Kazakhstan and Ukraine had been liquidated), space systems, basic command centres and systems, and almost the whole navy. In December 1991 the largest Soviet combat ship, the aircraft-carrier *Admiral Kuznetsov*, was transferred from Sevastopol, Ukraine, to Severomorsk, Russia.

Illustrating Russia's commitment to maintaining its place at the nuclear top table, the Tupolev Tu-160 'Blackjack' fleet has been swelled through the addition of new aircraft, plus those bought back from the Ukraine. This is the 121st TBAP flightline at Engels.

Russian air bases

Armed Forces of the Russian Federation

The Russian armed forces operate under two principles. The first is to prevent a possible aggressor from attacking Russia by means of strategic nuclear power capable of destroying such an aggressor. The present military doctrine reserves for Russia the right of first use of nuclear weapons, which applies equally to an enemy which has no such weapons (this is unlike the Soviet policy, under which the USSR guaranteed officially that it would not employ first use of nuclear power). The second principle is to hold back any regional or local aggressor by maintaining sufficient combat abilities of general-purpose forces.

Recently, some items of the Russian military budget were disclosed for the first time. According to these data, in 2004 the Russian armed forces will spend 51.9 billion roubles on research and development work, 54.9 billion roubles on purchasing new equipment, and 30.9 billion roubles on repairs to military technology. Of the budget total of 137.7 billion roubles, 11.1 billion roubles will be spent on air technology. ($US1 = 28 roubles)

Russia maintains a credible strategic bomber force, complete with tanker support. However, like US strategic forces, the 37th Air Army is modernising to provide a sophisticated conventional capability while maintaining and modernising its nuclear assets. Here an Il-78 'Midas' tanker demonstrates the refuelling technique with a Tu-95MS 'Bear-H'.

Structure of the armed forces

The Verkhovnoye Glavnoye Komandovaniye (VGK, Supreme High Command) is responsible for preparing and carrying out war campaigns and strategic operations. The Supreme Commander-in-Chief is the President of Russia, Vladimir Putin. The Ministerstvo Oborony (MO, Ministry of Defence) is charged with daily administration of the armed forces. The Minister of Defence is Sergey Ivanov (born in 1953), a former intelligence officer. The Generalnyi Shtab (GSh, General Staff) wields direct control over services of the armed forces and combat arms of the services. The Chief of General Staff and First Deputy Minister of Defence is Army General Anatoliy Kvashnin (born in 1946).

Currently, there are three services of the armed forces: Sukhoputnye Voyska (Ground Troops), Voyenno-Vozdushniye Sily (Air Forces), and Voyenno-Morskoy Flot (Navy). There are also three combat arms of the services: Raketnye Voyska Strategicheskogo Naznacheniya (Strategic Rocket Troops), Kosmicheskiye Voyska (Space Troops), and Vozdushno-Desantnye Voyska (Airlanding Troops). Additionally, there are rear services and other auxiliary formations.

The **Sukhoputnye Voyska** (SV) or Ground Troops (i.e., the army) is numerically the biggest service, with an effective strength of 321,000 officers and servicemen. The Ground Troops include motorised, armoured, rockets and artillery, and anti-aircraft troops, as well as special troops (reconnaissance, communications, electronic warfare, engineering, radiological, chemical and bacteriological protection services, transport service, and others). The forces are equipped with more than 20,000 battle tanks, 27,000 armoured vehicles, and more than 21,000 artillery pieces. The Ground Troops are organised into six Military Districts (Voyennyi Okrug, VO): Moskovskiy, Leningradskiy, Privolzhsko-Uralskiy, Severo-Kavkazskiy, Sibirskiy and Dalnievostochnyi. The Commander-in-Chief of the Ground Troops (and Deputy Minister of Defence) is Army General Nikolay Kormiltsev (born in 1946). Until 2003, the army had its own aviation assets, comprising over 2,000 aircraft and helicopters. Recently, the whole air component of the Ground Troops was moved to the command of the Air Forces (see below).

The **Voyenno-Vozdushniye Sily** (VVS) or Air Forces (literally, Military Air Forces) were created in their present form in 1998 from the merged Air Forces (VVS) and Air Defence Troops (PVO). Their present effective strength amounts to 184,600 officers and servicemen. The Air Forces have 5,100 fixed-wing aircraft and 2,000 helicopters, including long-range bombers, tactical bombers, fighter aircraft, attack aircraft, transport aircraft, and special role aircraft (reconnaissance, electronic warfare, airborne command posts, and others). After consolidation with the Air Defence

Glossary of unit abbreviations

AB – Aviatsionnaya Baza, Air Base
ABON – Aviatsionnaya Baza Osobogo Naznacheniya, Air Base for Special Purposes
ABSDRLO – Aviatsionnaya Baza Samolyotov Dalnego Radiolokatsionnogo Obnaruzheniya, Air Base of Long-range Radiolocation Detection Aircraft (airborne early warning aircraft)
ADON – Aviatsionnaya Diviziya Osobogo Naznacheniya, Air Division for Special Purposes
AG – Aviatsionnaya Gruppa, Air Group
APIB – Aviatsionnyi Polk Istrebiteley-Bombardirovshchikov, Fighter-Bomber Air Regiment
APON – Aviatsionnyi Polk Osobogo Naznacheniya, Air Regiment for Special Purposes
ARZ MO RF – Aviatsionnyi Remontnyi Zavod Ministerstva Oborony Rossiyskoy Federatsii, Aircraft Repair Plant of Ministry of Defence of Russian Federation
BKhRBLA – Baza Khraneniya i Remonta Bespilotnykh Letatelnykh Apparatov, Storage and Repair Base for Unmanned Air Vehicles
BKhUAT – Baza Khraneniya i Utilizatsii Aviatsionnoy Tekhniki, Base for Storage and Utilisation of Air Technology
BRS – Baza Rezerva Samolyotov, Aircraft Reserve Base
BRV – Baza Rezerva Vertoliotov, Helicopter Reserve Base
DPVO – Diviziya PVO, PVO Division
DShD – Desantno-Shturmovaya Diviziya, Landing-Assault Division
GLITs – Gosudarstvennyi Lyotno-Ispytatelnyi Tsentr, State Flight-Test Centre
IAP – Istrebitelnyi Aviatsionnyi Polk, Fighter Air Regiment
IBAP – Instruktorskiy Bombardirovochnyi Aviatsionnyi Polk, Instructor Bomber Air Regiment
IISAP – Issledovatelskiy Instruktorskiy Smeshannyi Aviatsionnyi Polk, Research Instructor Composite Air Regiment
IIVP – Issledovatelskiy Instruktorskiy Vertoliotnyi Polk, Research Instructor Helicopter Regiment
ITs – Ispytatelnyi Tsentr, Test Centre
IVTAE – Instruktorskaya Voyenno-Transportnaya Aviatsionnaya Eskadrilya, Instructor Military Transport Air Squadron
IVTAP – Instruktorskiy Voyenno-Transportnyi Aviatsionnyi Polk, Instructor Military Transport Air Regiment
KPVO – Korpus PVO, PVO Corps
LO VTA – Lyotnyi Otriad Voyenno-Transportnoy Aviatsii, Air Detachment of Military Transport Aviation
OAE – Otdelnaya Aviatsionnaya Eskadrilya, Independent Air Squadron
OAPSZ – Otdelnyi Aviatsionnyi Polk Samolotov-Zapravshchikov, Independent Air Regiment of Tanker Aircraft
ODRAO – Otdelnyi Dalniy Razvedyvatelnyi Aviatsionnyi Otriad, Independent Long-range Reconnaissance Air Detachment
OIAP – Otdelnyi Istrebitelnyi Aviatsionnyi Polk, Independent Fighter Air Regiment
OIT APON – Otdelnyi Ispytatelno-Trenirovochnyi APON, Independent Test-Training APON
OKIAP – Otdelnyi Korabelnyi Istrebitelnyi Aviatsionnyi Polk, Independent Shipborne Fighter Air Regiment
OKPLVE – Otdelnaya Korabelnaya Protivo-Lodochnaya Vertoliotnaya Eskadrilya, Independent Shipborne Anti-Submarine Helicopter Squadron
OKPLVP – Otdelnyi Korabelnyi Protivo-Lodochnyi Vertoliotnyi Polk, Independent Shipborne Anti-Submarine Helicopter Regiment
OKVE PSS – Otdelnaya Korabelnaya Vertoliotnaya Eskadrilya Poiskovo-Spasatelnoy Sluzhby, Independent Shipborne Helicopter Squadron of Search-Rescue Service
OMRAP – Otdelnyi Morskoy Raketonosnyi Aviatsionnyi Polk, Independent Naval Missile-carrying Air Regiment
OMSAP – Otdelnyi Morskoy Smeshannyi Aviatsionnyi Polk, Independent Naval Composite Air Regiment
OMShAE – Otdelnaya Morskaya Shturmovaya Aviatsionnaya Eskadrilya, Independent Naval Attack Air Squadron
OMShAP – Otdelnyi Morskoy Shturmovoy Aviatsionnyi Polk, Independent Naval

Troops, the Air Forces now also include anti-aircraft missile units and radio technical units. The 2003 absorption of army aviation gave the Air Forces combat, combat-transport and transport helicopters. The Commander-in-Chief of the VVS is Colonel General Vladimir Mikhailov.

The **Voyenno-Morskoy Flot** (VMF) or Navy (literally, Military Sea Fleet) presently has 171,500 officers and servicemen; the Commander-in-Chief of the VMF is Admiral of the Fleet Vladimir Kuroyedov (born in 1944). The main component of the Russian Navy is 13 strategic nuclear submarines armed with 320 ballistic missiles (as of 31 July 2003). The Navy has one aircraft-carrier (the *Admiral Kuznetsov*), seven heavy nuclear cruisers, five missile cruisers, 14 large ASW ships, 10 frigates, 35 multipurpose submarines (with nuclear and diesel propulsion) and about 600 smaller ships. Naval aviation has about 600-650 aircraft and helicopters.

The Russian Navy is organised into four fleets: Severnyi Flot (Northern Fleet), Tikhookeanskiy Flot (Pacific Fleet), Chernomorskiy Flot (Black Sea Fleet), and Baltiyskiy Flot (Baltic Fleet), as well as one Kaspiyskaya Flotilla (Caspian Sea Flotilla).

The **Raketnye Voyska Strategicheskogo Naznacheniya** (RVSN) or Strategic Rocket Troops (literally, Missile Troops of Strategic Purpose) make up less than 10 per cent of total personnel yet control 75 per cent of the nuclear intercontinental ballistic missiles (ICBM). Their objectives are the nuclear deterrence of possible aggression and the attack of enemy strategic objects with concentrated, grouped or isolated strikes. In 2002, the rank of the Strategic Rocket Troops in the armed forces' structure was reduced from a service of the armed forces to a combat arm of the service. The commander of the RVSN is Colonel General Nikolay Solovtsov.

The **Kosmicheskiye Voyska** (KV) or Space Troops were formed on 1 June 2001. Their purpose is to warn of missile attack against Russia and to undertake anti-ballistic missile defence of Moscow, as well as launching and controlling various spacecraft (military and civil). Since March 2004 the KV commander is Colonel Lieutenant Vladimir Popovkin (born 1957). The Space Troops consist of an outer space control system, an early-warning system for ballistic missile attack (nine long-range ground-based radars, 12 medium-range radars, and warning units installed in space ships), a system for the anti-ballistic missile defence of Moscow (100 missiles of long and medium range), and spaceports in Plesetsk and Baykonur.

The **Vozdushno-Desantnye Voyska** (VDV) or Airlanding Troops are destined for aeromobile operations, such as airlanding and fighting behind the front line. The VDV has about 45,000 officers and servicemen in four divisions, one brigade and several independent regiments. The commander of the VDV is Lieutenant General Aleksandr Kolmakov (born in 1955).

Other agencies

Other government organisations in Russia with military components include: the Ministerstvo Vnutrennikh Del (MVD) or Ministry of Interior Affairs, which has 649,000 personnel including 186,300 Interior Troops; the Federalnaya Sluzhba Bezopasnosti (FSB) or Federal Security Service – 66,200; the Federalnaya Pogranichnaya Sluzhba (FPS) or Federal Border Guard Service (which in 2003 was incorporated into the FSB) – 165,000; the Federalnoye Agentstvo Pravitelstvennoy Svyazi i Informatizatsi (FAPSI) or Federal Government Communications and Information Agency – 38,500; the Ministerstvo po Cherezvychaynym Situatsiyam (MChS) or Ministry of Emergency Situations – 22,100; the Gosudarstvennaya Pozharnaya Sluzhba (GPS) or State Fire Service (recently incorporated into the MChS) – 73,000; the Federalnaya Sluzhba Okhrany (FSO) or Federal Protection Service – 11,500; the Zheleznodorozhnyye Voyska (ZhV) or Railway Troops – 48,000; and others.

The Russian Ministry of Defence has 1,132,000 military personnel including 1,400 general officers, and 867,000 civil personnel (as of 1 January 2004). In total, military-orientated organisations in Russia have about 3.2 million military and 4.5 million civilian personnel.

Russian naval airpower includes a small air group for its only aircraft-carrier. Carrier-capable fixed-wing aircraft are from the Sukhoi stable, comprising the Su-25UTG, Su-33 and Su-33UB, all of which are represented here during a slow, hooks-down flypast.

Attack Air Regiment
OPLAP – Otdelnyi Protivo-Lodochnyi Aviatsionnyi Polk, Independent Anti-Submarine Air Regiment
OPLVP – Otdelnyi Protivo-Lodochnyi Vertoliotnyi Polk, Independent Anti-Submarine Helicopter Regiment
OSAE – Otdelnaya Smeshannaya Aviatsionnaya Eskadrilya, Independent Composite Air Squadron
OSAP – Otdelnyi Smeshannyi Aviatsionnyi Polk, Independent Composite Air Regiment
OTAE PSS – Otdelnaya Transportnaya Aviatsionnaya Eskadrilya Poiskovo-Spasatelnoy Sluzhby, Independent Transport Air Squadron of Search-Rescue Service
OTBVP – Otdelnyi Transportno-Boyevoy Vertoliotnyi Polk, Independent Transport-Combat Helicopter Regiment
OVDB – Otdelnaya Vozdushno-Desantnaya Brigada, Independent Airborne Brigade
OVE – Otdelnaya Vertoliotnaya Eskadrilya, Independent Helicopter Squadron
OVE REB – Otdelnaya Vertoliotnaya Eskadrilya Radio-Elektronnoy Bor'by, Independent Helicopter Squadron of Electronic Warfare
OVO – Otdelnyi Vertoliotnyi Otriad, Independent Helicopter Detachment
OVP – Otdelnyi Vertoliotnyi Polk, Independent Helicopter Regiment
OVP BU – OVP Boyevogo Upravleniya, OVP for Battle Control
OVTAE – Otdelnaya Voyenno-Transportnaya Aviatsionnaya Eskadrilya, Independent Military Transport Air Squadron
OVTAP – Otdelnyi Voyenno-Transportnyi Aviatsionnyi Polk, Independent Military Transport Air Regiment
RAP – Razvedyvatelnyi Aviatsionnyi Polk, Reconnaissance Air Regiment
RGNIITsPK – Rossiyskiy Gosudarstvennyi Nauchno-Issledovatelskiy Institut Tsentr Podgotovki Kosmonavtov, Russian State Scientific-Research Institute Centre for Cosmonauts Training
SAD – Smeshannaya Aviatsionnaya Diviziya, Composite Air Division
SAP – Smeshannyi Aviatsionnyi Polk, Composite Air Regiment
TBAD – Tiazholaya Bombardirovochnaya Aviatsionnaya Diviziya, Heavy Bomber Air Division
TBAP – Tiazholyi Bombardirovochnyi Aviatsionnyi Polk, Heavy Bomber Air Regiment
TsBPiBP – Tsentr Boyevoy Podgotovki i Boyevogo Primeneniya, Centre for Combat Training and Combat Application
TsBPiPLS – Tsentr Boyevoy Podgotovki i Pereuchivaniya Lyotnogo Sostava, Centre for Combat Training and Flight Personnel Training
TsBPiPLS AA – TsBPiPLS Armeyskoy Aviatsii, TsBPiPLS of Army Aviation
TsBPiPLSChBS – Tsentr Boyevogo Primeneniya i Pereuchivaniya Lichnogo Sostava Chastey Bespilotnykh Samolyotov, Centre for Combat Application and Training of Personnel of Unmanned Aircraft Units
TsPAT – Tsentr Pokaza Aviatsionnoy Tekhniki, Centre for Demonstration of Air Technology
TsPLI – Tsentr Podgotovki Lyotchikov-Ispytateley, Centre for Test Pilots' Training
TsSFP DA – Tsentr Spetsialnoy Fizicheskoy Podgotovki Dalney Aviatsii, Centre for Special Physical Training of Long-range Aviation
UAP – Uchebnyi Aviatsionnyi Polk, Training Air Regiment
UTs – Uchebnyi Tsentr, Training Centre
UTsBP – Uchebnyi Tsentr Boyevogo Primeneniya, Training Centre for Combat Application
UVP – Uchebnyi Vertoliotnyi Polk, Training Helicopter Regiment
VA – Vozdushnaya Armiya, Air Army
VAI – Voyennyi Aviatsionnyi Institut, Military Aviation Institute
VAII – Voyennyi Aviatsionnyi Inzhenernyi Institut, Military Aviation Engineer Institute
VDD – Vozdushno-Desantnaya Diviziya, Airborne Division
VNITs – Vozdukhoplavatelnyi Nauchno-Ispytatelnyi Centr, Aerostatic Scientific-Test Centre
VTAD – Voyenno-Transportnaya Aviatsionnaya Diviziya, Military Transport Air Division
VTAP – Voyenno-Transportnyi Aviatsionnyi Polk, Military Transport Air Regiment
ZATO MO RF – Zavod Aviatsionnogo Tekhnologicheskogo Oborudovaniya Ministerstva Oborony Rossiyskoy Federatsii, Aircraft Plant for Manufacturing Equipment of Ministry of Defense of Russian Federation

Voyenno-Vozdushnye Sily – Air Forces

Headquarters: Balashikha near Moscow

The Russian Air Forces (Voyenno-Vozdushniye Sily, VVS) were created in their present form in 1998 as the result of the consolidation of the then-VVS and part of the Air Defence Troops (Voyska Protivo-Vozdushnoy Oborony, PVO). PVO units attached to the Air Forces included surface-to-air missiles (SAM), interceptors, and radio technical units (the radar system used to monitor airspace). The remaining elements of the PVO, including a warning system against ballistic missile attack, Moscow anti-missile defence corps and a network of space control systems, were allocated to the Strategic Rocket Forces (Raketnye Voyska Strategicheskogo Naznacheniya, RVSN); in March 2001, these elements were re-assigned to the newly created Space Forces (Kosmicheskiye Voyska).

The main objective in merging the VVS and the PVO was the reduction of maintenance costs of the armed forces. Personnel numbers have been reduced 20 per cent to 180,000, air regiments are down to 70 from 100, and 37 anti-aircraft missile regiments remain. The command system for the Air Forces is based on that of the former PVO, enabling automatic operation and combat command (before consolidation, in the VVS only Long-range Aviation had an automatic control and command system).

Alterations to the Air Forces' structure are ongoing. In 2001, the 5th Independent Corps of VVS and PVO was transformed into the 5th Army of VVS and PVO. In October 2002, Moscow's District of VVS and PVO became the Special Purpose Command.

On 1 January 2003, army aviation, which had belonged to the Ground Troops, was subordinated to the Air Forces. Russian officials insist that this change had been planned for some time, but, in reality, it was the result of the shooting down in Chechnya on 19 August 2002 of an Mi-26 heavy transport helicopter and the deaths of 119 people onboard. The subsequent inquiry disclosed much negligence and many weak points in the command system of army aviation within the ground forces. There are no official indications of additional transformations, but it is known that the VVS is insisting on also taking over the Tu-22M3 bombers belonging to the Navy (about 50 aircraft). The Air Forces argue that concentration of all Tu-22Ms in one service will improve their effectiveness and reduce maintenance costs.

Army General Vladimir Mikhailov was appointed to the post of Commander-in-Chief of the VVS on 21 January 2002. Mikhailov, born on 6 October 1943, had been the commanding officer of the 4th Air Army in Rostov-on-Don (1991-1998) and deputy Commander-in-Chief of the VVS (1998-2002), and was awarded with the title of Russian Federation Hero. In October 2003, President Vladimir Putin gave permission for Mikhailov's military service to be extended by two years, until October 2005 (in Russia, active military service ends at the age of 60; the President may extend it, but only to age 65). The Deputy C-in-C of the Air Forces, and Chief of Aviation, is Lieutenant General Aleksandr Zelin (born in 1953), former commanding officer of the 4th Army of VVS and PVO in Rostov and commander of the 14th Army of VVS and PVO in Novosibirsk.

The High Command and Main Staff of the Russian Air Forces are now located in Balashikha near Moscow (the former headquarters of the PVO prior to reorganisation in 1998).

The following elements are subordinated to the Air Forces:

- direct reporting units
- the 37th Air Army of the High Supreme Command (Strategic Purpose), which oversees the long-range aviation assets of Tu-95MS and Tu-160 strategic bombers, as well as Tu-22M continental bombers
- the 61st Air Army of the High Supreme Command (Military Transport Aviation), which has 210 Il-76 transport aircraft and a dozen or so each of An-124s and An-22s; smaller transport aircraft such as the An-12 and An-26 belong to VVS and PVO armies rather than to the 61st Army
- the Special Purpose Command (until October 2002, known as the Moscow District of VVS and PVO), which is responsible for the defence of the capital
- the 4th Army of VVS and PVO in Rostov-on-Don, which bears the burden of fighting in Chechnya
- the 5th Army of VVS and PVO in Yekaterinburg at the border of Europe and Asia (an internal region), which has a relatively small force
- the 6th Army of VVS and PVO in St Petersburg, which protects the European part of Russia from the north
- the 11th Army of VVS and PVO in Khabarovsk, which operates in the Russian far east and on the Pacific coast
- the 14th Army of VVS and PVO in Novosibirsk, which has a huge zone of responsibility: 3500 km (2,175 miles) from west to east and more than 5000 km (3,105 miles) south to north; the Air Forces are concentrated in the southern part of this region near the borders with China, Mongolia and Kazakhstan.

Direct-reporting units

8th Krasnoznamionnaya Aviatsionnaya Diviziya Osobogo Naznacheniya
8th Air Division for Special Purposes awarded with Red Banner – Chkalovskiy

The division was formed on 6 October 1942 at Khodinka airfield in Moscow. In spite of several reorganisations and renamings, the division has always served the supreme command of the armed forces, as well as the state administration in wartime. (Unlike in many other states, in peacetime the President of Russia and the supreme authorities of the state do not use military aircraft; they fly on the state-owned airline Rossiya). The division is equipped with Il-80 (strategic) and Il-22 (operational) airborne command posts, as well as Tu-134 and Tu-154 executive aircraft, among other types. The division consists of two regiments and an air base, plus the 223rd LO (see below). The CO of the division is Colonel Belayev (2002).

Based at Chkalovskiy near Moscow, the 8th ADON operates the force of airborne command posts, although it rarely flies. The mainstay of the fleet is the Ilyushin Il-22, an example of which is seen here in the latest antenna configuration.

223rd Lyotnyi Otriad
223rd Air Detachment – Chkalovskiy

Services performed by military transport aviation are a major source of income for the Russian Ministry of Defence. The 223rd LO has been separated from the 8th ADON and registered as an air company. The unit operates various types of aircraft, including Il-18, Il-62, Tu-134 and Tu-154 airliners. In a similar manner, the 224th LO (heavy transports) has been separated from the 61st Air Army.

1st Gvardeyskiy Krasnogvardeyskiy ordena Lenina dvazhdy Krasnoznamennyi ordena Kutuzova Instruktorskiy Bombardirovochnyi Aviatsionnyi Polk
1st Red Guard Instructor Bomber Air Regiment of Guards awarded with Lenin Order, Red Banner (twice) and Kutuzov Order – Lebiazhye (Mikhaylovka)

The regiment was created in 1953 in Tambov from the 1st Bomber Air Regiment of High

The 8th ADON's most prized asset is the Il-80 national airborne command post.

Above: Presidential transport is handled by two Ilyushin Il-96s assigned to the state-owned airline Rossiya.

Below: A number of Tupolev Tu-134s remain in use for staff transport duties.

Right: The Beriev/Ilyushin A-50 AWACS fleet flies from Ivanovo under the auspices of the 2457th Air Base.

VVS direct-reporting units

UNIT (1ST LEVEL)	UNIT (2ND LEVEL)	BASE	INVENTORY/NOTES
8th ADON		Chkalovskiy	Il-22, Il-80
	353rd APON	Chkalovskiy	
	354th APON	Chkalovskiy	
	206th ABON	Chkalovskiy	
	223rd LO	Chkalovskiy	An-12, Il-18, Il-62, Il-76, Tu-134, Tu-154
1st IBAP		Lebiazhye (Mikhaylovka)	Su-24
764th IAP		Bolshoye Savino (Sokol)	MiG-31, MiG-25PU
2457th AB SDRLO		Ivanovo	A-50
929th GLITs		Akhtubinsk	
	1338th ITs	Chkalovskiy	
	unknown ITs	Nalchik	
	unknown ITs	Feodosiya	seaplanes
	386th OSAE	Nalchik	An-12
	VNITs	Volsk	balloons and airships
	267th TsPLI	Akhtubinsk	
4th TsBPiPLS		Lipetsk	
	968th IISAP	Lipetsk	MiG-29, Su-24, Su-25, Su-27
	4020th BRS	Lipetsk	storage base with MiG-23, MiG-27, MiG-29, MiG-31
344th TsBPiPLS AA		Torzhok	
	696th IIVP	Torzhok	Ka-50, Mi-8, Mi-24, Mi-26
	92nd IIVE	Sokol (Vladimir)	Mi-8, Mi-24
924th TsBPiPLSChBS		Yegoryevsk	unmanned air vehicles
RGNIITsPK		Zviozdnyi Gorodok	
	70th OIT APON	Chkalovskiy	Il-76K and other
185th TsBPiBP		Astrakhan	
	116th UTsBP	Astrakhan	MiG-23, MiG-29
	42nd UTsBP	Ashuluk	anti-aircraft missile systems
unknown BKhRBPLA		Tunoshna (Yaroslavl)	
2881st BRV		Totskoye-2	storage base with Mi-8, Mi-24
4215th BRS		Dmitriyevka	storage base with MiG-23, MiG-27, Su-17, Su-22, Su-24
118th OVE		Dmitriyevka (Cheben'ki)	Mi-8, Mi-6

Courses of Night Training for Bomber Pilots, as a training regiment. In the 1960s, it was armed with MiG-17 fighters. Later, it was a fighter-bomber regiment and was equipped with MiG-27s until 1992, when it converted to Su-24 bombers from the disbanded 3rd BAP. The regiment undertakes many experimental operations within the domain of bomber aviation tactics, and its crews take part in combat actions in Chechnya. The commanding officer is Colonel Sergey Shatalov (2003).

764th Istrebitelnyi Aviatsionnyi Polk

764th Fighter Air Regiment – Bolshoye Savino (Sokol) near Perm

Created in November 1952 with MiG-15 fighters, the regiment subsequently operated a succession of MiG fighter types between 1971 and 1992: MiG-17, MiG-19 and MiG-25. In 1992, the regiment equipped with MiG-31s and also operates a few MiG-25PU trainers. For the time being, this Regiment is subordinated directly to VVS High Command, but it will probably be included in the 5th Army of VVS and PVO depending on where it is stationed.

2457th Aviatsionnaya Baza Samolyotov Dalnego Radiolokatsionnogo Obnaruzheniya

2457th Air Base of Long-range Radiolocation Detection Aircraft (airborne early warning aircraft) – Ivanovo

The base was created at Ivanovo airfield after A-50 early-warning aircraft were ferried there in October 1998. It replaced the 81st VTAP, which disbanded at the same time at Ivanovo.

929th Gosudarstvennyi ordena Lenina Krasnoznamionnyi Lyotno-Ispytatelnyi Tsentr imeni V.P. Chkalova

929th State Flight-Test Centre awarded with Lenin Order and Red Banner, named after V.P. Chkalov – Akhtubinsk

This institute carries out stage B of state acceptance tests for aircraft. Stage A is undertaken at a similar civil test centre of aircraft industry at Zhukovskiy near Moscow, under which aircraft are evaluated as platforms (i.e., their flying properties are assessed). In stage B, tests are carried out to evaluate the aircraft as a weapon. Since 1920, 390 aircraft types have been tested by the 929th GLITs, 280 of which were accepted into service. Tactical combat aircraft stationed in Akhtubinsk were used in Chechnya, particularly between 1994 and 1996. Lieutenant General Yuriy Tregubenkov (2003) is in charge of the GLITs.

The centre came into being on 21 September 1920 as the Opytnyi Aerodrom (Test Aerodrome) at Khodinka airfield in Moscow. In 1922 it was renamed the Nauchno-Opytnyi Aerodrom (Scientific-Test Aerodrome) and in October 1926 it became the Nauchno-Ispytatelnyi Institut VVS (NII VVS) or the Scientific-Test Institute of the VVS. In 1932, the NII VVS was moved to a new and larger airfield, Chkalovskiy outside Moscow. In June 1944, the institute was renamed as the Gosudarstvennyi (State) NII VVS and later (date unknown) the number 8 was assigned.

Lipetsk is the Russian equivalent of Nellis, the based unit conducting operational evaluation and advanced tactics instruction for the tactical aircraft force. The resident 4th TsBPiPLS operates examples of the current tactical combat types, such as the MiG-29, MiG-31, Su-24M (left), Su-25 and the Su-27 (above).

In 1947 in Vladimirovka village (which in 1959 was renamed as Akhtubinsk), construction began of a new base for another NII VVS institute – the 6th, dealing with aircraft weapons testing. In 1960, the 8th NII VVS moved from Chkalovskiy to Akhtubinsk and absorbed the 6th NII VVS. In 1989, the institute received its current designation of the 929th GLITs. The centre has its own huge testing ground that reaches several hundred kilometres into the neighbouring state of Kazakhstan. The 267th Tsentr Podgotovki Lyotchikov-Ispytateley (Test Pilots' Training Centre) has been active since 29 October 1971 as part of the institute.

The 929th GLITs's detachment at Chkalovskiy (the 1338th ITs) carries out tests of transport aircraft, helicopters, basic trainers, parachutes, airlanding equipment, etc. Another detachment of the institute at Nalchik tests aircraft, unmanned air vehicles and weapons in mountainous conditions. Also at Nalchik is the 386th OSAE squadron that belongs to the institute.

The next branch of the 929th GLITs is the Vozdukhoplavatelnyi Nauchno-Ispytatelnyi Tsentr (VNITs) or Aerostatic Scientific-Test Centre at Vol'sk near Saratov. The VNITs was created in 1957 for testing balloons and airships. Recent work carried out in Vol'sk between 1999 and 2001 included tests of the Aerostatika-2 airship, as well as tests of USU-85M Mezhbrov'ye and UAN-400 Bars captive balloons. The commanding officer of the VNITs is Colonel Aleksandr Golovenkin (2002).

Under the terms of an agreement with Ukraine, Russia still uses the GLITs's testing ground and detachment in Feodosiya for assessing naval aircraft and helicopters. For example, in 2001 a modernised Su-30KN fighter took part in experimental firings of Kh-31A anti-ship missiles. This detachment has its origins in the Morskaya Ispytatelnaya Baza (Sea Test Base) of the NII VVS, which was created on 4 March 1932 in Sevastopol and moved to Feodosiya in 1948. For many years, Russia has planned to establish a new base on Russian territory, at Gelendzhik, but no practical steps have yet been taken.

4th Krasnoznamionnyi ordena Lenina Tsentr Boyevogo Primeneniya i Pereuchivaniya Lyotnogo Sostava imeni V.P. Chkalova

4th Centre for Combat Application and Training of Air Crew awarded with Red Banner and Lenin Order and named after V.P. Chkalov – Lipetsk

In March 1923, the 2nd Vysshaya Shkola Krasnykh Voyennykh Lyotchikov (2nd High School of Red Military Airmen) was established at Lipetsk. It was renamed many times until in 1960 it received its present designation of the 4th TsBPiPLS. Between 1925 and 1933, a German Air School was stationed in Lipetsk to train future Luftwaffe pilots (after World War I, Germany was forbidden to have an air force).

The centre is tasked with preparing air tactics as well as the combat training of pilots. It has fighter, reconnaissance, attack and bomber (tactical) departments, in addition to departments for flying control, mathematical modelling, aerodynamics and combat application. The centre includes the 968th IISAP mixed air regiment and the 4020th BRS storage base. The commanding officer of the 4th TsBPiPLS is Major General Aleksandr Kharchevskiy (2004).

Until 1998, the Savastleyka-based 148th TsBPiPLS of the PVO and the 54th IAP regiment were subordinated to the 148th Centre and equipped with Su-27, Su-30 and MiG-31 fighters. In 1998 the 148th TsBPiPLS became a detachment of the 4th TsBPiPLS, and in 2002 both it and the 54th IAP regiment were disbanded.

968th Sevastopolskiy Krasnoznamionnyi ordena Suvorova Issledovatelskiy Instruktorskiy Smeshannyi Aviatsionnyi Polk

968th Sevastopol Research Instructor Composite Air Regiment awarded with Red Banner and Suvorov Order – Lipetsk

The regiment was created as the 402nd IAP Osobogo Naznacheniya (402nd Special Purpose Fighter Air Regiment) on 25 June 1941, a few days after the German invasion of the USSR. It was equipped with MiG-3 fighters flown by the most experienced test pilots. During World War II, the regiment made 13,511 sorties and shot down 810 enemy aircraft. On 5 February 1949, the unit was renamed as the 968th IAP. It was stationed in Germany until May 1992, and upon its return to Russia was based in Lipetsk and subordinated to the 4th TsBPiPLS. Presently, the regiment is equipped with Su-27 (1st squadron), MiG-29 (2nd squadron), Su-24 (3rd squadron), and Su-25/-25T aircraft (4th squadron). The commanding officer is Aleksandr Ishchenko (2002).

344th Tsentr Boyevogo Primeneniya i Pereuchivaniya Lichnogo Sostava Armeyskoy Aviatsii

344th Centre for Combat Application and Training of Personnel of Army Aviation – Torzhok

On 1 November 1979, two months before the Soviet invasion of Afghanistan, the 344th TsBPiPLS was created from the 696th OVP and several smaller units. The 696th OVP helicopter regiment, stationed in Torzhok, was the first unit to master operation of new helicopter types, such as the Yak-24, Mi-4, Mi-6 and Mi-8. From 1947 to 1955 it had been the training regiment of airlanding gliders.

Helicopter operations in the Afghan War was the main subject of tests carried out over the next 10 years by the centre, including tactics of various combat actions, helicopter self-defence

The 344th TBPiPLS AA at Torzhok is the equipment and tactical evaluation centre for army aviation. Among its mixed inventory are the first production examples of the Ka-50, which were sent to Chechnya in late 2000 for combat evaluation.

Bell AH-1

It can be a 'Snake' or a 'Viper' or a 'Skid' or a 'Whiskey'. Soon it will be a 'Zulu'. Bell's near-immortal AH-1 Cobra family has had many names and many forms. It was born in haste during the Vietnam War and was only ever supposed to be a stop-gap aircraft. Many of its intended replacements have come and gone, while the Cobra stays fighting. Operators around the world still fly the single-engined AH-1S Cobra, while others have the twin-engined AH-1W SuperCobra to call on. The US Marine Corps has just taken its AH-1Ws to war in Iraq, for the second time – and now the Marines are looking to the latest AH-1Z, almost a new aircraft, to take their aviators well into the 21st Century.

Above: Cobra in its element – an AH-1W from HMLA-367 cruises along the southern California beach near its base at Camp Pendleton. This sprawling base, just north of San Diego, houses four of the six active-duty USMC Cobra squadrons, the training unit and a reserve squadron.

Right: Now fully replaced by the Apache in US Army service, the single-engined Cobra continues to play an important part in the operations of several air arms around the world. This view of an AH-1F highlights the flat-pane canopy which was introduced by the Production AH-1S (Step 1, AH-1P) as a means of reducing the helicopter's glint signature.

The grunt's friend – Marine Corps Cobras are an extension of the ground forces, and operate very closely with them to secure the flanks and to scout ahead of the lead elements.

Bell's AH-1 Cobra was born in war. Like so many of the best aircraft designs, it was conceived to answer an urgent operational requirement. It was also the first of a whole new class of aircraft. The AH-1 was the first true combat helicopter, the first real gunship. Today, Cobras remain in service in the United States, Bahrain, Pakistan, Iran, Israel, Japan, Jordan, Taiwan, Thailand and Turkey. With some of these operators it is coming close to the end of its operational career, but with others the AH-1 is about to embark on a whole new lease of life. The US Marine Corps is gearing up to take its next-generation AH-1Z SuperCobras into service later this decade. The AH-1Z re-invents the Cobra design and is sure to find a home with other export users. The fact that the Marines' 180 AH-1Zs are being rebuilt from their 20-year old fleet of AH-1W warhorses (some of which, in turn, were converted from even older AH-1Ts) proves the soundness of the Cobra's design. It also underlines Bell's claim to have consistently delivered the most affordable and effective package of attack helicopter capability available anywhere.

The Cobra has never been a perfect aircraft. With each new variant the problems of the preceding versions have been tackled and solved, to some degree – but the Cobra has still been a clever compromise. The current benchmark service version, the AH-1W SuperCobra, is light years away from the original AH-1G – although still very recognisable as the same basic aircraft.

The early years

The pre-history of the AH-1 begins in Korea where helicopters found themselves truly at war for the very first time. Korea was not really the birthplace of the armed helicopter, however. There are stories of how US Army OH-13s (Bell Model 47 Sioux) were rigged to carry weapons on the front-line – some Army Aviation legends talk about bazookas strapped to the skids of these fragile machines – but there is no hard documentary evidence that this ever happened. Most helicopter historians agree that helicopters began to shoot back at the enemy during the French conflicts in Algeria that flared into life from 1954 onwards. There, the French forces did arm their Bell 47Gs and H-21C Shawnees, fitting them with machine-guns and 20-mm cannon for use against the rebel forces of the Armée de Libération Nationale, dug into mountain strongholds.

Armed helicopters in the 1950s were not combat helicopters in the modern sense – they had weapons for self-defence. They did show that the helicopter could be a major battlefield asset and, in the United States, work progressed on arming Army helicopters. Early work with the OH-13 was led by Colonel Jay Vanderpool, who championed the Aerial Combat Reconnaissance platoon concept (known to some as 'Vanderpool's Fools'). The OH-13 was always too small to be of practical use, but the ACR philosophy was tested with Sikorsky H-19s, H-34s and Vertol H-21s. By the early 1960s, H-34s had even fired Bullpup air-to-surface missiles and the birth of the attack helicopter was just around the corner.

Hand-in-hand with the development of the Bell AH-1 is the story of Bell's UH-1 – 'grandaddy' of the modern helicopter (and helicopter industry), engineering marvel, Vietnam icon and still one of the most numerically important helicopter types in service today. Bell's XH-40 prototype flew on 22 October 1956 and it entered production in 1959 as the HU-1A for the US Army. It was the 'HU' designation that spawned the eternal 'Huey' nickname, although

the re-alignment of US service designations in 1962 transformed it into the more familiar UH-1. The UH-1 made the theory of air mobile warfare practical. The new tactics for 'air cavalry' called for US forces to be highly mobile across a wide area. They would not stand and fight pitched battles, and they would not stay and hold static positions. Instead, troops would range across the countryside, carried by fleets of Hueys, to fight the enemy at times and places of their own choosing. They would win with overwhelming firepower and then fly out again to secure areas. That, at least, was the plan.

It immediately became clear that the unarmed UH-1 troop transports – the 'Slicks' – were not able to make unopposed troop drops in the landing zones (LZs). Some heavy firepower was needed to clear the Viet Cong and NVA enemy out of the way first. By 1962 a handful of armed HU-1As (soon to be UH-1As) were escorting H-21 (CH-21) troop transports in and out of the LZs (strict rules of engagement at the time meant that the gunships could not fire until fired upon, which did little for their combat effectiveness). By November 1962 UH-1B gunships had arrived in Vietnam, factory-fitted with four 7.62-mm M60 machine-guns mounted on pylons outboard of the main cabin. The US was sliding deeper and deeper into the pages of history in Vietnam, and the massive expansion of the American military presence there ushered in a whole new era of war from the air. Helicopters were the linchpin of US Army tactics and the protection of those helicopters became a new and vital role.

Birth of the Cobra

By December 1962 Bell had stepped in with a company-funded venture to provide a purpose-built gunship to fight in Vietnam. The role of this new helicopter would be to protect the troopships, but also to wield a fully-fledged combat capability of its own. Strangely, Bell's first design was built around the Model 47 (albeit a much modified one) leading to the small and sleek Model 207 Sioux Scout that first took to the air in July 1963. As the Sioux Scout was being put through its paces, Army Aviation was still refining its concept of helicopter operations. The Army was not clear on what its next steps needed to be, and it was certainly not clear that there was any requirement (or any budget) to support the kind of dedicated armed helicopter that Bell was proposing.

A glance back at the Sioux Scout design shows how prescient it was. All the key features of a modern combat helicopter were there – a tandem cockpit, stub wings for weapons and a chin-mounted gun turret. Beginning in January 1964, the Sioux Scout was evaluated by the 11th Air Assault Division, and it shone. The Army was impressed by what it saw but equally it believed that the Sioux Scout was too small, too underpowered, too unsophisticated and too fragile to be of real practical use. Unfortunately for the Army, its solution to this 'problem' was to launch the Advanced Aerial Fire Support System (AAFSS) competition.

It was the AAFSS requirement that would give birth to the infamous Lockheed AH-56 Cheyenne – a heavyweight battlefield helicopter that would prove to be over-ambitious, over-complex and massively over-budget before being cancelled in 1972. The 10-year Cheyenne programme was a splendid failure. It developed some fascinating technology and demonstrated some impressive performance, but it was never made to work as a functional combat helicopter. The Cheyenne experience did serve to underline the one unbreakable rule of the combat helicopter – survival on the battlefield would be ensured only by the right mix of speed, agility and weapons.

Meanwhile, despite the Army's preference for the AAFSS (for which Bell Helicopter was not selected to compete), Bell had the vision and perseverance to stick with its own smaller, lighter gunship ideas. In January 1956, it stumped up $1 million to proceed with its next-generation Model 209 design. It would prove to be a very wise investment.

Taking the proven elements of the UH-1 (transmission, rotor system, T53 turboshaft) and mating them with the design philosophy of the Sioux Scout, Bell's engineers (headed by J. R. Duppstadt) produced a design classic. The slender, sharp-nosed Model 209 looked every inch a thoroughbred – an aircraft built for fighting. In Vietnam, events were also conspiring in the Model 209's favour. Vietnamese attacks on US forces were increasing, and by 1965 the US response had escalated into the Rolling Thunder bombing campaign. By the end of June 1965 there were 50,000 US

In 2001 the US Army retired the AH-1F from service, although numbers remain available for onward sale to export customers or use by other government agencies such as the Forest Service. Conversion of active-duty units to the sophisticated AH-64 Apache and the OH-58D Kiowa Warrior took over 10 years, with the last Cobras leaving Europe in 1996 and Korea in 1997. Cobras saw out their last days of service with the National Guard. In 1990 the Army still flew over 1,000 Cobras, but by 2000 the force was down to 346 (many of which were unflyable), operating with Army National Guard units like California's 1-18 Cav (below). Under the US Army's force modernisation roadmap, most of the Cobra units were scheduled for eventual conversion to the AH-64 Apache and (now cancelled) RAH-66 Comanche, but in the interim converted to the OH-58A Kiowa, itself slated for retirement by October 2004. A massed flypast over New York was planned for the AH-1F's retirement at the end of September 2001, but the events of the 11th of the month put an end to that.

This view of an AH-1F highlights the M197 cannon which was introduced by the Up-Gun AH-1S. This aircraft was allocated to 1-7 Cav at Fort Hood, one of the last active-duty squadrons. The final active-duty user of the AH-1F was the 1st Battalion, 25th Infantry Division, based at Wheeler Army Airfield, Hawaii, which retired its last aircraft on 15 March 1999. When the AH-1F was retired from Army National Guard service in 2001, all survivors were gathered together at Fort Drum, New York, for storage in the first of a three-phase retirement plan. In the second phase aircraft were released for export sale, use by other government agencies (including the USMC which took some for spares recovery), or donation to museums and schools. The final phase covers the destruction of remaining aircraft at the end of FY04.

ground troops in Vietnam. 1965 was also supposed to be the deadline for AAFSS selection, but the programme was mired in technical difficulties and political in-fighting. Suddenly, the US Army needed an interim gunship to carry the load in Vietnam and it asked five companies to provide a rapid solution. Submissions came in for armed variants of the Boeing-Vertol CH-47A, Kaman UH-2, Piasecki 16H Pathfinder, Sikorsky S-61 and the Bell 209. While the others were noble attempts at delivering a 'fast-track' capability, none of these offers came near the level of capability promised by the Model 209.

On 3 September 1965 Bell rolled out its prototype Model 209. Four days later, on 7 September, it made its first flight. It had taken just eight months from go-ahead to maiden flight. The Army was not persuaded out-of-hand. Bell still had to face up to an evaluation of all the rival helicopters, but the 209 waltzed through the competitive fly-off held in November-December. By April 1966 the US Army had signed the first production contract for 110 aircraft.

The first AH-1G HueyCobras were delivered in June 1967. Originally, the Army designated the aircraft as the UH-1H, seeing it as an offshoot of the existing Huey family. However the 'A' for attack designation was soon adopted, and when the improved UH-1D became the UH-1H, the HueyCobra became the AH-1G. Between 1967 and 1973 Bell built 1,116 AH-1Gs for the US Army. The aircraft gave sterling service in Vietnam, where they had a rapid baptism of fire and chalked up something like one million operational flying hours. The basic AH-1G weapons fit featured an undernose Emerson M28 turret fitted with an M134 7.62-mm Minigun and an M129 40-mm grenade launcher. The stub wings could carry additional 7.62-mm gun pods and 2.75-in rocket pods. The M28 turret could also be fitted with twin M134s or twin M129s.

Cobras for the Corps

As it turned out, the real future of the Cobra would not remain in US Army hands. Army Cobras would stay in service with regular aviation units until 1999 – and even longer with the Army National Guard – while equivalent aircraft still fly with several export operators. However, it was the twin-engined, upgunned SeaCobras fielded by the Marine Corps that would fully exploit the potential of the AH-1 design. It is Marine Cobras that make up today's US front-line AH-1 fleet, and it is their new aircraft that will be in service for many years to come.

When the US Marine Corps came to assess the AH-1 in 1967 it recognised that the helicopter would be much more effective and reliable with two engines. This was an essential requirement for over-water operations. The Corps also saw that the basic gun armament fit was too light and lacked range. As thoughts turned to a twin-engined Cobra there was an obvious solution already in place. Pratt & Whitney Canada had developed its PT6T Twin Pac configuration, under a jointly funded US/Canada programme for the Model 212 Twin Huey (UH-1N). It was a logical step to swap out the AH-1G's single Lycoming T53L-13 and replace it with the PT6T-3. However, the Marines' ambitions were not matched by their wallets. The Corps did not have the funding to adopt and support an entirely new engine, so it was forced to retain the T53 in its single-engined AH-1Js.

Like the HueyCobra before it, the navalised AH-1J SeaCobra was thrown in at the deep end in Vietnam. In 1968 the Tet Offensive took US forces by complete surprise and the requirement for the AH-1J (set at a modest 28 aircraft in the FY68 budget) soared. More importantly, it was decided to dump the compromise T53 configuration and go for the full-spec twin-engined layout. The new AH-1J was also armed with the much more effective three-barrelled M197 cannon. On 10 October 1969 the first SeaCobra was rolled out at Fort Worth. By early 1971 the first aircraft were undergoing combat evaluation in Vietnam.

At this stage it was becoming very clear that the AH-56 Cheyenne was a failed project and that there would be no immediate replacement for the 'interim' AH-1. The Cheyenne programme had essentially died in 1969 and survived only as a research project, but with no future. Bell saw this as a good time to introduce a developed version of the AH-1, the twin-engined Model 309 KingCobra. This was a scaled up AH-1 with a more advanced set of mission systems than the AH-1G (drawing heavily on equipment developed for the Cheyenne) that would be a true day/night-capable combat helicopter. The KingCobra would prove to be a dead-end (two prototypes were built), but it did point to what the basic AH-1 design was capable of, and underlined the superiority of the twin-engined layout.

All Cobras up to this stage had been daytime-only aircraft, with very limited navigational aids and simple optical sights. American attention would soon turn away from Vietnam and refocus on the Soviet threat in Europe, where it was becoming clear that the ability to fight at night would be crucial for NATO forces. This would be especially true for combat helicopters, which were now taking on a whole new role with a whole new weapon. The Cheyenne and the KingCobra were designed to use the Hughes BGM-71 TOW (Tube-launched, Optically-tracked, Wire-guided) missile, now a product of Raytheon Missile Systems. For the first time, TOW gave helicopters the ability to kill armoured vehicles, up to and including main battle tanks, to a distance of 3750 m (4,100 yards). The French had demonstrated the first air-to-surface missiles on helicopters many years previously with SS.10- and SS.11-armed Alouettes, but the TOW was a next-generation weapon that was more effective – and more practical – than anything that had come before it. As the Cheyenne and KingCobra were smothered by the post-Vietnam draw-down, the AH-1 once again stepped up to take on the job of the US military's *de facto* battlefield attack helicopter.

Work on integrating the TOW with a handful of UH-1B test beds began in 1972, to pave the way for its adoption on dedicated combat helicopters. When North Vietnamese forces slashed across the DMZ and into the South's Quang Tri province in March 1972, they fielded heavy armour for the first time in the war. The NVA was operating out-dated but still effective Soviet vehicles, including T-34 and T-54 tanks, plus the PT-76 light tank. Six days after the order to deploy, on 14 April, TOW-armed Hueys fitted with the experimental XM26 launch system arrived in Vietnam. In the right place at the right time, they played an important part in pushing back the NVA forces, particularly in the fighting around Kontum. The first combat use of a TOW fired from an Army helicopter came on 2 May 1972 (this was also the first ever launch of an American-made guided missile by American troops).

Missile-armed HueyCobras did not fight in Vietnam, but by 1973 the Improved Cobra Armament Program (ICAP) was delivering TOW-capable AH-1Gs to the US Army. By January 1974 Bell was delivering the follow-on AH-1Q, designed from the outset to carry eight TOWs. It had the (now production standard) Hughes M65 optical sight in a nose turret, known as the telescopic sight unit (TSU), and uprated 1,800-hp (1343-kW) Lycoming (now AlliedSignal) T53-L-703 engines. All service AH-1Qs were rebuilt from existing AH-1Gs. This was the first purpose-built TOW-armed variant for the US Army but, despite its better engine, the AH-1Q was still lacking in performance and aircraft rarely carried the maximum load of eight missiles (anything between two and six being common).

AAH competition

The AH-1's shortcomings were recognised and by 1972 the US Army had already begun the search for another new attack helicopter. Out of the ashes of the Cheyenne programme (finally cancelled in August 1972) came the new Advanced Attack Helicopter (AAH) requirement. The AAH competition would prove to be almost as controversial as the AAFSS, but it did finally deliver the Hughes (later McDonnell Douglas, now Boeing) AH-64A Apache. The operational driver behind the AAH (and ultimately the Apache) was the ability to fight at night on the threat-heavy European battlefield of the future. At the same time, the US Army had a substantial fleet of well over 600 valuable AH-1s (becoming AH-1Qs) that could not just be thrown away. In 1976 a new modernisation effort for the AH-1Q was launched, the ECAP (Enhanced Cobra Armament Program). The product of the ECAP was the AH-1S, a fully TOW-armed aircraft with a more powerful engine, an uprated transmission and a higher gross operating weight.

There were several standards of AH-1S as new modifications were made to progressive production blocks of aircraft. The initial Production standard (known as the Step 1) introduced the trademark seven-pane flat-plate canopy, improved main rotor blades, AN/APR-39 radar warning system and a curved, semi-circular exhaust suppressor. The first examples entered service with the 82nd Airborne in August 1977.

Over-water safety was the prime driver behind the USMC's requirement for a twin-engined Cobra, which initially resulted in the Pratt & Whitney Canada PT6T (T400)-powered AH-1J SeaCobra. The AH-1T had an uprated T400, whereas the current AH-1W has two General Electric T700s.

Marine helicopter units routinely practise operations from austere locations.

Four Middle East countries fly the Cobra, including Jordan, which received 24 AH-1Ss in 1985 (plus a top-up batch in 2000/01). The aircraft fly with two squadrons based at Al Matar, near the capital Amman. The other Cobra users in the region are Bahrain, Iran and Israel.

Thailand had ambitious plans to establish an attack helicopter force equipped with the Cobra. Four AH-1Fs were acquired in 1990 to form an evaluation cadre, around which it was planned to create a force of at least squadron size. At various points in the 1990s further procurement was planned of either new-build AH-1Ws or ex-US Army AH-1Fs. As Thailand's 'tiger' economy suffered dramatically in the late 1990s, and the defence budget was prioritised towards more pressing needs, so plans for any further attack helicopters expired. One of the quartet was lost in a crash, but the remaining three continue to serve with the army aviation centre at Lop Buri.

Beginning in 1978 the Production AH-1S standard gave way to the Up-Gun AH-1S (AH-1S Step 2). On these aircraft, the outdated M28 turret was replaced by an M197 20-mm cannon mounted in the General Electric-designed M97A1 Universal Turret that had already been adopted by Marine Corps Cobras. The M128 helmet sight system (HSS) was another feature of these aircraft. Linked to the Universal Turret, it allowed crew-members to point the gun at a target just by looking at it.

The final standard was the 'Modernized AH-1S' (AH-1S Step 3). Produced between 1979 and 1981, these aircraft were fitted with the M143 air data sensor (located on the forward fuselage, starboard side) that fed information to the new M26 fire control system. Rocket-firing capability, not available to the previous AH-1S models, was returned on the Step 3 aircraft with the inclusion of the M147 rocket management system (RMS). A laser rangefinder was added to the M65 sight and an AN/AAS-32 laser tracker was fitted in a bullet fairing in front of the rotor housing. The crew was also provided with the improved M136 HSS.

Better self-defence measures included the AN/ALQ-144 infra-red jammer (mounted above the rear engine housing), along with a new upwards-angled tubular exhaust shroud. The cockpit was fitted out with the M76 HUD, a new Doppler navigation system, the APX-100 IFF and KY-58 secure radios. NVG-compatible instruments were also added for the first time. These aircraft were much easier and safer to fly at low-level (thanks largely to the HUD), thus making them more survivable. The layout of the cockpit was rearranged to make key instruments and switches more accessible (although the Cobra cockpit was far from being a masterpiece of man-machine interface design). The AH-1S upgrade programme was completed by 1985. A total of 530 aircraft was fielded (387 AH-1G upgrades and 143 new-build).

Further enhancements were made to several hundred late-model aircraft, including the integration of the Cobra Night Vision (C-Nite) system that turned the AH-1 into a night-capable aircraft for the first time, and the improved TOW 2 missile. In 1988 the US Army redesignated all of its Cobra variants – leaving a trail of confusion that has inhabited the AH-1 story ever since. The Step 1 (Production standard) AH-1S became the AH-1P. The Step 2 (Up-Gun) AH-1S became the AH-1E. The Step 3 (Modernized) AH-1S became the AH-1F.

From the late 1980s onwards, the spread of AH-64s into Army Aviation saw the withdrawal of the AH-1 from front-line units and into the hands of the Army National Guard (ArNG). The AH-1s that remained in use with regular units shifted into scout/gun teams flying alongside OH-58C Kiowas, often operating for Cavalry units. Both types were largely replaced in that role by the OH-58D(I) Kiowa Warrior. The last AH-1Fs in the regular US Army were retired in March 1999. On 15 March, eight AH-1Fs of the 1st Battalion, 25th Aviation Regiment (Attack), 25th Infantry Division (Light) launched for a ceremonial flypast from their base at Wheeler Army Airfield, Hawaii, to close the book on 32 years of front-line Cobra flying. The unit would become the last Army Aviation battalion to transition to the OH-58D(I), in May 2000.

On 4 April 2000 the Army released its Aviation Force Modernization Plan, which specified that the AH-1F would be retired by 30 September 2000. This date was subsequently extended to 16 December, and then again to 30 September 2001, the Army coming under pressure from Adjutant Generals in the affected state ArNGs who were concerned about the readiness of their forces. On 3 May 2001 the US Army Aviation Logistics School graduated its last class of Cobra armament technicians, and on 30 September the AH-1F was formally retired. A number were passed on to existing overseas customers and the Forest Service.

Marine developments

The Cobra story is certainly not over for the Marine Corps. The development of the Cobra with the Marines has followed a much different path. In the mid-1970s the Marines jumped from the twin-engined AH-1J to the AH-1T Improved SeaCobra. This version had an extended fuselage and tail boom, and could operate at even higher weights than the J model (a sizeable jump to a 14,000-lb/6350-kg maximum take off weight). The new weight limits meant the AH-1T was also far better able to operate with a full TOW load. In 1979 another step forward came when Bell began fitting General Electric T700 turboshafts to an AH-1T. A demonstrator aircraft flew in April 1980 and represented a highly attractive proposal because it offered powerplant commonality with the bulk of the modern twin-engined US helicopter fleet (AH-64, UH-60, SH-60). The improved Cobra was offered to the Marines as the AH-1T Plus (AH-1T+), and by 1982 attrition in the Marines' Cobra fleet made the purchase of new aircraft almost inevitable.

Bell made a cut-price offer to build the first batch of 44 AH-1T Plus aircraft at $15.2 million dollars each (unit cost), equivalent to a 'flyaway' price of $7.1 million. A deal was done and the new Cobras were ordered for delivery from March 1986 onwards. A definitive T700-GE-401-powered prototype flew on 16 November 1983 and the new up-engined and fully marinised aircraft became the AH-1W.

The AH-1W SuperCobra, often referred to as the Whiskey Cobra, is the most powerful, well-armed and best equipped evolution of the Cobra family to enter service. The first aircraft began operations with the US Marine Corps in 1986 – at around the same time that the AH-1S started to give way to the Apache in the US Army's inventory (the first Apaches were deployed to Europe in 1987). The USMC acquired a mix of converted (from AH-1T) and new-build AH-1W airframes.

AH-1Ws are fitted with the Night Targeting System (NTS) – a modification of the original M65 – that was developed from 1987 by Kollsman, in the US, from an Israeli design produced by IAI's Tamam Division. The NTS comprises a FLIR and CCD TV sensors (with a video recording function), a laser designator/rangefinder and automatic target-tracking capability. Its introduction gave the Marines a significant leap forward in capability over the outdated M65. As part of the NTS integration effort, the AH-1W received the canopy cockpit modification (CCM) that replaced all the existing canopy, nose section and the front cockpit (co-pilot/gunner) instrument panel. Along with the NTS came a new tactical navigation system (TNS) that, from 1996, incorporated the AN/ASN-163 GPS/INS.

For improved self-defence the AH-1W received the ECP-1674 electronic warfare suite. This integrated the AN/AAR-47 missile warning system with the AN/ALE-39 countermeasures dispenser. The AAR-47 is an optical sensor that gives the crew an aural warning when it detects the plume from a surface-to-air missile launch. The launch direction is shown on a cockpit display and chaff/flares are automatically deployed by the ALE-39. During Operation Iraqi Freedom Cobra crews found that the AAR-47 was responsive enough to alert them to rocket-propelled grenades (RPGs) being fired at their aircraft. The AH-1W is also fitted with the AN/AVR-2 laser warning receiver that detects and warns if any pulsed laser energy (such as laser rangefinders or weapons designators) are being aimed at the aircraft. The AVR-2 is even sensitive enough to detect office laser pointers. Finally, the AN/APR-39A(V)2 radar warning receiver monitors ground and airborne radars and alerts the crew if any become threatening by tracking or locking on to the aircraft.

Cobra weapons

The basic weapons set of all AH-1s in service today comprises the M197 20-mm cannon, the BGM-71 TOW and 2.75-in (70-mm) rockets – typically in LAU-61 or LAU-68 pods. This is the weapons fit first fielded on the AH-1J and Up-Gun AH-1S. The M197 cannon can carry a maximum load of 750 20-mm shells and has a maximum rate of fire of 650 rounds per minute. The gun is normally fired in nine-second, 100-round bursts. For years now, the M197 has been acknowledged to be prone to jamming and this problem came to the fore again during the 2003 war in Iraq. The gun still remains a vital weapon and the gunners

Japan is the only country so far to build the Cobra under licence, Fuji Heavy Industries being responsible for building 89 AH-1S Step 3 aircraft (similar to the US Army's AH-1F) following the delivery of two Step 2 (AH-1E) aircraft from the US. The later production machines were fitted with the C-Nite system, adding night capability. Japan's Cobras serve in five front-line attack helicopter units, each of two squadrons, plus a single training squadron. The front-line units also operate OH-6D/OH-1 scouts. Production was originally to have included a number for attrition, but no Cobras were lost during the first years of service and the eight-aircraft batch was cancelled. In February 2004 the JGSDF suffered its first AH-1S losses when two aircraft collided during a training mission. Thankfully all four crew survived.

air-to-air missile, in the shape of the AIM-9 Sidewinder. The Marines qualified the AGM-122 Sidearm, an anti-radiation missile based on the Sidewinder airframe (and using the common LAU-7 launcher), for their Cobras too.

While the AIM-9 was a useful capability to possess, it was a compromise weapon that is too large and too unwieldy to be used effectively on the AH-1W. While the Corps has always been mindful of the air-to-air role for its attack helicopters, it has been unable to spend much money on developing more suitable weapons. The AH-1W has never adopted the air-to-air Stinger (ATAS), as found on US Army Kiowa Warriors, and recent US Army Apache trials with the British-built Helstreak (Starstreak) AAM have passed the AH-1W by. The AIM-9 remains an option for the AH-1W (the AIM-122 is now out of the inventory) and Marine pilots still do some air-to-air training, but the Sidewinder is no longer seen as a practical weapon. One big disincentive for carrying Sidewinder is that it precludes the loading of any other stores on the remaining wing station.

Over the years Marine aviators have gazed across at the Army's Apaches and wished they could afford them. Several attempts have been made to introduce the Apache into USMC service but, always the branch of the US military that has had to do the most with the least, the Marines have never been able to fund an Apache purchase. Paying for a marinised Apache variant would not come cheap and the USMC was likely to be the sole customer. This is why the AH-1W (and soon the AH-1Z) remains the core combat helicopter for the Corps. While this has often been viewed as a second-best situation for the USMC, much of the Apache's gloss has worn off in recent years, and today the Marines actually find themselves with what the numbers show to be the most survivable combat helicopter in US service.

Although the IDF/AF now operates a sizeable fleet of Petens (AH-64s), the Zefa (AH-1) remains an important part of the Israeli military machine, seeing widespread action throughout its career. In recent times it has been spotted regularly during news footage of attacks on targets within the Palestine Authority, and some analysts suggest that it is preferred over the Peten for attacks where pinpoint accuracy is essential, such as missile strikes against individual houses where terrorist subjects are believed to be hiding and where collateral damage has to be kept to the absolute minimum. The use of Spike has been reported in such attacks, although there is no official confirmation.

have learned how to manipulate the ammunition feed, keeping pressure on and off the ammunition belt, to minimise the risk of jams.

The AH-1W went a step further than any previous AH-1 variant, adding the laser-guided AGM-114 Hellfire missile that was developed originally for the AH-64A Apache. In order to provide some air cover for troops 'on the beach', the AH-1W also became the first US helicopter to field an

Hellfire for the Cobra

What the Cobra has always lacked was hitting power, hence the Hellfire for the AH-1W. The Marines knew that if their SuperCobras were going to have to fight their way off the beaches against serious opposition, they would need some serious additional firepower. The TOW's 3750-m (12,300-ft) range is insufficient to out-reach current battlefield air defence systems. Neither can its warhead defeat the most modern tank armour. These deficiencies were already recognised during the mid-1970s when (what was then) Rockwell International was contracted to develop the

Right: A 'Southern Squadron' Zefa flies at typically low altitude during a training sortie. Israeli Cobras are very well protected, this aircraft being fitted with radar/missile warning receivers, infra-red countermeasures turret and large flare dispensers.

AGM-114 Hellfire. This new laser-guided missile was far more flexible than the TOW (offering lock-on before launch or lock on after launch guidance modes), had a warhead capable of wiping out anything in its path and came with an effective range of more than 8000 m (26,247 ft). Hellfires can be fired from either seat in the AH-1W but the controls for the laser are in the front cockpit only.

When the Marines went to war in Iraq in 2003 they brought the latest version of the laser-guided Hellfire II (AGM-114K), the newly-developed blast/fragmentation variant (AGM-114M) and the (hitherto) completely unknown AGM-114N missile with its thermobaric warhead. Thermobaric weapons (from the Greek *therme*, meaning heat and *baros*, meaning pressure) use intense heat and blast pressure to destroy targets. They are often equated to fuel air explosives (FAE) but modern thermobaric (TB) explosives have a far more concentrated and direct effect. The Marine Corps has led the development of TB weapons in the US, with an eye on the considerable range of operational systems already tested and fielded in Russia (and elsewhere).

In late 2001, spurred on by the terrorist atrocities of 11 September, the Indian Head weapons facility began serious work on new TB weapons, drawing on its existing development efforts with the BLU-118/B TB bomb for the USAF. The Marines wanted an equivalent weapon to use for fighting in caves and mountains. It was decided to adopt the Hellfire, and replace the PBX-109 explosive fill of the AGM-114N with a new TB compound, PBX N-12. By mid-2002 the development team was getting ready for the first live fire tests from a helicopter. The aim was to produce a weapon capable of completely destroying, for example, six rooms inside a building over a penetration range of 60 ft (18.3 m). Two AGM-114Ns had been successfully fired before the Iraqi conflict and a development batch of 60 missiles was rebuilt from existing AGM-114Ks.

During Operation Iraqi Freedom Marine Cobras typically flew with a mix of Hellfires, TOWs and rockets, which meant that they were always operating at or beyond their maximum weight limits. One pilot who fought in Iraq recalled, "the Cobra has a NATOPS limit of 14,750 lb [6691 kg] all-up weight but we always flew above that – because it was war. You'd be carrying four TOWs, four Hellfires, seven HE rockets and seven flechette rockets, plus 300 rounds for the gun and you'd be operating at over 15,000 lb [6804 kg], always." The AH-1W can carry a maximum ordnance load of 1,500 lb (680 kg).

Israeli Zefa

In Israel, the AH-1 has a special role to play within the ranks of the Israel Defense Force/Air Force (IDF/AF). Israel has been a Cobra operator since the late 1980s and currently operates an uncertain number of around 50 AH-1S (AH-1F) aircraft, known locally as Zefas (vipers). When the Cobras entered service they formed part of a high/low mix together with the IDF/AF's TOW-armed Hughes 500MD Defenders. The AH-1s were Israel's sophisticated attack helicopters while the 500MDs were the affordable force multipliers. As the Defenders were retired from use and the AH-64A began to enter IDF service in increasing numbers, the status of the Cobra force was reversed, but they remained invaluable.

Operated by Nos 160 and 161 Sqns, Israel's Cobras were armed with a range of TOW variants including the BGM-71A TOW (P'teel, chord), BGM-71C I-TOW (Peeg'yon, dagger), BGM-71D TOW 2 (Pit'yon, bait) and the subse-

Left: Wearing the badge of the 'Northern Squadron', an IDF/AF AH-1F fires a Spike-LR missile during tests for the Polish army. This weapon uses fibre-optics relaying seeker imagery directly back to the Cobra, resulting in very high accuracy. It is believed that the Spike has been used operationally by the IDF/AF in pinpoint attacks against urban targets.

Above and above left: A Zefa fires a BGM-71 TOW at a target during a firepower demonstration. The TOW uses a wire link between the missile and the aircraft, through which guidance commands are relayed.

The IDF/AF's two Zefa units regularly practise operating from deployed locations, like these 'Northern Squadron' AH-1Fs.

Pakistan is one of the countries to have used the Cobra in action. In March 2004 the gunships supported a large army force as it moved into mountainous areas along the border with Afghanistan to surround a force of al-Qaeda terrorists. For many years Pakistan was embargoed by the US owing to the country's nuclear policy. However, Pakistan's response in the wake of the 9/11 attacks, and its improving relations with India, means that is viewed in a new light by the US. The result is the resumption of arms deliveries, and a large batch (possibly up to 40) surplus AH-1Fs are expected to be delivered.

quent TOW 2A and TOW 2B variants. Israel also developed its own second-generation anti-tank guided missiles, including the TAAS (later IMI) MAPATS, which seems to have been based on the TOW, but is improved and refined.

During the 1990s the wraps came off a new fibre optic-guided missile system developed by Rafael. The Spike series of weapons was developed as the NT family (Neged Tank, anti-tank) and three variants were produced under the codenames Gill, Spike and Dandy (NT-G, NT-S, NT-D). In 2002 Rafael gathered all these weapons under the Spike banner. The original Spike variant (now known as Spike-ER) was the first helicopter-launched version to be developed. It was a 13-kg (28.6-lb) missile with a range of around 4000 m (13,123 ft). This was followed by the much larger Dandy (now known as Spike-LR) which had an 8000-m (26,247-ft) range.

All of these missiles use an innovative (and so far unique) man-in-the-loop guidance system. After launch, the fibre-optic cable that connects the missile to the launch platform relays seeker imagery back to the operator, who can see exactly where the missile is going and correct its course accordingly. This makes the Spike a very accurate

In the 1990s Turkey used the AH-1P/S (illustrated) and AH-1W in action in its fight with PKK Kurdish separatists, a campaign which drew international criticism. In early 2004 the long-running saga to procure new AH-1Z KingCobras for the Türk Kara Kuvvetleri appeared to be nearing an inconclusive end. Earlier, Bell had also courted Romania, which had ambitious plans to build 96 AH-1Ws under licence as the AH-1RO Dracula. As may prove to be the case in Turkey, these plans were eventually overtaken by economic reality.

weapon, more accurate than any laser-guided missile, for example. The Spike also has an autotracking fire-and-forget capability and the combination of these two very different control modes makes it a unique weapon. The Spike has been integrated on Israel's Cobras – in fact, they are thought to be the only operational Spike platforms in service. The weapon gives IDF/AF Cobras an unparalleled precision attack capability. There have been persistent rumours that Spike, or a modified version of the missile with a smaller warhead (for lower collateral damage), has been used on Israel's targeted 'assassination' air strikes during 2003 and 2004.

IAI has developed another air-launched weapon that is probably also in service on IDF/AF Cobras. This is the LAHAT, known as the SkyBow in its air-launched form. SkyBow is a small laser-guided missile, originally designed to be fired from tanks or self-propelled guns. It is therefore a very compact and lightweight weapon suitable for use on small helicopters or even UAVs. IAI is now offering the LAHAT/SkyBow for export – usually (although not always) a sign that the system is already in Israeli service. There has been a suggestion from some Israeli sources that this missile, claimed to be accurate to within 72 cm (2.4 ft), has been used in the anti-personnel strikes in the Occupied Territories.

Cobras at war

Post-Vietnam, Israel's Cobras were the first to be used in combat. Press reports place their operational debut in May 1979, when TOWs were fired at targets in the Halil refugee camp, near Tyre. Their first major combat experience was during the 1982 invasion of Lebanon (Operation Peace for Galilee). This does not appear to have been an unqualified success, as the terrain and tactics used did not suit the aircraft. At least two Cobras are believed to have been shot down by Syrian air defences, while several others were damaged. However, the Cobras were used to engage Syrian armour and fired over 100 TOWs, claiming kills on T-72 tanks (later hotly disputed by other branches of the Israeli military who also claimed credit).

While Israel's Cobras were on the front line through the 1980s, Iran's substantial force of AH-1Js fought long and hard against Iraqi forces during the first Gulf War of 1980 to 1988. The Islamic Republic of Iran Army made very effective use of its 'International' AH-1Js, which fought Iraqi ground and air forces over the course of the conflict. There is good evidence to support several reported air-to-air kills against Iraqi helicopters using both TOW and the M197 gun. Iranian sources even claim three air-to-air kills of Iraqi MiG-21s by IRIAA Cobras. At the same time Iranian helicopter losses were high and only about half the original fleet of 202 AH-1Js remained available at the end of the war.

The Cobras fought in several key offensive and defensive engagements, supporting advances through the Iraqi front lines in 1982, 1984 and 1986, while fending off the Iraqi offensive of 1988 and claiming kills on T-72 tanks. Not all of Iran's Cobras were TOW-capable and, towards the end of the conflict, Iranian technicians began to integrate alternative weapons on the AH-1Js (several such efforts were undertaken, including attempts to fit Hawk and Standard SAMs on F-14s and F-4s). A few Iranian Cobras were armed with AGM-65 Mavericks and the small supply of available missiles was reportedly used with success against static positions. The IRIAA also tried to arm its Cobras with Sidewinders, but this project failed.

In the early hours of 20 December 1989, US forces invaded Panama to overthrow the regime of General Manuel Noriega. Operation Just Cause was launched from the existing American bases in the Canal Zone and was supported by additional troops flown in from the United States. Helicopters played a key role in the brief – but bloody – action, which saw the first combat missions undertaken by female US Army pilots (piloting UH-60 Blackhawks). Five US Army Task Forces were deployed against various targets in Panama, and AH-1s delivered fire support for these operations. This included Task Force Bayonet's capture of Fort Amador, the air assault of El Renacer prison, and the crucial H-hour battle for the Cerro Tigre army barracks. The fighting around Cerro Tigre saw the first combat use of NVGs by US Army aviators.

Guns and rockets remain essential arrows in the Cobra's quiver. Marine AH-1s operate very closely with the ground forces, and they may be regularly called upon to clear small pockets of resistance being faced by the infantry. At close quarters the 2.75-in rocket and 20-mm cannon remain highly effective means of countering 'soft' targets, while the carriage of TOW and Hellfire allows the Cobra to handle 'hard' targets such as tanks and bunkers at greater ranges. Here a pair of AH-1Ws from the USMCR squadron HMLA-775 fire rockets on the range at Camp Pendleton, California.

Short-range MANPADS (man-portable air defence systems – shoulder-launched weapons) are one of the main threats to the attack helicopter, and a sophisticated and fast-reacting defensive suite is required. Here an HMLA-775 Cobra releases flares while firing 'on the run'.

This HMLA-367 AH-1W is releasing flares while in the hover, not a typical Marine operating procedure – the Corps favouring to keep moving at all times during combat operations as a stationary helicopter is considered a much easier target for gunners than a moving one. The flares are dispensed from ALE-39 boxes mounted above the wingtips, triggered automatically by the AAR-47 missile warning system.

Although the 26th MEU(SOC) was on hand aboard USS **Kearsarge** *(LHD 3) during the Kosovo war in 1999, the Cobras played no direct part in Operation Allied Force. However, to aid the policing actions which followed the NATO occupation, AH-1s were deployed ashore. Here an aircraft allocated to the 26th MEU(SOC)'s HMM-263 lands at Camp Bondsteel in 2002.*

During the 1990s Turkey's Cobras were used to fight a high-intensity but low-profile war against Kurdish insurgents along the southern border region with Iraq. In 2004 Pakistan's AH-1S force found itself fighting a similar action as part of the March campaign to hunt down suspected al-Qaeda figures hiding out in the border regions of South Waziristan. Between 5,000 and 7,500 troops were involved in fighting with perhaps 400 armed tribesmen.

Operation Desert Storm

Both US Army and US Marine Corps AH-1s went to war in Iraq during 1991, but the 100-hour ground campaign of Operation Desert Storm gave the crews relatively little operational exposure. Army Apaches grabbed the headlines when Task Force Normandy made the first strike across the Iraqi border in the early hours of 17 January. Eight AH-64As of the 101st Airborne Division fired the first shots of the war in an attack that destroyed key Iraqi radar positions. According to the official record the US Army took 145 AH-1s (and 277 AH-64As) to war in Iraq. The Cobras played a supporting role to the Apaches – in what was to be the high-point of the AH-64's operational career. While the Apaches took on the deep strike mission, the Cobras flew constant security and reconnaissance missions, providing over-watch and flank security for the advancing Army units.

The Marines deployed 48 AH-1Ws for the first war in Iraq, or about 20 per cent of the available attack helicopter force. Yet these aircraft flew more than 50 per cent of the total attack force flight hours. The AH-1Ws also turned in a mission readiness rate of 92 per cent. The Marine Cobras were heavily involved in the fighting around the Saudi town of Khafji, between 29 January and 2 February, which was subject of an Iraqi incursion before being retaken by Saudi and US troops. After the ground war had begun, Marine Cobras operated in their classic roles of armed reconnaissance, escort and close air support. The speed of the Allied advance, the bad weather and smoke from burning oil fields made life hard for the helicopter crews. The AH-1Ws operated in direct support of the 1st Marine Division, and as part of the Task Force Cunningham manoeuvre element (alongside AV-8B Harriers).

On G Day +2 (26 January) Marine Corps Cobras, flying with a single UH-1N, supported Task Force Ripper in its battle with the Iraqi 3rd Armoured Division. Along with Task Force Papa Bear, TF Ripper was one of the two main assault forces of the 1st Marine Division into Iraq. Fitted with a more modern FLIR system than the AH-1s, the UH-1N led units through the smoke and under power lines into battle positions where they could engage Iraqi tanks with Hellfires. The Huey even designated targets for the Cobras' Hellfires. The Cobras were also very active with the two flanking elements (TFs Grizzly and Taro).

In the years that followed the liberation of Kuwait, Marine Corps Cobras supported expeditionary deployments around the world and saw active service in Somalia, Bosnia, Haiti and Liberia. The AH-1W is a key component of the Marine Expeditionary Unit (Special Operations Capable), or MEU(SOC), concept.

When the Cobra went back to war against Iraq in 2003, this time the Marines' AH-1Ws were on their own – the Army had only Apaches. The aircraft that the crews flew in 2003 were essentially unchanged from those of 12 years previously, but the nature of the war was dramatically different. In 1991 Marine Corps troops had pushed almost as far as Baghdad in about four days but were not attempting to take over the country. This time US forces would cross virtually all of Iraq and, while the combat phase of the campaign was relatively brief, the scale and ferocity of the fighting bore no resemblance to that of Desert Storm.

Operation Iraqi Freedom

The experience of HML/A-269 (REIN) in Operation Iraqi Freedom (OIF) was typical. During peacetime, Marine Light Attack Helicopter Squadron 269 (HML/A-269) 'Gunrunners' is one of two AH-1W squadrons based at MCAS New River, North Carolina. New River is itself attached to the huge

The mechanised units of TF Tripoli arrived at Tikrit on 13 April and within 24 hours the town had been taken by US forces. HML/A-269 flyers recall that on the first day they were taking heavy fire from Iraqi positions on Tikrit South airfield, but by 14 April they held the airfield which had been turned into an operational FARP, named for pilot Ben Sammis who had been killed on 4 April. The Cobra crews were particularly pleased with the footage of their aircraft, dripping with Hellfires, that appeared on CNN's coverage of the 'Taking of Tikrit'.

One final event that several crews all remember was the discovery and destruction of a huge ammunition dump near Tikrit. One pilot described the scene, "We found this expanse of camouflage netting and we thought 'what the hell's under there?' – because those dudes didn't camouflage jack. I pushed the rotor blades towards and under the netting, to blow it all back, and we found about 350 or 400 Roland missiles, all just sitting there. That was the last target to be attacked in Iraq. We took seven aircraft and fired about 100 PGMs into that place, plus it was hit by 25 or 30 fixed-wing aircraft too. On NVGs it was like *Die Hard 3*. There were huge explosions reaching up to 12,000 and 15,000 ft."

By 20 April the fighting was done and units had already begun to pull out of the Jalibah FOB (Riverfront) to the USS *Saipan*. MAG-29 helicopter assets, including six AH-1Ws, moved to Blair Field, at Al Kut (a FARP named for a Marine killed in action near An Nasiriyah). The Cobras and other helicopters flew security patrols over the areas still occupied by TF Tarawa until 15 May. On 29 May the first batch of HML/A-269's AH-1Ws embarked aboard the USS *Kearsarge* to begin the return journey home. Unfortunately for the Marines the squadron was then reassigned to support operations in Liberia, where civil strife had broken out. Eventually, the Cobra crews got home in June.

During the war HML/A-269's Cobras – it took 18 to Iraq in total – fired 334 Hellfires, 345 TOWs, 5,665 2.75-in rockets and 64,106 rounds of 20-mm ammunition. The squadron's nine UH-1Ns disposed of 107,787 0.50-cal bullets and 119,891 7.62-mm rounds. The squadron claimed 697 tactical targets destroyed, including 47 T-55s, nine T-62s and 34 T-72s, 77 APCs, 47 S-69 AA guns, 65 ZU/ZPU AA guns, 25 missile positions (SAMs), 82 artillery positions, 20 mortar sites, 156 cars/trucks/'technicals', plus buildings, bunkers and storage sites. About 30 or 40 per cent of those targets were manned, says Hewlett. "While we were out there we were duelling with 37-mm SPAAGs (self-propelled anti-aircraft guns), we killed FROGS and Ababils (short-range surface-to-surface missiles), we hit SA-2s and SA-3s and nailed a couple of helicopters on the deck. I spotted an Su-25 on the ground at Tikrit but when I went back for it, it had gone. The Iraqi air force was hiding in the streets of northern Baghdad. We saw at least four MiG-29s under cover, but we were not allowed to engage them."

Cobra tactics

Marine Cobra tactics were to fly in two-ship formations (sections) or four-ships (divisions). Typically the East Coast crews flew sections while the West Coasters flew divisions. Hewlett's view was, "I didn't see any targets that required four Cobras. It's a bad use of (scarce) assets, and if we go away (to rearm and refuel) we'll only be gone an hour. We didn't have to cover the entire country, we were only needed where the fighting was, so we conquered Iraq within 200 m (650 ft) of the MSRs (main supply routes)." From 20 March to 13 May HML/A-269's 18 AH-1Ws flew a total of 1,957 combat hours, while the nine UH-1Ns flew 951 hours. The squadron always had a minimum of 12 AH-1Ws and six UH-1Ns available throughout the conflict (an 80 percent mission capable rate). The Cobras averaged 36 flight hours per day. Throughout all that time the unit had no personnel losses to enemy action or mishaps.

Each day that they went out, the Cobra crews had to load up weapons with their best guess of what the threat would be. Major Streeter explained, "TOW gives you a

As has become traditional in wartime, many AH-1Ws committed to operations in Iraqi Freedom sprouted nose art, as displayed by these HML/A-269 aircraft. The squadron was one of the principal participants in the operation, dispatching 18 aircraft (out of a total Cobra force of 54). During the operation three AH-1s were lost, but these losses were lighter than could be expected given the type of operations in which the Cobras were involved, and compared favourably with the US Army's experience with its better equipped AH-64s. The answer, it would appear, lay in the tactics involved, the Marines attacking their targets while moving at speed. In a battlefield where guns and even RPGs represented as big a threat as SAMs, hovering was off the agenda as far as the Marines were concerned.

Above: Nearly 40 years after it entered service with the USMC the AH-1 Cobra is still a highly potent vehicle, as displayed by its 'star' appearance in Iraqi Freedom. The conflict represented an ideal environment for the current warfighting doctrine of the Marines in general, and the Cobra in particular: a fast and fluid battlefield with a mix of 'hard' and fleeting 'soft' targets, many of which would pop up unexpectedly to threaten the advancing forces, requiring immediate action from the escorting helicopters.

Above right: For several years Marine AH-1Ws have had a meaningful anti-aircraft capability thanks to the AIM-9 Sidewinder, although no other ordnance can be carried on the wing when it is fitted. The missile is carried on an LAU-7 launch rail which could also launch the AGM-122 Sidearm anti-radiation derivative when it was still in the inventory.

close-in warm and fuzzy feeling, but Hellfire gives you stand-off. In practice we always kept the TOWs and Hellfires on there because it's too much trouble to reboresight the stations. Ninety-nine per cent of the time the debate was over rockets – do you carry HE, or flechettes or both? We fired more PGMs as the war progressed. In the early days we shot more from the hip – rocket, rocket, shoot, run. The gun has reliability issues. In my experience it jammed about 35 per cent of the time, which is not good. In fact, the more you shoot the gun the more problems it has. We had problems with the rockets too – just one of those things I guess. The ground guys would always fix it, but it took a lot of trouble-shooting. It was very rare to go out with everything working. I had a go/no go line – if 80 per cent of my helicopter was working I was still in the fight. "

The threat of NBC (nuclear, biological, chemical) warfare was a constant for all the deployed coalition forces. The Cobra crews were not always happy about their ability to cope with an NBC battlefield. The directives on NBC posture were widely felt to be confusing and the effect of wearing protective gear in the desert environment had dangerous heat and fatigue implications. Pilots were already carrying a large amount of equipment into the cramped cockpits of the Cobras, and to this had to be added NBC protective gear for the air and for the ground. If the Cobras did find themselves flying through contaminated areas no-one quite knew what the effect of bringing a 'dirty' aircraft into a 'clean' FARP might be – and everyone knew that the NBC detection equipment available to them was substandard.

And yet, the AH-1W had a good war. The aircraft and crews gave exemplary service and always fulfilled their mission of flying and fighting with their troops on the ground. One pilot commented, "The Cobra is great to fly and so durable. If a target needs shooting, you can shoot it in a Cobra. The whole air/ground thing works so well in the Marines. We are so close to the grunts, we understand fire support. We live, eat and breathe it." The Cobras were often called on to fly ISR (intelligence, security and reconnaissance) missions – in other words, scout flying – but this did not always prove to be a good idea. Although it is a classic Cobra role, in Iraq the crews knew that if their helicopters were pushed too far forward of the lead ground elements they became vulnerable to hostile fire.

Maintaining the bubble

Lt Col Hewlett had no doubts about what the Cobras needed to do, "Our job was to maintain a 2-km bubble around the troops as they headed up the MSRs. Outside that it was Indian country. Don't go there. I should stress that that was true for the day only. At night things were much better and we didn't need to fly down low where it gets dangerous." Pilots knew never to fly in a straight line, never to follow linear features and never fly predicable routings. Because of the pace at which CAS missions were being flown, hot-reloading skills became essential and the Cobras will now spend more time practising this at home than ever before. New skills in urban CAS planning have to be developed and new tactics to work out the split between Section and Division flying will have to be considered.

The war in Iraq also highlighted the deficiencies that everyone knows exist in the AH-1W, and that everyone is waiting to be fixed in the AH-1Z. Chief among these are the aircraft's archaic sensor fit and its compromised payload/range capability. Even when conditions are at their best it can take valuable time and some real operator skill to locate and identify targets using the Cobra's NTS sight. Also, because of the gross weight limitations on the AH-1W, pilots are frustrated by their inability to carry all the weapons that they should, in theory, be able to hang on their helicopters.

Lt Col Hewlett explained, "Our number one problem is the (FLIR) sensor in the nose. Right now we have to close to small arms range to visually ID a BMP, theirs, or a LAV, ours. With the Zulu we will have an 8- to 10-km stand-off range instead of the 800- to 1000-m that we have now. In Iraq a vast amount of the ID work we did was visual, using NVGs. You'd look with the goggles, mark the position and then try with the FLIR again. If it's a nice night then maybe you can break out an image by 800 m [2,625 ft]. By day the TV camera is fine. I can look out to 6 or 7 km and ID things. But at night it's no good. When the Zulu enters service we'll have a lot more fuel and a third-generation FLIR. At the moment an AH-1W can carry enough fuel to stay airborne for 1.8 hours. With the Zulu that will climb to 2.7 hours. All the wing stations have a universal interface, the Whiskey only has two stations for PGMs (the outboard hardpoints, Nos 1 and 4). With the Zulu we'll be able to carry 16 Hellfires, although you'd probably never want to."

Ir

Bel

1 Fou
2 All-
blac
3 Pitc
4 Fina
angl
5 Ang
6 Tail
7 Ang
gear
8 Rea
ante
9 Tail
10 End
11 Elev
12 Vent
13 Elev
14 Con
15 Tail
strin
16 Tail
17 Deta
18 Sha
coup
19 Rea
port
20 Elec
exte
21 100-
exte
22 100-
napa
23 BGM
surfa
miss
24 TOW
latch
25 Four
carri
26 250-
frag
27 Low

Cobra cockpits

***Above and left:** The front cockpit of the AH-1W is dominated by the binocular weapons sight for the Night Targeting System. Behind the sight is a small screen for displaying sensor imagery.*

***Below:** The rear cockpit of the AH-1W is topped by a weapons sight for firing rockets, and a display for the RWR.*

The cockpits of the AH-1F differ significantly from those of the W. The instruments in the pilot's cockpit (above) are laid out differently, while in the front (left) the co-pilot/gunner does not have a screen display.

Missile armament

Despite its age, the BGM-71 TOW wire-guided missile remains very useful for attacks in certain scenarios, especially against targets which require a side-on attack. The missiles are tube-launched, the tubes being quickly fitted into the launcher modules (left), of which two or four are carried under each outboard pylon. The laser-guided AGM-114 Hellfire (right), is carried on rails on a four-round launcher, also on the outboard pylon. Cobras usually carry a 2.75-in (70-mm) LAU-61/69 or -68 rocket pod on the inboard pylon.

Sighting systems and cannon

At the heart of the AH-1W's weapon system is the Night Targeting System (left), a modification of the standard daytime M65 sighting system which is used in the AH-1F (right). NTS incorporates a FLIR, TV sensor, laser rangefinder and laser designator, allowing the Whiskey Cobra to be fully night-capable and to designate for AGM-114 Hellfire missiles. Some AH-1Fs, including export aircraft, have been fitted with the C-Nite system which adds a FLIR to provide night capability. The standard gun armament of both variants is the M197 three-barrelled 20-mm cannon – unturreted in its AH-1F installation. This gun has replaced the six-barrelled Minigun found on earlier versions.

AH-1Z details

Compared with the AH-1W, the Zulu introduces a four-bladed main rotor mounted above a reprofiled fairing (right). The blades are bearingless and have prominent cuff sections to which are attached the actual blades themselves. The tail rotor is also a four-bladed unit (left). The new stub wing (below) retains two underside hardpoints, but also allows the carriage of air-to-air missiles on the tips. The pylon itself contains fuel, significantly improving the Zulu's endurance. This test aircraft is covered with photo-calibration marks.

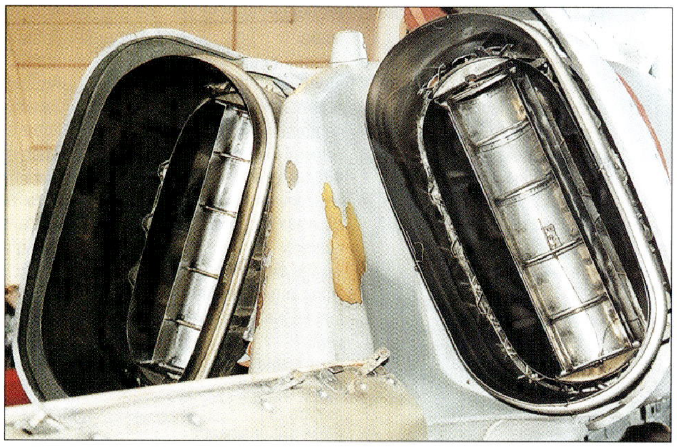

The engines exhaust through flattened nozzles which suppress IR signature by mixing cool air with engine exhaust.

Another OIF veteran agreed, "Right now we have 1980s technology and a borderline Gen II FLIR. At night, on a good night, you might be able to see a tank at 5 km, but those perfect conditions just don't exist and in Iraq we were lucky if we could see things at 2 km. We would fix the FLIR to look straight ahead, set the rad alt [radar altimeter] at 300 ft and just go."

Cobra pilots are also looking to the Zulu to fix the cockpit interface problems that have built up in the AH-1W over the years. One pilot spoke for them all, saying "It has very poor human engineering, you have to toggle too many switches and really be a contortionist to get ready for the fight. For example, you can't actually see the switch for the defensive aids system. You have to reach over and feel for it. Forget to flip another switch and that's it – no weapons. There are too many black boxes in the back seat. There's the AIM-9 controls, the RWR, the HUD – it all blocks your view. The forward visibility is also impaired, so instead of flying anywhere nose on, you have to weave so that you can see to get there." While every AH-1W pilot loves their aircraft, all are waiting for the AH-1Z and it can't come too soon.

AH-1Z: the 'Zulu Cobra'

For an aircraft that entered service as an 'interim solution', the AH-1 still has a big future ahead of it. Built-up from the current baseline twin-engined AH-1W configuration but incorporating a host of new technology, the AH-1Z – the Zulu – is the Cobra for the 21st Century. The AH-1Z is at the heart of the US Marine Corps' aviation rotary-winged modernisation plans (along with the similarly upgraded UH-1Y) and it is also on offer to several prospective export customers, including Turkey and Taiwan. Despite the radical differences between it and its predecessors, the Zulu will share the same name as the AH-1W. To the Marine Corps it will be the AH-1Z SuperCobra.

Colonel Doug Isleb, the Marine Corps H-1 Upgrade Program Manager, described his new helicopter thus, "It will be a much more effective warfighting platform. We will do all the same missions, but we'll do them better. With the sensor, we are going from one of the worst fielded to what is probably the best. We can carry more weapons and payload so we can find the opposition, identify them from further away and shoot more of them. The biggest problem with the AH-1W is the cockpit, there's too much management and not enough fighting. With the Zulu's digital integrated systems and helmet-mounted sight, day or night the crew is going to be heads-up and fighting."

Designs for an improved aircraft that embodied many of the features of today's AH-1Z have been around for a long time, and have had many names. Back in December 1979 Bell flew its Model 249 – essentially an AH-1S airframe fitted with the four-bladed rotor system that had been developed for the Model 412. This aircraft, the Cobra II demonstrator, made its international debut at the 1980 Farnborough air show. The Cobra II would have offered full Hellfire capability with a new targeting system and improved engines. Bell suggested that the changes could be incorporated in the ongoing AH-1S upgrade programme but the Army decided instead to invest its resources in the LHX competition that spawned the RAH-66 Comanche. The Cobra II was then offered for the Advanced Scout Helicopter requirement but was beaten by the OH-58D.

There was also a Cobra 2000 proposal that had some key elements of the Zulu design, such as T700 engines and the four-bladed rotor. The Marine Corps was attracted by the performance offered by a four-bladed Cobra, but again there was no funding to support the programme. Bell persevered and offered the Cobra to Germany as a rival to the PAH-2 design that eventually gave birth to the Eurocopter Tiger. The Model 249 remained in service as a test aircraft only, but plans for a four-bladed Cobra never went away.

In 1993 the UK's Army Air Corps issued a tender for a new attack helicopter. Bell, together with UK prime contractor GEC-Marconi Avionics (now BAE Systems), offered the British an aircraft very similar to what became the AH-1Z – the CobraVenom. Based on the AH-1W airframe, the CobraVenom would have been a fully modernised, twin-engined, new-build aircraft. It had redesigned stub wings with four 'smart' pylons that were

A new generation of Cobra – the AH-1Z completely revamps the AH-1 concept, turning the aircraft into a digital warrior capable of operating in the information technology-based warfare of the 21st century. The equipment upgrade is matched by airframe/ transmission improvements which significantly increase performance in all areas while retaining the AH-1W's T700-GE-401 engines.

This view of the first AH-1Z highlights the differences between the Zulu and the standard AH-1W behind. Most obvious are the four-bladed main and tail rotors, and the taller and reshaped rotor mast fairing. Less obvious, but just as important to the crew, is the complete replacement of the avionics system, with all-digital systems and a 'glass' cockpit/helmet-mounted sighting system which allows the crew to navigate and communicate with much reduced workload, and to fight 'heads-up'.

At present the gun of the AH-1Z remains the same A/A49E-7(V4) system with M197 20-mm cannon, despite its acknowledged jamming problems. The weapon can be operated in three modes: in the fixed mode the gun fires directly ahead, aimed by the pilot manoeuvring the helicopter; in IHSS mode the gun is slaved to the helmet-mounted sight; and in TS/Gun mode the gun is slaved to the sight's auto-tracking system. The gun fires at 650 rounds per minute, and can fire a maximum of 450 rounds in a single burst, after which it must cool for six minutes.

each capable of carrying Hellfires or TOWs, along with the Brimstone anti-armour missile. It was hoped that the new fully-integrated digital avionics suite designed by GMAv would also be adopted by the US Marines, as part of an integrated weapons system upgrade then being contemplated for the AH-1W. In June 1995 the CobraVenom design was revised to include the four-bladed rotor system, but in July the AH-64D was announced as the winner of the UK competition. The CobraVenom proposal had many fans within the AAC and the work done on that design translated directly into the AH-1Z programme, which was officially launched in 1996.

Right: The third NAH-1Z poses with its maximum missile load of 16 AGM-114 Hellfires plus two AIM-9 Sidewinders.

Below: Designated NAH-1Z, the no. 1 Zulu is seen in the hangar at Patuxent River. This aircraft lacks the avionics suite. Note the test air data probe in front and temporary low-speed air data boom.

The Marines had long been looking at how they could field a credible battlefield helicopter fleet into the 2010 timeframe and beyond. The answer lay in a joint upgrade path for both the AH-1W and the UH-1N. On 15 November 1996 the USMC signed the contract with Bell Helicopter that launched its H-1 Upgrade Program. The aim was to remanufacture the existing fleet of AH-1Ws and UH-1Ns into 180 AH-1Zs and 100 UH-1Ys. The USMC inventory, circa 2001, included 194 AH-1Ws and 96 UH-1Ns. The USMC will get 'zero-time' airframes (good for 10,000 hours), with integrated avionics systems and glass cockpits, along with four-bladed rotors and upgraded drive trains that are common to both aircraft. The AH-1Z (and UH-1Y) is expected to deliver dramatic performance improvements, including increased range, payload and speed, plus increased ballistic tolerance and crash survivability. The two upgraded aircraft are expected to have 84 per cent identical major components (said to be identical down to the part numbers), including their T700-GE-401/C engines and transmission, tail unit, composite main rotor system, hydraulic and fuel system, integrated avionics and software, and crashworthy seats.

Looked at in the round, most people would describe the AH-1Z as virtually a new aircraft. Identifiable components taken from existing AH-1Ws include the (modified) tail section, engine doors and cowling, (modified) combining gearbox and the (modified) forward fuselage. Items that are found on the Whiskey but which have been newly built for the Zulu include its tail fin (plus elevators and tail stinger), a redesigned 90° offset gearbox and magnesium-free gearbox casting, improved crashworthy landing gear (capable of withstanding a 12 ft/3.6-m per second impact), the uprated rotor driveshafts, the revised upper fuselage cowling and redesigned stub wings. Neither does this list include the completely new equipment such as the rotor system and cockpits.

Key Performance Parameters (AH-1Z/UH-1Y)

KPP	Threshold	Objective
Payload (lb)	2500/2800	3500/4500
Cruise Speed (kt)	140/140	165/165
Mission Radius (nm)	110/110	200/200
Manoeuvrability (*g*)		
AH-1Z	-0.5 to +2.5	-0.5 to +2.5
UH-1Y	-0.5 to +2.5	-0.5 to +2.5

By the end of 2003 the AH-1Z had demonstrated a maximum airspeed of 222 kt (411 km/h) and 160 kt (296 km/h) in the cruise, while the UH-1Y had reached a maximum of 190 kt (352 km/h) and was cruising at 166 kt (307 km/h).

There are five development aircraft in NAVAIR's joint test flight programme, underway at NAS Patuxent River.

Airframe	MSN	Reissued BuNo.	Former BuNo.	First Flight
AH-1Z (Z1)	59001	166477	162549	7 December 2000
AH-1Z (Z2)	59002	166478	163933	4 October 2002
AH-1Z (Z3)	59003	166479	162532	26 August 2002
Aircraft Z3 was the first production AH-1W				
UH-1Y (Y1)	55001	166475	160446	7 July 2002
UH-1Y (Y2)	55002	166476	159193	20 September 2002

From 1996 to 2003 the AH-1Z programme was essentially a research and development effort. The first aircraft to fly (Z1, at Fort Worth) had the airframe and drive train of a Zulu, but none of the advanced avionics. The programme's early years were hit with problems and delays that brought cost overruns and schedule slippages. Difficulties encountered by the H-1 team included cracking in the vertical stabiliser and issues with the main rotor blade, yoke, hydraulic system, manufacturing tooling and the integrated avionics systems. Bell also had to re-assess the amount of material that would be reused in each rebuild, while still trying to keep to the target per helicopter. The original plan called for an LRIP start in FY03 (meaning before October

2003) with aircraft in operational service by FY06. By May 2001 the low-rate initial production date had been pushed back by a year. Then, the number of aircraft in the EMD (engineering manufacturing and development) phase was cut by three to try and balance the budget. To smooth things along all five aircraft moved from Bell facilities to NAVAIR's home at NAS Patuxent River, Maryland, where the test effort is being supported by a combined contractor and government team.

By the second half of 2003, the programme was in a 'design reconfiguration' phase. Only one example (Zulu 1) was in flight test, at that time evaluating a new moving stabiliser. The two additional test AH-1Zs and UH-1Ys had been stripped down to be fitted with the new avionics suites. The OT-IIA and OT-IIB operational assessment phases, to include live-fire trials, were set for completion in time for operational evaluation in late 2004. On 23 October 2003 the Defense Acquisition Board approved Lot 1 low-rate initial production (LRIP) for the H-1 Upgrade Programme. This was a major triumph for the project (worth about $6.2 billion in its entirety) which had come close to cancellation on several occasions. The DAB gave the go-ahead for the remanufacture of three AH-1Ws to AH-1Z standard during FY04, along with six UH-1Ys. The Lot 2 LRIP batch, another three AH-1Zs and six UH-1Ys, should enter production in FY05. Full rate production is expected to commence with the subsequent Lot 3 order.

Service entry schedule

The Marines hope to begin training operations with the first batch of upgraded H-1 aircraft in 2006 using LRIP aircraft delivered to HMT-303 at MCAF Camp Pendleton, California. Initial operating capability (IOC), defined as the first group of six AH-1Zs and three UH-1Ys capable of independent operational deployment, is hoped for 2008/09. Those aircraft will be based either at Camp Pendleton, or MCAS New River, North Carolina. Peak production rate requires 24 AH-1Ws to be upgraded annually, with final deliveries in FY14 for a goal of continued operations beyond 2020. Any export deliveries could begin around 2005.

The AH-1Z is built by Bell at Fort Worth, Texas, using new airframe assemblies that come from the company's Amarillo plant and teardown kits that have been produced from disassembled helicopters at the Cherry Point NAD. The most obvious difference between the AH-1W and the AH-1Z is the new four-bladed main rotor. This uses all-composite blades with an elastomeric bearingless (no lubricated components) rotor head that incorporates a semi-automatic folding system. The composite head assembly takes up all the normal lead/lag and feathering functions of a helicopter rotor system, and is ballistically tolerant to a 23-mm shell impact.

The General Electric T700-GE-401 engines that power the AH-1W are retained, but they are coupled to an entirely new main transmission (and a higher output tail rotor system) that can cope with and deliver increased power levels. One Whiskey pilot describes the experience of flying his current aircraft thus, "the aircraft has power but you have to treat it gently. The engines will perform but the transmission can't handle them. You have to be very careful not to torque it out. You end up flying with an invisible hand that holds the collective down." The Zulu should solve these problems. It will also come with a new Hamilton Sundstrand APU that has its own independent gearbox. The Zulu offers an empty weight of 11,900 lb

BuNo. 162549 was the first remanufactured AH-1Z, and is seen here flying near the test centre at 'Pax' River.

The main sensor of the AH-1Z is the AAQ-30 Hawkeye Target Sighting System (TSS) turret, which completely replaces the M65/NTS of the AH-1W. This combines the very latest in FLIR technology with eye-safe lasers and a colour low-light TV, and incorporates a highly accurate auto-tracker. It can be controlled by hand grip or by the helmet-mounted sight. It can provide seeker cues for AIM-9 seeker heads.

The AH-1Z can carry the current range of Cobra ordnance, but the wingtip stations allow it to carry Sidewinders as well as other stores.

Developed by Thales, the TopOwl helmet-mounted display can project FLIR imagery as well as flight information.

'Zulu 1' prepares for a flight on 14 June 2001 with deputy commandant for Marine Aviation Lieutenant General Fred McCorkle in the front cockpit.

(5398 kg) and a maximum gross weight of 18,500 lb (8392 kg). Simply expressed, on any given mission it will offer crews either twice the range or twice the payload of the AH-1W. The AH-1W has a combat radius (out and back range) of only 70.5 km (43.8 miles).

The airframe uses a mix of conventional metal assemblies (aluminium, steel and titanium) with some composites. The tail unit is taken directly from the AH-1W, but fitted with a new 'stinger' and some structural modifications to cope with the higher loads and sink-rates that the Zulu will experience. The test programme has suffered several problems with the tail, not least the tendency for the (deflected) exhaust stream to heat-damage it. A similar problem has been faced by the Eurocopter design team. The engines are able to consistently deliver an extra 500-600 lb (2.22-2.67 kN) extra thrust, thanks to the new transmission, and this has forced the AH-1Z test team to scab on a doubler to thicken the tail skin. For the longer term an exhaust redesign is under consideration to divert the hot air flow away from the sensitive areas.

The Zulu's new longer span wing stubs look simple but are very different to those on the AH-1W. Each of the AH-1Z's pylons has a 'smart' interface (MIL-STD 1760 bus) and each wing also carries 50 US gal (189 litres) of internal fuel. The new wing assembly weighs the same as the old unit, but the centre-section airframe has been reinforced to cope with the increased loads that will be hung on the wing.

The Zulu cockpit is a radical shift away from anything that Cobra pilots have known before. Thirteen Marine pilots and an aircrew system advisory panel worked for 18 months in simulators and mock-ups to create an efficient man-machine interface. Development of the integrated avionic system (IAS) is now in the hands of Northrop Grumman, ever since NG took over the original system developer Litton. Several delays to the H-1 upgrade programme were due to problems with this very advanced mission fit, which was common to Australia's SH-2G(A) SuperSeasprites – and caused similar difficulties for that programme. The IAS operates on a MIL-STD 1553 databus, linked to a central mission computer and all the communications, navigation and mission systems. The two crew stations are dominated by a pair of 6 x 8-in (15 x 20-cm) high-resolution (640 x 480 pixel) colour liquid crystal MFDs (multi-function displays), each with a digital moving map capability. The MFDs are backed up by a smaller (4.2 x 4.2-in/10.6 x 10.6-cm) dual function display (DFD). The DFD also serves as the standby flight display if the main avionics system should fail.

TopOwl helmet

The Zulu comes with a Honeywell dual embedded GPS/INS navigation fit and a brand new helmet-mounted sight (HMS) in the shape of the French-built Thales Avionics TopOwl system. The TopOwl is a modular HMS that integrates day and night sensors. It uses an electromagnetic head-tracking system that requires a magnetic 'map' to be produced for each aircraft cockpit so that the position of the crewmember's head can be accurately plotted. The system is already in operational service and is being fielded on at least four helicopter types (Tiger, NH-90, Rooivalk and AH-1Z) in more than 10 countries.

The basic protective helmet incorporates radio communications systems, while an add-on display module projects sensor imagery onto the visor, using integrated night vision sensors (image intensifier tubes) or aircraft FLIR, video and/or synthetic symbology. The Marines are integrating a 'virtual HUD' function on their helmets which will display flight symbology to the pilots when they are facing straight ahead, but which will 'declutter' when their heads move from side-to-side. A pointer on the display indicates where the other crewmember is looking at any given time.

At the moment the Marines have decided not to display sensor imagery on the helmet visor, seeing it as a possible safety issue. Instead they will use the helmet's own 'plug in' night vision system and keep sensor imagery on the MFDs. The HMS is also slaved to the gun and the TSS. The HMS is binocular with a 40° field of view (FoV). It is lightweight (2.2 kg/4.8 lb) and its own integral image intensifying system (that replaces conventional NVGs) provides a valuable dual-sensor capability for night flying. Thales delivered a pre-production batch of 16 TopOwls to the Marines in 2002/03 and production of the full batch of 560 helmets to support both the AH-1Z and UH-1Y began in 2004. The first flight with a TopOwl was made by a UH-1Y on 9 October 2003.

Unlike the Whiskey, the Zulu can be flown and fought from either cockpit – only the position of a few circuit breakers sets the two crew stations apart. The slick new design will come as a relief to one experienced HML/A-269 flyer who noted, "as the years have passed the ergonomics (in the AH-1W) have become very bad. The systems are not well integrated and so much stuff has been stuck in, over time, that the cockpit is packed out with all this bulky kit and there's just no room." The flight controls are now fully hands-on-collective-and-stick (HOCAS) configured, and there are no longer any switches located behind the crew members. The aircraft is mush easier to fly thanks to a four-axis automatic flight control system, which has functions such as hover hold and height hold that all reduce crew workload.

Every pilot who flew in Iraq said they needed a better sensor on the aircraft. There was praise for the embedded

GPS/INS fitted to the AH-1W, which was a great targeting aid. Pilots could 'lase' a target and then send its co-ordinates (accurate to a 10-digit grid) over the secure radio net. However, the existing NTS sight system badly needs replacing. On the AH-1Z the primary mission sensor is the Lockheed Martin Missile and Fire Control-built target sight system (TSS), the AN/AAQ-30 Hawkeye. Gone are the direct view optics of Cobras past, replaced by a three-chip colour TV and FLIR system. The AAQ-30 incorporates several off-the-shelf components, housed in the same L-3/Wescam Model 20 turret fitted to US Navy P-3Cs. The Hawkeye TSS also uses elements of the USAF's AN/AAQ-33 Sniper XR targeting pod, and even the electro-optical targeting system now under development for the Joint Strike Fighter. The Hawkeye has a FLIR sensor, CCD colour TV camera, laser rangefinder/designator (LRFD), inertial measurement unit (IMU), boresight module and electronics unit. Each sensor is auto-boresighted to the others and to the AH-1Z's dual-embedded GPS/INS. The third generation (Gen 3) FLIR has a relatively large (81/2-in/21.5-cm) aperture, and uses a 640 x 512 staring array (indium antimonide) that operates in the mid-range IR spectrum (3 to 5 µm). The FLIR has already demonstrated its ability to pick up targets with "TV quality" at ranges of 13 miles (21 km). The integrated avionics system can direct the FLIR (or any of the sensors) to look at or look ahead to any designated point on the digital map.

A Sony-built TV camera covers the visible and near-IR wavebands with three CCD detector arrays. The camera is fitted with a x2.5 extender lens and provides continuous zoom up to x18 magnification. Two TV FoVs are matched to two of the four offered by the FLIR, permitting easy switching between either system. The LRFD has been adopted from the AN/AAQ-14 LANTIRN targeting pod, with a selectable eye-safe function. It is the same system fitted to US Navy SH-60s, giving them Hellfire capability too. The TV system is capable of tracking 13 targets simultaneously, with three held in the current FoV and another 10 previously selected targets being tracked by the sight's own IMU. The operator has just to cycle between the different target sets to keep them updated and tracked. The system is capable of launching Hellfires to engage these targets every eight seconds. The weapons system is augmented by a new air data sensor, developed by BAE Systems (formerly GEC-Marconi) and fitted to a Marine Cobra for the first time. The TSS EMD phase was completed by March 2003 and the first TSS-equipped AH-1Z was scheduled to enter Operational Evaluation in October 2004.

Longbow radar

In 2001 it was announced that Longbow International had teamed with Bell to develop the Cobra Radar System (CRS), consisting of a pod-based millimetre wave radar (as used on the AH-64D Apache Longbow) that could be mounted on a wingtip or stores position of the AH-1Z. This would give the AH-1Z access to the RF-guided AGM-114L Longbow Hellfire missile that currently equips only the Longbow Apache. The USMC has not made any decision on whether or not to acquire the CRS and it remains largely an export option. Marine operators say that you cannot rely on a radar alone to deliver positive IDs on the battlefield, and a visual sensor is also needed, but they view the Longbow 'enhancement' as desirable when it comes to target classification or the ability to cue the FLIR to unseen targets.

The AH-1Z retains the AH-1W's ability to carry the AIM-9 Sidewinder and not just on the wingtip stations, but on any of the hardpoints. The Zulu's higher operating weights make the missile a more viable weapon than it is on the Whiskey. The M197 cannon also has an air-to-air role and can fire M50-series rounds designed specifically for anti-

Above: The wider wings and taller rotor mast fairing are readily apparent in this view of an NAH-1Z. The main rotor can be semi-automatically folded, with one blade folding forwards and another backwards so that the aircraft takes up a similar amount of space as the two-bladed AH-1W. The Zulu continues the Cobra's tradition of presenting a low visual signature, especially when viewed from the front. The canopy presents little glint, and the paint is an IR suppressant. The four-bladed rotors have a much lower noise signature than the familiar 'slap' of the two-bladed Cobras.

Above left: This cutaway highlights (the shaded portions) the main airframe/powerplant elements of the AH-1Z that are common with the UH-1Y Yankee that will serve alongside the Zulu in HMLA squadrons. Other elements, such as portions of the avionics system, are also common, combining to greatly ease the logistic supply lines and maintenance requirements for the two types.

Reserialled as BuNo. 166477, NAH-1Z no. 1 hover taxis at Patuxent River with four dummy Hellfire missiles carried on the outboard pylon. Zulu trials are being overseen by HX-21, the US Navy's rotary-wing test squadron. The aircraft is fitted with a new moving tailplane which lacks the endplate fins on the tailplanes originally installed on the prototypes.

By late 2003 only the no. 1 NAH-1Z aircraft was flying while the other two 'prototypes' had new avionics installed. In October the go-ahead was given for six LRIP aircraft to be converted, these machines being earmarked for service with the training unit (HMT-303) at Camp Pendleton from 2006. Full production is currently slated to cover 180 aircraft, all conversions of existing AH-1Ws, to be delivered by 2014/15.

aircraft use. The Zulu team is also trying to address gun reliability problems with the M197, which is unchanged on the new aircraft. A new linkless feed system has been considered but at the moment this remains an "unfounded deficiency" in the words of one programme official.

The Zulu comes with a big boost in its self-defence capabilities. Four ALE-47 countermeasures dispensers are now fitted (compared to the AH-1W's two ALE-39s). The APR-39B(V)2 radar warning system has been improved, and integrated on an individual databus to read out on the cockpit MFDs. The AAR-47(V)2 combined optical warning system has now taken on the role of the earlier AVR-2 laser warning system. One item that is missing from the current AH-1Z EMD plan is the AN/ALQ-144 infra-red jammer found on the AH-1W. The ALQ-144 is seen as old technology by many, and is only effective against uncooled IR seekers, but the Marines are still pushing to have the ALQ-144 reinstated. The Zulu uses its inherent design and specialised exhaust shrouds to minimise its IR signature but the crews, especially those who have been shot at recently, are not convinced. Says one "Are you nuts?! Put that thing on my aircraft." The Zulu was supposed to have been fitted with an active countermeasures system like DIRCM, but this has not yet been funded.

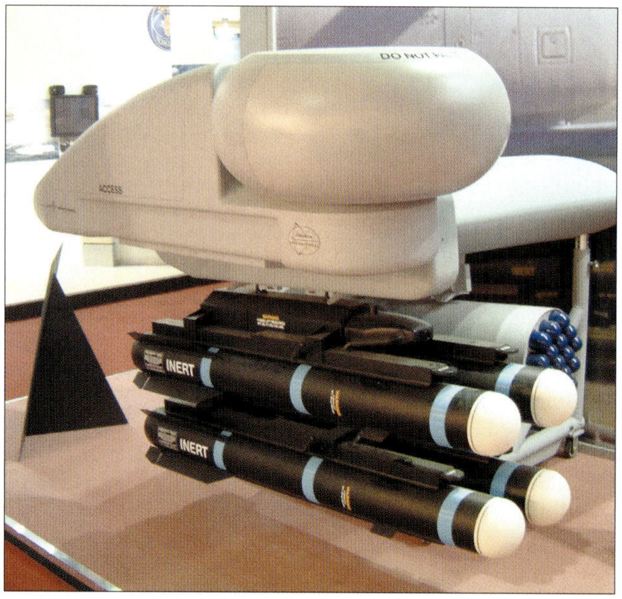

Right: Currently an option for the AH-1Z is the Longbow millimetre-wave radar, as employed in a mast-mounted position on the AH-64D. In the Zulu the radar would be mounted in a wingtip station, as illustrated by this mock-up, or possibly in an underwing pod. The radar can detect targets that would otherwise remain unseen by the FLIR, and can be used to cue other sensors. The radar would allow the support of AGM-114L Hellfire missiles.

There are more misgivings about other items of equipment that will be left off the AH-1Z, under current plans. The most significant of these, in the eyes of many, is the loss of the TOW missile. As a 'legacy' weapon the TOW has no place on the all-digital Zulu, but combat experience in Iraq proved that the TOW – or a weapon like it – was invaluable. Because the TOW is directly controlled by an operator, using wire guidance, it can be flown at a target in a number of different ways. A TOW does not have the range and the killing power of a Hellfire, but because Hellfire is a top attack weapon only, a TOW can reach targets that Hellfires cannot. The most common examples of this encountered in Iraq were armoured vehicles hidden under bridges, overpasses or any kind of top cover. TOWs could be flown under or around the cover, to hit the target. "We need a trajectory shaping missile," says one OIF veteran "we need TOW on the Zulu. It looks like they are going to give us APKWS, which will be a flat trajectory weapon and useful, but Hellfire just can't kill the tank hidden under an underpass. We're going to get the Joint Common Missile (JCM) too – well OK, but that's going to be expensive."

Another OIF pilot agreed, "there will be no replacement for TOW and that's bad. In Iraq we shot more TOWs than Hellfires because TOW is much more easily employable in urban environments." A member of the AH-1Z test team concedes, "There will be no TOW and we have not demonstrated Zunis (rockets). We want the APKWS to take care of the TOW 'issues' and there's always the JCM, but that is an expensive way to blow up tanks."

APKWS and JCM

The APKWS (Advanced Precision Kill Weapons System) – also known as LCPK (Low Cost Precision Kill) – is a new laser-guided 70-mm (Hydra 70) rocket system that the US is developing to give its helicopters (primarily) a new affordable precision-guided weapon capability. The weapon is being developed by General Dynamics and BAE Systems, and the development programme was launched in February 2003. APKWS will fill the large capability gap between the Army's 2.75-in (70-mm) rockets and the AGM-114 Hellfire. The Marines are also very interested in the system and the Zulu will be one of the first types to be equipped with the APKWS.

The Joint Common Missile (JCM, formerly known as the Common Modular Missile) is another 'big ticket' future weapons programme that is intended to replace the AGM-65 Maverick and AGM-114 Hellfire on a range of rotary- and fixed-wing platforms across the US Army, Navy Air Force and Marines. Three competing bids from Boeing/Northrop Grumman, Lockheed Martin and Raytheon answered the JCM request for proposals in September 2003 and a decision on which to take forward is expected before the end of 2004. The JCM is slated to enter service in 2009 on the AH-1Z, which has been designated as the 'pilot platform' for the missile. The JCM is supposed to incorporate a direct fire mode, to replace the TOW capability that Cobra crews crave.

AH-1Ws pilots are supremely enthusiastic about the AH-1Z but they still retain an element of professional cynicism. As one pilot put it, "Well, we look at the Hornets and their experience. Only now, with the E/F, are they doing everything they promised for the A/B. They tell us that the Zulu will do everything. We'll wait and see."

With a positive LRIP decision secured for the Marines, Bell Helicopter's attention has now firmly switched to potential export customers for the AH-1Z. It was one of the bidders for the Australian Army's Air 97 requirement, that was won by the Eurocopter Tiger. Bell offered a variant dubbed the 'Viper' that was essentially identical to the Marines' AH-1Z. It became embroiled in a legal challenge to the Australian selection process when it failed to make

Below: Fitted with Hawkeye turret, an NAH-1Z taxis along the HX-21 ramp at 'Pax'. In the background is a prototype of the UH-1Y, which shares many of the features of the Zulu. The bulk of avionics test work is to be undertaken by Zulus 2 and 3, leading to an operational evaluation in late 2004.

Bell has been trying to sell its AH-1Z to Turkey for some years, and despite being chosen as the winner of the ATAK competition in 2000, a conclusion to the deal seems as far away as ever in early 2004. Turkey currently operates both single- and twin-engined Cobra variants (AH-1W above). The AH-1Z KingCobra on offer for the ATAK requirement has been illustrated with a wheeled undercarriage (below), an option for AH-1Z customers.

the shortlist in 1999. The following year Bell was allowed to resubmit its bid but it was ultimately unsuccessful.

Bad news in Australia was offset by some good news in Turkey where Bell was selected to supply up to 145 new attack helicopters worth up to $4 billion. A variant of the AH-1Z, the AH-1Z KingCobra, was chosen by the Turkish Army to meet its ATAK combat helicopter requirement. Turkey was already a satisfied customer for the AH-1S and AH-1W. The plan for the KingCobra, in July 2000, called for three production batches to be built in Turkey by Tusas Aerospace Industries (TAI), beginning in November 2003 (delayed from an the intended date of 2002). Since then, the Turkish deal has been mired in controversy and dispute – most concerning pricing and industrial co-operation issues – and it seems to have largely disintegrated. Hopes that the deal would get back on track following the 2003 DAB approval of the AH-1Z programme in the US proved to be over-optimistic.

While Bell will make little comment on the saga of the renegotiation of the renegotiations, the latest press reports in March 2004 suggest that the whole programme will now be cancelled (along with others in Turkey) as a result of budgetary restrictions. If this happens it would leave Taiwan as the only near-term prospect for the AH-1Z.

Taiwan is already an AH-1W customer, but beginning in mid-2003 it evaluated both the AH-64D Longbow Apache and the AH-1Z as possible future attack helicopters. Taiwan had hoped to launch its procurement plan by the end of that year, leaving the AH-64D as the only available candidate, but this schedule has now slipped well into 2004. No funding has yet been reserved for the programme. There is a requirement for up to 75 new helicopters to equip three aviation brigades in Taiwan, although most observers expect an initial order of around 25 helicopters to equip just one unit and perhaps a final buy of only 45. A decision in Taiwan is now expected to emerge in mid-2004. Bell's chances may be improved by the prospect of a deal with Taiwan's AIDC (rather than TAI) to supply tail boom assemblies for the US AH-1Z programme. If the AH-1Z is selected by Taiwan, Bell and AIDC may also strike a local production deal for the whole aircraft. There is also the possibility that Bell could find a role in Korea's still emerging Korean Multi-role Helicopter (KMH) project.

Robert Hewson

As the chance of selling new Cobras to Turkey wanes, Taiwan – which already operates the AH-1W (illustrated) – has become the brightest hope for an export order for the AH-1Z. Export capacity theoretically becomes available in 2005, fitting well with Taiwan's nascent plans. Bell is also dangling the 'carrot' of sub-contract work on the USMC's AH-1Z programme, and also the possibility of licence production if Taiwan chooses the AH-1Z over the rival Boeing AH-64.

AH-1 Cobra operators

Bahrain

The Bahrain Amiri Air Force, which was established in 1976 as a small paramilitary air wing, was expanded considerably in capacity from the mid-1980s. In the aftermath of the Iraqi invasion of Kuwait in August 1990, Bahrain was offered eight AH-64A Apaches plus 450 associated AGM-114 Hellfire missiles in October, but this offer was rejected in favour of a cheaper option. This covered eight AH-1E Cobras and six TAH-1P combat trainers. The Cobras were delivered from 1994. Another batch of 16 AH-1Es followed, deliveries of which ran until July 1997. The Cobras were allocated to Nos 8 and 10 Squadrons at Rifa'a. It is thought that 12 of the AH-1Es remain in service, and all of the TAH-1Ps. In 2001 Bahrain announced its intention to upgrade the aircraft. However, the BAAF is now looking at a batch of 17 ex-US Army AH-1Fs, of which 14 would be made operational with the remainder acting as spares sources.

Iran

Between 1975 and 1978 202 specially-developed twin-engined AH-1J 'Internationals' were delivered to the Shah's Imperial Iranian Army Aviation corps, in both TOW-capable and non-TOW versions. Flying with the post-revolutionary Islamic Republic of Iran Army, the Cobra won a solid combat reputation during the war against Iraq in 1980-1988. Both the armed forces and the Iranian Revolutionary Guard Corps have expressed a keen interest in keeping the AH-1 fleet in active service, and improving the aircraft where possible. About half of Iran's Cobra fleet was destroyed in the war with Iraq. The number of currently operationally available Cobras is unclear, but is probably in the region of 50 or 60, if not more.

Iran has kept a sizeable number of its AH-1Js operational by the reverse-engineering of spare parts. The type serves with both regular army aviation (IRIAA) and the Revolutionary Guards aviation (IRGC).

The Tehran-based Iran Helicopter Support & Renewal Company (IHSRC) is the primary overhaul source for civil and military helicopters in Iran. With such a diverse helicopter fleet in Iranian military service, the IHSRC has developed a wide range of expertise on types from the JetRanger and the Mi-17, to the Chinook and RH-53. However, Iran's large number of pre-revolutionary aircraft is ageing, and hit by international embargoes on parts supply. As a result, an innovative series of helicopter upgrade and reverse-engineering projects have sprung up in Iran.

Overhaul and maintenance support for the AH-1 force is undertaken by the IHSRC and by the country's primary aviation umbrella organisation, HESA (the Iran Aircraft Manufacturing Industry Co.). HESA has exhibited an upgrade for the AH-1 that is understood to be flying on at least five AH-1Js. During the Iran Air Show of 2002 the fourth Cobra upgrade prototype, painted as 'HESA P4', participated in both the static and flying displays. This aircraft had been fitted with a range of new avionics systems, although other items such as the engines and weapons are believed to remain unchanged. The aircraft did, however, sport new flat-pane glazing, similar to that fitted to the AH-1F.

In the forward (gunner's) cockpit a new rectangular MFD screen with soft-touch controls has been fitted in the main panel. This screen was demonstrated during the show to display flight-related data – including engine status (torque, temperature etc.) and basic navigational and positional information. The new MFD completely dominates the space in the cockpit previously given over to the TOW targeting system for the nose-mounted M65 sight, and so it must be assumed that the MFD is directly linked to the TOW weapons system. Mention was made by one HESA spokesman of a new FLIR system for the AH-1, which was not fitted to this aircraft (the M65 sight is a daytime-only system). A Garmin GPS receiver has also been fitted on top of the front coaming of the forward cockpit, with the relevant satellite antenna fitted above the Cobra's nose. Below the aircraft a new radome corresponding to a radar altimeter has been fitted.

The upgraded Cobra has also been equipped with a new RWR system, with four receiver antennas mounted on the nose and tail to provide 360° coverage. The operator's panel for this system is located in the rear cockpit.

Robert Hewson

Above and below: Under 'Project 2091' Iran's HESA has upgraded at least five AH-1Js with flat-pane canopies and a new front cockpit.

ISRAEL

Israeli interest in the attack helicopter concept was accelerated by lessons from the Yom Kippur War. The fundamental issue was how to compensate for an initial 10:1 imbalance of force between the regular armies of Israel and its surrounding Arab nations until such time as Israel's reserves could be mobilised and deployed to the front. Until the October 1973 Yom Kippur War, Israeli generals were sure that any sign of Arab aggression could be detected in time to mobilise the Israeli reserves, but the surprise Egyptian and Syrian attack on 6 October 1973 shattered that belief. Moreover, it became apparent that the Israel Defence Force/Air Force (IDF/AF) was unable to fill the void, for four main reasons:

■ during the first 24 to 48 hours of a war the IDF/AF was committed to the air superiority battle, and could only spare a meagre force of less capable fighters to support the hard-pressed regular ground forces during that critical period.

■ the air superiority battle which opened the Yom Kippur War was a complete disaster for the IDF/AF, which failed to destroy the Egyptian and Syrian Air Defence Force (ADF) and consequently had to operate within hostile ADF-protected areas, in turn restricting the ability of the IDF/AF's fighter-bombers to properly support the ground forces.

■ lack of accuracy or precision strike capability.

■ lack of real-time intelligence.

Interestingly, NATO was facing similar a scenario in the event of a surprise Soviet attack in Europe. The US answer was to modify the attack helicopter concept, which had gained momentum in Vietnam as a suppressive fire platform in heliborne operations, into a Main Battle Tank (MBT) hunter. The attack helicopter promised to cure the four shortcomings in IDF/AF Close Air Support (CAS) during the first 24 to 48 hours of a war:

■ the attack helicopter force mission would be CAS, therefore these helicopters would not participate in the air superiority battle and would be dedicated to the support of IDF troops.

■ operating around the flanks and using Nap Of Earth (NOE) flying, it was hoped that the attack helicopters would be able to operate within ADF-defended zones.

■ compared with 1970s fighter-bomber technology, the attack helicopter armed with Anti-Tank Guided Missile (ATGM) was a true pinpoint attack platform.

■ as the attack helicopters would operate in direct contact with brigade-level forces, the problem of lack of real-time intelligence would be eliminated.

The IDF/AF's first Cobra unit formed at Tel Nof with AH-1Qs. It subsequently standardised on the AH-1S, moved to Palmachim and became known as the 'Northern Squadron'.

Israel's first Cobras were six AH-1Gs, which served with the Cobra Evaluation Flight (badge left) at Tel Nof between 1975 and 1977.

Initially, Israel purchased six Bell AH-1G Cobra helicopters, essentially still heliborne assault escorts without any ATGM capability. The Cobra was allocated the Hebrew name Zefa (Viper), initiating the IDF/AF tradition of naming its attack helicopters after snakes. An evaluation unit was formed at Tel Nof air base during 1975, staffed by a mix of helicopter pilots, fighter pilots and fighter navigators. Less than a year after the arrival of the six AH-1Gs in April 1975, the decision was made to upgrade the flight to squadron status, to convert the AH-1Gs to AH-1Q standard (which was capable of launching TOW ATGMs), and to purchase additional helicopters.

The six AH-1Gs were shipped to the US in May 1977, the flight was disbanded and a squadron was formed. The return of the six AH-1Qs in August 1978 prompted an extensive campaign of co-operation exercises with IDF armoured brigades. The drill was to allocate a formation of attack helicopters to a certain brigade, and it was up to the brigade commander to decide how to utilise the re-enforcement. The most common variations were to allocate an individual sector to the attack helicopters – usually on the flanks of the brigade's sector where the helicopters enjoyed freedom of operation, or to co-ordinate the attack helicopters with a regiment's sector, where they would work closely with the MBT crews.

As tactics were refined the concept of scouts was also introduced, at first using Bell 206s and then Hughes 500M-Ds. During 1979 the squadron added six Zefa helicopters to its inventory and moved from Tel Nof to Palmachim. The new Zefa helicopters were to AH-1S ECAS standard, with 20-mm cannon and flat-pane canopy to reduce sun reflections that revealed the presence of the attack helicopters to hostile forces. Thus, with 11 Zefa helicopters and four 500M-D scouts, the squadron was tested during the June 1982 Lebanon War. The scenario was not quite like the one for which the attack helicopters were purchased, as in this case Israel was on the offensive. The squadron flew 62 sorties, launched 72 TOW ATGMs, claimed the destruction of 51 targets including T-62 and T-72 MBTs, and lost two helicopters.

The first of 30 AH-1S Modernized Cobras were delivered to Israel in 1983, the new variant being equipped with an improved weapon-aiming system including a laser range-finder, composite rotor blades, and a large black nozzle to cool the hot air emission rather than just pointing it upwards, as in earlier Cobra

The 'Southern Squadron' also has a snake motif as part of its badge, as well as a fuselage marking. It is co-located with the 'Northern' unit at Palmachim.

variants. In 1985 a second Zefa squadron was formed, with the first unit becoming known as 'The Northern Squadron' while the new unit became 'The Southern Squadron' to reflect the respective location of the two units within Palmachim air base.

Following the service entry of the McDonnell Douglas AH-64A Peten (Python) in 1990, the IDF/AF turned its attention to a 'night-time operational capability gap closure' programme, aimed at giving the Zefa fleet the option to operate at night with much the same capability as the Peten. The primary elements of this programme were the Elbit Yarchon (Journal) Night Vision Goggles/Head Up Display (NVG/HUD) system that was introduced in 1992, and the replacement of the original daytime targeting system with the IAI Tamam Night Targeting System (NTS), a system identical to the NTS installed in the US Marine Corps AH-1W Super Cobra and known in IDF/AF service as Reshafim (Flashes). Deliveries commenced in 1992, with a formal service entry date in June 1993. This resulted in a vastly expanded operational envelope, as well as certain operational changes. The use of air-to-ground rocket pods was stopped in 1993 as a result of the shift towards precision strike with minimal collateral damage, while the installation of a small Multi-Function Display (MFD) in the front cockpit just above the NTS sight caused a change in the gunner's working methods. Instead of placing the gunner's head close to the sight and virtually ignoring the outside world during the launch of a TOW missile, the MFD (displaying either a daytime TV image or night-time Forward Looking Infra Red/FLIR image), offers the gunner the option to retain situational awareness even when launching a TOW. The original sight has been retained, however, and pre-NTS era gunners tend to use it as they did in the old days. However, all of the younger gunners use the MFD.

Once Zefa crews got accustomed to the Yarchon NVG/HUD system they demanded the introduction of a similar daytime system. The El-Op Zaharon (Noon) – basically a helmet-mounted HUD – entered service in 1994. Further improvements included a Data Transfer Equipment (DTE) system to ease the loading of the navigation system, GPS receiver, missile launch debrief system and a self-protection system. The most recent addition in the ongoing modernisation of the Zefa fleet is the installation of the Elbit Nadiv (Generous) D-Map digital map display, while Zefa aircrews hope that the introduction of the digital map will be followed by the installation of a data link that will allow them, for example, to receive an image of their designated target in real-time for an Unmanned Air Vehicle (UAV).

The Zefa is currently the IDF/AF 'entry-level' attack helicopter platform. The delivery of 14 AH-1Es in 1996 enabled the IDF/AF to introduce the Zefa as the IDF/AF Academy's attack helicopter pilot trainer, while a further 15 US Army surplus Cobras delivered in 2003 are a source of spare parts. A capable platform, the Zefa is also elderly and, as such, is an obvious candidate for an upgrade project. IAI has proposed a General Electric T700 re-engining project, while Bell is promoting the AH-1Z, including the demonstration of a cockpit in Israel in December 2002. However, the IDF/AF is currently committed to an AH-64D purchase programme so attack helicopter force funding is tight at the present. Furthermore, there is an opinion that the operational gap that the attack helicopter was created to fill no longer exists. Combat aircraft field Precision Guided Munitions (PGM), the ADF no longer poses such a threat as it did during the 1970s and a battlefield littered with MBTs is a vision of the past. Reflecting this attitude, a committee headed by former IDF/AF commander (1996-2000) Eitan Ben-Eliyahu recommended in December 2003 to cut the number of IDF/AF attack helicopters by 20 per cent.

Though Soviet MBTs flooding Europe or Syrian MBTs flooding the Golan Heights may be a vision of the past, the attack helicopter has diversified considerably in its operational repertoire since the 1970s, when it was viewed primarily as an MBT hunter. It is fair to assume that the familiar sound of the AH-1's two-bladed rotor will continue to be heard over Israel for at least another decade.

Shlomo Aloni

IDF/AF gunship crews learn their trade on the AH-1Es of the IDF Academy (badge above) attack helicopter training squadron. Although the aircraft have NTS, and some have IR suppressing nozzles, they do not have IRCM or chaff/flare dispensers.

Not all of Japan's AH-1s have yet been fitted with wire-cutters, but the work to fit them fleet-wide continues. The Cobras wear their unit assignment in black letters behind the pilot's cockpit, in this case 'IIIATH' for the 3rd ATH at Metabaru.

JAPAN

In the late 1970s the Rikujo Jiei-tai (Japan Ground Self Defence Force, JGSDF) began to study the potential of attack helicopters seriously. Shortly before this, some JGSDF UH-1B/Hs had been equipped with rockets and tested during some exercises, but they were ineffective against heavily armoured vehicles such as main battle tanks, and also inaccurate. Furthermore, they had no sophisticated sighting systems. In 1977, the JGSDF decided to purchase a dedicated attack helicopter to fully evaluate this capability, and the Bell AH-1S was selected as a matter of course. The AH-1S was the only such aircraft in existence at that time.

The first two AH-1Ss (s/n 73401 and 73402) were imported from the US, and were in the Step 2 configuration. In the US this was redesignated as the AH-1E but the JGSDF continued to use the name of AH-1S. The two aircraft were assigned to the Kyoiku Shien Hiko-tai (School Support Squadron, SSS) at Akeno AAB in 1979. The main role of the SSS was to establish the standards of education and training for aircrew, but they also flew the aircraft on research, development, test and evaluation tasks. The evaluation period for the AH-1S took almost three years, after which the JGSDF confirmed the firepower effectiveness that the dedicated attack helicopter would bring. A plan to establish the Tai Sensha Herikoputa-tai (Anti-Tank Helicopter Unit, ATH) was settled in 1982 and procurement of the AH-1S was authorised.

Fuji Heavy Industries (FHI) had a long relationship with Bell Helicopters through maintenance and modification programmes, and had manufactured Model 204s and 205s under the license. Therefore, FHI was named as the licence-manufacturer of the AH-1S. In the initial plan, enough helicopters for 5½ squadrons were to be produced. The regular complement of a squadron is 16 aircraft, so 5½ squadrons equated to 88 AH-1Ss. A further eight aircraft were to be manufactured to cover for attrition, for a total of 96. In the event, the plan was revised to 90 AH-1Ss due to the lack of attrition, and finally cut to 89. In the late 1990s the JGSDF embarked on a follow-on attack helicopter programme (AH-X), ending procurement of the AH-1S. The final order for the 89th aircraft (73492) was made in FY1998 and delivered to the JGSDF on 14 December 2000.

These licence-built AH-1Ss were all in the Step 3 version, which subsequently became the AH-1F in US Army parlance. However, again the JGSDF elected to retain the AH-1S designation for its aircraft.

AH-1S in service

Five front-line ATHs have been formed since 1986. The JGSDF divided Japan into five Homen-tai (Regional Army) by geographic areas. They are Hokubu Homen-tai (Northern Army), Tohoku Homen-tai (Northeastern Army), Tobu Homen-tai (Eastern Army), Chubu Homen-tai (Central Army) and Seibu Homen-tai (Western Army). Each Regional Army has a Homen Koku-tai (Regional Army Air Group, for example Hokubu Homen-tai has Hokubu Homen Koku-tai = Northern Army Air Group) and one ATH was assigned to the control of each Regional Army Air Group. Each ATH consists of two squadrons called Dai 1 Hiko-tai (1st Squadron) and Dai 2 Hiko-tai (2nd Squadron), and each squadron was equipped with eight AH-1Ss and two OH-6Ds for scout duties (in recent years, conversion to the Kawasaki OH-1 from the OH-6D has been under way).

Dai 1 Tai Sensha Herikoputa-tai (1st ATH): the 1st ATH was formed on 25 March 1986 at Obihiro AAB in Hokkaido as the first ATH in the JGSDF. The 1st ATH started to receive the AH-1S on 30 September 1985. 1st ATH is assigned to Northern Army Air Group.

Dai 2 Tai Sensha Herikoputa-tai (2nd ATH): the 2nd ATH was formed at Hachinohe AAB on 25 March 1989. The unit received its first AH-1S on 25 March 1988. 2nd ATH is under the control of Northeastern Army Air Group.

Dai 3 Tai Sensha Herikoputa-tai (3rd ATH): the 3rd ATH was formed at Metabaru AAB on 29 March 1990 and is assigned to Western Army Air Group.

Dai 4 Tai Sensha Herikoputa-tai (4th ATH): the 4th ATH was formed at Kisarazu AAB on 31 March 1992 and is assigned to Eastern Army Air Group. The first two AH-1Ss which were imported were eventually assigned to this unit although they were withdrawn from use in 2001.

Dai 5 Tai Sensha Herikoputa-tai (5th ATH): the 5th ATH was formed at Akeno AAB as the final ATH in the JGSDF on 28 March 1994. Because of procurement budget restrictions, the 5th ATH was established with only half of its strength with only one squadron initially assigned. When the final AH-1S was delivered to the JGSDF, the 5th ATH reached full strength. The 5th ATH is under the control of Central Army Air Group.

Koku Gakko Honko (Headquarters School): this unit at Akeno AAB handles training with a strength of eight aircraft.

Japan's AH-1S aircraft are essentially the same as those of the US Army. After the two earliest aircraft (73401 and 73402) were retired, all of the JGSDF's AH-1 fleet were to the same standard as the AH-1F. Standard weapons consist of eight BGM-71 TOWs and two M261 rocket pods (each containing 19 2.75-in/70-mm FFARs). It is possible to switch M261 and TOW launchers, so alternate configurations are 16 TOWs or four M261 pods. An M197 three-barrelled 20-mm cannon is fitted under the nose. In the extreme nose is an M65 telescopic sight unit for the TOW.

After the 73rd aircraft (73473, including the two imported AH-1Ss), the C-Nite FLIR system for night operations was introduced. In the JGSDF's C-Nite aircraft the transmitter for the radar jammer was moved to forward of the front cockpit, while the same system's receiver was moved to the main rotor mast fairing. A pair of wire-cutters was added to the top and bottom of the forward fuselage. These are also being retrofitted to the existing aircraft but the work is proceeding slowly.

At present, the JGSDF is to reorganise and to form a new unit called the Kuchu Kido Ryodan (Air Mobility Brigade, AMB). This unit will be a mixed aviation force comprising eight heavy transport helicopters (CH-47JA), eight utility helicopters (UH-60JA), five observation helicopters (OH-1) and eight attack helicopters. The attack helicopter for the AMB is the AH-X, for which the Boeing AH-64D Apache Longbow has been selected. Initially, there will be no effect on the current AH-1S fleet and the five ATH units, but when procurement of the AH-64D is half-complete, at least one ATH is due to disband.

Yoshitomo Aoki

These AH-1Ss are from the 5th ATH (right) at Akeno and the 4th ATH (below) at Kisarazu. Akeno was the first JGSDF base to host Cobras during the evaluation phase, and today houses the Cobra training school.

Jordan

The Royal Jordanian Air Force identified a requirement for an anti-armour helicopter in the late 1970s. Initially the HOT-armed Aérospatiale SA 342 Gazelle was considered, but this was rejected after the US offered a batch of 24 AH-1S Cobras with BGM-71 TOWs in 1981. A contract worth US$196 million was signed in September 1982, and the aircraft were delivered between January and December 1985, equipping Nos 10 and 12 Squadrons at Al Jafr (serialled 1001 to 1012 and 1201 to 1212). The Cobras were delivered in AH-1S Modernised configuration with heat-shields on the exhausts but lacking IRCM turrets. They were subsequently redesignated as AH-1Fs.

The two Cobra squadrons subsequently moved to Al Matar Air Base near Amman/Marka, from where they continue to operate the survivors (believed to number 20). In 2000/01 nine ex-US Army AH-1Fs were delivered to augment the fleet.

Jordan's Cobras are operated by No. 10 (top left) and No. 12 (top right) Squadrons, both based at Al Matar Air Base.

Republic of Korea

In 1978 the Republic of Korea Army Aviation received eight TOW-armed, twin-engined AH-1J Internationals. Experience with these led to the later adoption of the single-engined AH-1S as its principal attack helicopter. An initial order, for 21 TOW-armed AH-1Ss, was placed in 1985 with a value of US$178 million, including spares and support. By 1987 the contract had been expanded to cover 42 aircraft at a cost of US$260 million, to be purchased through regular commercial channels.

The first of them was delivered in 1988, and deliveries were maintained at around three per month. In 1990 an additional 20 were ordered through FMS (part of an option for 28), the sale being approved by the US Congress in November. The helicopters were delivered as Modernised AH-1S equipped with C-Nite, subsequently becoming AH-1Fs.

Augmentation and replacement of the RoKAA Cobra fleet has been a long saga, beginning in July 1992 when the service ordered 37 AH-64A Apaches (and 775 Hellfires) through FMS channels but did not complete the purchase. The requirement was revived in 1996, and in March 2000 Requests for Proposal for a new attack helicopter were issued to seven companies, including Bell. The competition was subsequently whittled down to the AH-1Z, AH-64D and Kamov Ka-52K. An initial order for 36 aircraft was expected, with an eventual requirement of around 60. Funding problems continue to postpone the final decision, and it is likely that the requirement will be rolled into a requirement for a smaller multi-purpose attack scout helicopter, for which the indigenous KMH is front-runner.

This AH-1S belongs to the RoKAA's 107th/1st Aviation Brigade.

Pakistan

Pakistan's Army Aviation Corps underwent a considerable expansion from the mid-1970s. An attack helicopter requirement resulted in competition between the Hughes 500MD and the Bell AH-1S, resulting in an order for the Cobra. Twenty were ordered (with an option for a further 20) and the first batch of 10 was delivered in late 1984. They officially entered service in March 1985. The second batch of 10 was delivered in early 1986. A revised US$89 million option for 10 more through FMS was due to be exercised, and formal notification was given in Washington in January 1990. However, the subsequent arms embargo placed by the US later that year over Pakistan's nuclear weapons programme ended any further deliveries. In late 1995 Washington was prepared to resume arms sales, resulting in the delivery of 18 C-Nite night targeting sets, 135 TOW 2 launchers and 16,720 2.75-in rockets for the 19 remaining Cobras.

Pakistan's Cobras (serialled 786-001 to 786-020, now designated AH-1F) were placed in service with Nos 31 and 32 Squadron at Multan, from where the survivors continue to operate. Both squadrons also operate a few Bell 206B JetRangers for communications and training. Serviceability of the Cobras was hampered by the arms embargo, but it is believed that 18 of the 20 are still in use.

In 2004 Pakistan was expecting to receive a large batch of ex-US Army AH-1Fs, thought to number 40 aircraft.

Pakistan has two squadrons of AH-1Fs at Multan. Since the end of Operation Enduring Freedom the fleet has seen sporadic action against Taliban/al-Qaeda insurgents operating in the mountains along the Afghan border.

TAIWAN

The RoCAA's AH-1Ws are a powerful counter to any aggression from the mainland.

Taiwan issued an initial requirement for attack helicopters in 1984, evaluating the MBB BO 105 and Hughes 500. Following the ironing out of export difficulties, by early 1992 the requirement had crystallised into a firm order for 42 Bell AH-1Ws. The aircraft were completed to Integrated Weapons System (Phase 1) specification, including the Tamam/Kollsman NTSF-65 night targeting system, compatible with the AGM-114A Hellfire missile. As part of the same package 12 Hellfire-capable OH-58D Kiowa Warriors were also ordered (with 14 options, later exercised). In May 1997 Taiwan placed an order for a further 21 AH-1Ws as part of a US$479 million contract.

Taiwan's 'Whiskeys' were built in six production lots, the first (serialled 501) being delivered in March 1993. Deliveries of the first batch of 42 ran until 1997. The aircraft are distributed between the Army Aviation Training Centre at Kuejien-Tainan and two Attack Helicopter Wings within the Republic of China Army Aviation. Each wing has two squadrons: No. 1 Attack Helicopter Wing (also known as the 601st Aviation Brigade) at Lungtan-Tao Yuan parents Nos 1 and 2 Attack Helicopter Squadrons, while No. 2 Attack Helicopter Wing (602nd Aviation Brigade) at Hsinshe includes Nos 3 and 4 Attack Helicopter Squadrons. The AH-1Ws work closely with the OH-58D force, which is divided into two helicopter Reconnaissance Squadrons, one of which is allocated to each wing.

In mid-2003 the RoCAA opened an evaluation for a new attack helicopter, for which the Bell AH-1Z and Boeing AH-64D are competitors. A decision to proceed with this requirement had not been taken by early 2004, but might be expected to result in an order for up to 75 aircraft.

THAILAND

Thailand's army aviation gained independence from the air force in 1952, and was established as the Kongbin Tha Han Bo – Royal Thai Army Aviation Division. The RTAAD was primarily equipped with transport types such as the UH-1H, but in the mid-1980s decided to add an attack helicopter capability.

In mid-1986 four Bell AH-1S Cobras were ordered to form an attack flight. They were delivered in 1990 as AH-1Fs (serialled 9996 to 9999), and were intended to form an operational cadre pending procurement of larger numbers of surplus US Army aircraft. However, plans for further procurement of AH-1Ws or more AH-1Fs have been shelved due to budgetary restraints in the light of Thailand's economic problems of the late 1990s. A minor upgrade for the three existing aircraft has been funded, 9999 having been written off in an accident on 23 July 2001.

TURKEY

The Türk Kara Kuvvetleri (Turkish army aviation) sought attack helicopters from the late 1970s. The TKK was offered six AH-1Ss through FMS in late 1983, the deal being worth US$50 million, including BGM-71 TOW missiles, but this deal lapsed. However, in July 1990 five AH-1Ws were delivered, diverted from Lot III production for the US Marine Corps. Another five were delivered in 1993, this time from Lot IV production. At the time it was planned that another 42 would be assembled locally. Attempts to purchase a further 10 AH-1Ws for US$145 million foundered in 1996 due to Turkey's disputes with Greece and over its continued use of attack helicopters against Kurdish insurgents.

The 10 AH-1Ws were delivered to a newly established Taaruz Helikopteri Taburu (attack helicopter battalion) at Ankara-Güvercinlik. The aircraft were subsequently augmented by 32 ex-US Army AH-1P/S aircraft delivered between 1993 and 1995. Included in this batch were four TAH-1P trainers. AH-1Ws serve with 1nci Filo, while the AH-1P/Ss are divided between 2nci and 3nci Filos.

The AH-1P/Ss were subsequently upgraded to a similar standard to the AH-1F, with GPS, three-barrelled 20-mm M197 cannon in place of the original 7.62-mm Minigun, radar warning receiver and night targeting system. One AH-1W was shot down by Kurdish guerrillas on 18 May 1997 during operations in northern Iraq.

In May 1997 the TKK requested proposals from Agusta, Bell, Boeing, Denel, Eurocopter, McDonnell Douglas Helicopters, Mil and Sikorsky for a new attack helicopter, outlining a US$3.5 billion requirement for an initial batch plus ensuing licence production for a total of 145. A last-minute proposal was also received from the MiG group, representing Kamov's Ka-50. The programme was repeatedly delayed by political changes and continuing international hostility to Turkey's handling of the Kurdish independence question. Eventually, in August 2000 the Bell AH-1Z King Cobra was selected, although continued wrangling over financing has resulted in further delays. Kamov's Ka-50-2 Erdogan with Israeli avionics technically remains in the running until a Bell deal can be concluded. The original 145-aircraft requirement is now considered too expensive, and current plans suggest 50 machines, with the first seven to be built at Fort Worth. However, in 2004 it was also considered likely that the whole attack helicopter requirement would be cancelled.

Based at Güvercinlik, the Taaruz Helikopteri Taburu has three squadrons of Cobras: one with AH-1Ws (below) and two with AH-1P/Ss (below left).

UNITED STATES MARINE CORPS

The US Marine Corps has operated the world's largest fleet of Cobras since 2001, when the US Army National Guard retired the AH-1F. Since receiving its first AH-1G HueyCobras in 1968 the USMC has operated four distinct variants of the Cobra, SeaCobra and SuperCobra that include the AH-1J, AH-1T and AH-1W. Although 186 AH-1Ws are currently in the inventory, Bell delivered 225 examples to the Marines between 1986 and July 1998. Production comprised a single prototype, and 180 new-build AH-1Ws. Additionally, 44 earlier AH-1Ts were updated to the later configuration and delivered between 1989 and 1992.

Whereas Marine Attack Helicopter Squadrons (HMA) originally operated the aircraft, today the AH-1W is in service with eight Marine Light Attack Helicopter Squadrons (HMLA) and a single Marine Helicopter Training Squadron (HMT). Six of the HMLA squadrons are assigned to the active component while the remaining two are components of the Marine Corps Reserve Force. Although the number of aircraft operated by the individual squadrons varies, each is nominally assigned 18 AH-1Ws and nine UH-1Ns. The reserve units are unique in being broken down into the parent squadron and a geographically separate detachment. While each of the parent squadrons is nominally assigned 12 AH-1Ws and six UH-1Ns, the detachments are equipped with six AH-1Ws and four UH-1Ns.

Super Cobras are also operated by three US Navy squadrons in support of test and operational evaluation efforts. The test pilots and support personnel within these squadrons include sailors and Marines.

Tom Kaminski

MCAS New River is 'home' to two Cobra/Huey units, HML/A-167 'Warriors' (above) and HML/A-269 'Gunrunners' (right). The New River-based squadrons officially retain the '/' in their unit designations, although it is deleted by West Coast squadrons.

2nd Marine Air Wing (MAW) – MCAS Cherry Point, North Carolina

HML/A-167 'Warriors' – 'TV'
Activated as Marine Light Helicopter Squadron One Six Seven (HML-167) on 15 March 1968, the 'Warriors' were equipped with the UH-1E. HML-167 was the last Marine helicopter squadron to operate in Vietnam and its final pair of helicopters departed on 15 June 1971. The squadron transitioned to the UH-1N in April 1972 and in January 1984 accepted its first AH-1Ts. Redesignated as HML/A-167 on 1 April 1986, the squadron transitioned to the AH-1W beginning on 11 August 1989. HML/A-167 is a component of Marine Air Group MAG-26 and is stationed at MCAS New River, North Carolina.

HML/A-269 'Gunrunners' – 'HF'
Initially equipped with the AH-1J, the 'Gunrunners' were activated as HMA-269 on 1 July 1971. Beginning in December 1977, the squadron transitioned to the AH-1T. It began operating the UH-1N alongside the Sea Cobra in 1983 and received its current designation on 1 April 1986. In 1990 the 'Gunrunners' exchanged the AH-1T for the AH-1W. HML/A-269 is assigned to MAG-29 and is stationed at MCAS New River, North Carolina.

3rd MAW Marine Air Wing (MAW) – MCAS Miramar, California

HMLA-169 'Vipers' – 'SN'
When activated as HMA-169 on 30 September 1971, the 'Vipers' were equipped with the AH-1G. The unit transitioned to the AH-1J in 1975 and the AH-1T followed in May 1978. On 1 October 1986 the squadron received its current designation and just 10 days later took delivery of its first AH-1W. HMLA-169 is stationed at MCAS Camp Pendleton, California, as a component of MAG-39.

HMLA-267 'Stingers' – 'UV'
The 'Stingers' were activated as Marine Observation Squadron VMO-5 on 15 December 1966 and were equipped with the OV-10A and UH-1E. Redesignated HML-267 on 15 March 1968, the unit operated both types until November 1971 when the Broncos were reassigned. The 'Stingers' took delivery of the first UH-1N in March 1976 and added the AH-1J in May 1983. The squadron received its current designation on 1 March 1987 and began its transition to the AH-1T later that same month. HMLA-267, which is assigned to MAG-39 and stationed at MCAS Camp Pendleton, California, took delivery of its first AH-1W in 1989

HMT-303 'Atlas' – 'QT'
Activated on 30 April 1982, Marine Helicopter Training Squadron HMT-303 is the fleet readiness squadron (FRS) for both the AH-1W and the UH/HH-1N. The squadron is a component of MAG-39 and is based at MCAS Camp Pendleton, California.

HMLA-367 'Scarface' – 'VT'
Stationed at Camp Pendleton, California, HMLA-367 is a component of MAG-39. Activated as VMO-3 on 1 August 1966 and equipped with the UH-1E, the unit was redesignated as HML-367 on 24 March 1968. It transitioned to the AH-1G in 1969 but later exchanged its Cobras for UH-1Es in 1971. In January 1976 the unit transitioned to the UH-1N, however a detachment of AH-1Js was added to the organisation in 1977. The first AH-1Ts arrived in 1985 and the

Above: HMLA-169 'Vipers'

Below: HMLA-267 'Stingers'

Above: HMLA-367 'Scarface'

Right: HMLA-369 'Gunfighters'

squadron received its current designation on 1 January 1988. It took delivery of its initial AH-1W on 11 August 1989.

HMT-303 'Atlas' acts as the training squadron for both the AH-1W Cobra and the UH/HH-1N Huey.

HMLA-369 'Gunfighters' – 'SM'
When activated on 1 April 1972, HMA-369 was equipped with the AH-1J. However, the squadron received a small number of UH-1Ns during 1984. The Hueys were permanently assigned in 1987 and the unit was redesignated as HMLA-369 on 15 September 1987. The squadron's first AH-1W was assigned on 30 October 1987 and the last AH-1J departed on 1 May 1988. HMLA-369 is currently stationed at MCAS Camp Pendleton, California, as a component of MAG-39.

4th Marine Air Wing (MAW) – NSA New Orleans, Louisiana

HMLA-773(-) 'Red Dogs' – 'MP'
Activated as HMA-773 on 1 September 1971, the 'Red Dogs' were initially equipped with the AH-1G but transitioned to the AH-1J in late 1978. HMA-773 accepted its first AH-1W on 8 October 1992 and the first UH-1N arrived in November 1993. The squadron, which is assigned to MAG-42 and stationed at NAS Atlanta, Georgia, received its current designation on 1 October 1994.

HMLA-773, Det. A 'Red Dogs' – 'MM'
On 1 October 1994 HML-767 was redesignated as HMLA-775 Detachment A, at NAS New Orleans, Louisiana. The unit, which was then a component of HMLA-775 at MCAS Camp Pendleton, was equipped with the AH-1W and UH-1N. During 2000 the unit was reassigned to HMLA-773 at NAS Atlanta, Georgia, and assumed its current identity.

HMLA-775(-) 'Coyotes' – 'WR'
Activated as HMA-775 on 7 January 1989, the 'Coyotes' were equipped with the AH-1J. The squadron received its first AH-1W in May 1992 and added the UH-1N to its inventory in June 1994. It received its current designation in October 1994. HMLA-775(-) is a component of MAG-46 and is stationed at MCAS Camp Pendleton, California.

Below left: HMLA-773. Below: HMLA-775

HMLA-775 Det. A 'Coyotes' – 'WG'
Originally established as HMLA-773 Detachment A, at NAS JRB Willow Grove, Pennsylvania, on 10 September 1997, this unit was equipped with AH-1W and UH-1N from the outset. During 2000 the unit was reassigned to HMLA-775 at MCAS Camp Pendleton, California, and assumed its current identity. On 3 March 2001 the detachment relocated to Johnstown-Cambria County Airport, Pennsylvania.

United States Navy

Commander Naval Air Systems Command (COMNAVAIRSYSCOM)

HX-21 'Rotary Wing'
Stationed at NAS Patuxent River, Maryland, HX-21 operates three AH-1Ws in support of research, development test and evaluation (RDT&E) associated with the AH-1W. Based at NAS Patuxent River, Maryland, and assigned to the Naval Air Warfare Center Aircraft Division (NAWC-AD), the squadron also supports the H-1 Upgrades programme.

VX-31 'Dust Devils'
VX-31 is tasked with conducting RDT&E in support of AH-1W weapons programmes. The squadron, which is based at NAWS China Lake, California, and is a component of the Naval Air Warfare Center Weapons Division (NAWC-WD), operates a single AH-1W.

Commander Operational Test & Evaluation Force (COMOPTEVFOR) – NS Norfolk, Virginia

VX-9 'Vampires' – 'XE'
Stationed at NAWS China Lake, California, the 'Vampires' operate three AH-1Ws in support of operational testing (OT) and tactics development efforts.

Trials and development work for the Cobra fleet is undertaken at the US Navy's two main test sites: China Lake and Patuxent River. Weapons and tactics trials are undertaken at the former by VX-31 and VX-9 (left). Systems evaluation is undertaken at 'Pax' by HX-21, which operates some AH-1Ws (right). The Zulu upgrade evaluation is also undertaken at 'Pax', using NAH-1Zs (above right).

Firefighting Cobras

The US Department of Agriculture Forest Service will soon begin testing a modified AH-1F as an aerial firefighting supervision platform. Known as the 'Firewatch Cobra', the aircraft is one of 25 retired AH-1Fs acquired by the Forest Service via the US Army's excess property programme in early 2003. The modified aircraft will initially be assigned to the Redding Air Attack Base in California during the 2004 fire season. In addition to serving as a lead plane and as an airborne command and control platform, the 'Firewatch Cobra' is capable of providing video and infrared mapping and downlinking the data to a receiver on the ground in real time. Although equipped with an EO/IR turret, the aircraft is not capable of carrying cargo or delivering retardant.

Three former US Army AH-1Ps that are operated by the Florida Division of Forestry Aircraft Operations Division are, however, capable of conducting firefighting operations. The helicopters can carry a 320-US gal (1211-litre) bucket, or a foam/retardant insertion system can be installed in the Cobra's forward ammunition bay. The system is equipped with a 360-US gal (1363-litre) fixed tank capable of delivering water/retardant or foam. Developed by Garlick helicopters Inc., the aircraft are known as B-209 'Firesnakes'.

AH-1F	N109Z	69-16422	USDA Forest Service
AH-1P	N130FC	76-22694	Florida Division of Forestry Aircraft Operations division
AH-1P	N131FC	76-22570	Florida Division of Forestry Aircraft Operations division
AH-1P	N132FC	77-22747	Florida Division of Forestry Aircraft Operations division

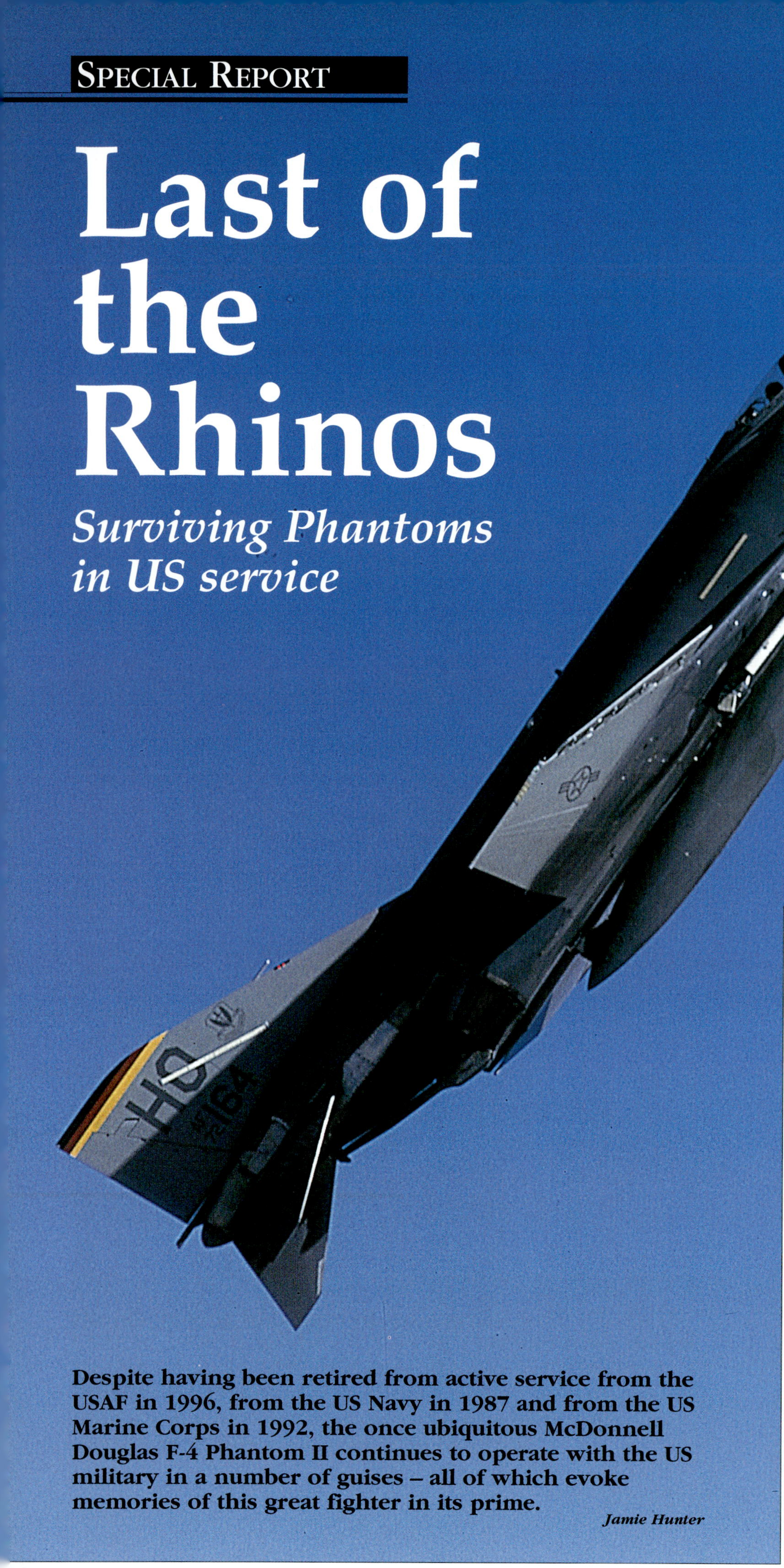

Last of the Rhinos

Surviving Phantoms in US service

Despite having been retired from active service from the USAF in 1996, from the US Navy in 1987 and from the US Marine Corps in 1992, the once ubiquitous McDonnell Douglas F-4 Phantom II continues to operate with the US military in a number of guises – all of which evoke memories of this great fighter in its prime.

Jamie Hunter

Left and below: For the foreseeable future the QF-4 (especially in its QF-4G version) will provide the US forces with a Full-Scale Aerial Target capability. The USAF's 82nd ATRS, which operates the Phantoms from Tyndall and Holloman, will assume FSAT duties for US Navy tests when VX-30 retires its last QF-4s. The yellow van parked in front of the QF-4G at left is the mobile control post for handling the take-off and landing of the drones in unmanned (NULLO) configuration.

Left: Holloman AFB in New Mexico is the last bastion of the Phantom in USAF service, with two F-4 units based there. Alongside the drones of the 82nd ATRS Det 1 are the F-4F ICE aircraft of the 20th Fighter Squadron. Attached to the 49th Fighter Wing, which operates the F-117 as its primary mission equipment, the 20th FS will train Luftwaffe Phantom pilots until 2004. In the Luftwaffe the F-4F is to be replaced by the Eurofighter Typhoon, and the long-standing requirement for new Phantom aircrew has finally come to an end. Note the German flag fin-stripe.

Below: In June 2004 US Navy Phantom operations come to an end, the last few aircraft serving with VX-30 at Point Mugu. The QF-4S+ aircraft carry out a variety of weapon-related test work in both manned and unmanned configurations.

82nd ATRS/53rd WEG – Air Force drones

The USAF retired its last active Phantoms when the F-4Gs of the 561st FS at Nellis AFB, Nevada, were followed into the boneyard by the fellow 'Wild Weasels' of the 190th FS, Idaho ANG, from Boise on 20 April 1996. These aircraft went directly to AMARC at Davis-Monthan AFB, but have steadily re-emerged for conversion as QF-4 Full Scale Aerial Targets (FSAT), or 'drones', by BAE Systems Flight Systems (formerly Tracor Flight Systems) at Mojave, California. The Phantom is the latest in a long and distinguished line of FSAT types (F-86, F-100, F-102 and F-106), which similarly saw service long after retirement from the front line. The first QF-4E joined the 82nd Aerial Targets Squadron on 31 October 1995.

The QF-4 conversion involves the installation of remote control equipment and proximity indication devices for missile and other weapons test programmes. An initial batch of 10 pre-production QF-4Es and QRF-4Cs for the USAF was completed from 1991, with a total requirement for up to 330 QF-4s for the USAF. Indeed, 2003 saw the award of the latest contract for 24 conversions. Contrary to popular opinion, these QF-4s – along with their US Navy FSAT counterparts – are equally important in manned test roles. However, their ultimate aim is to be used in an unmanned (NULLO – Not Under Live Local Operator) capacity for destructive missile testing – most recently becoming victims to AIM-120 AMRAAMs or AIM-9Xs.

The USAF QF-4s are 'owned' by the 53rd Weapons Evaluation Group (WEG) at Tyndall AFB, Florida. This unit is the umbrella for four squadrons and one detachment: the 81st Test Support Squadron, the 82nd Aerial Targets Squadron ('TD' tailcode, far left), the 83rd Fighter Weapons Squadron – all located at Tyndall, the 86th Fighter Weapons Squadron at Eglin, and Det 1 82nd ATRS at Holloman AFB, New Mexico ('HD' tailcode, below). The group's trials programme fulfils the USAF's Air-to-Air Weapon System Evaluation Program, known as Combat Archer (COMACC Plan 85), and the USAF Air-to-Ground Weapon System Evaluation Program, known as Combat Hammer (COMACC Plan 90). It also supports the Weapons Instructor Course air-to-air formal training syllabus under COMACC Plan 92.

Holloman's Detachment 1 provides QF-4s for weapons testing at the White Sands Missile Range, adjacent to Holloman. This detachment is commanded by Lieutenant Colonel Charlie 'Tuna' Hainline (left), who assumed command on 10 December 2003 having previously flown a variety of types, including the F-117. The Holloman unit operates up to 15 QF-4s, with the current fleet mainly comprised of QF-4Gs. Alongside Lt Col Hainline, one further military and four civilian contract pilots fulfil the manned QF-4 mission requirements and generally fly five missions each week. Recent trials have seen the QF-4 community heavily involved in trials with the USAF F/A-22 Raptor from Edwards AFB, California.

20th FS/49th FW – Luftwaffe training

As well as its QF-4 association, Holloman AFB is also home to a further Phantom unit in the form of the German AF F-4F training squadron (1st GAFTS) – also known by its USAF designator as the 20th FS 'Silver Lobos'. The 'Silver Lobos' are part of the 49th FW and provide the Luftwaffe with aircrew combat instruction on the F-4F. The squadron conducts both initial qualification training for new pilots and weapon systems officers (WSOs) and the Fighter Weapons Instructor Course for experienced F-4 crews. The squadron comprises both USAF and Luftwaffe instructors and the aircraft are 'loaned' to the USAF for these training duties. They are operated under USAF procedures and guidelines by Air Combat Command.

On 1 December 1972, the 20th Tactical Fighter Training Squadron (TFTS) stood up at George AFB, California, as part of the 35th Tactical Fighter Wing (TFW). The German AF training programme began in 1975 with early Luftwaffe F-4Fs. As USAF Phantom operations drew down at George the squadron transferred to Holloman in June 1992 and became the 9th FS. The unit again assumed the 20th FS mantle on 30 July 1993 as the F-117 joined the 49th FW and took on the 7th, 8th and 9th FS numberplates.

The 'Silver Lobos' have flown a mixture of F-4Es and F-4Fs, with the F-4Es purchased to free up assets for the front-line Luftwaffe units back home. By the late 1990s, the F-4Es had been retired and the unit operated 24 F-4Fs, with most wearing the USAF Egyptian One scheme. By late 2003 the unit had shrunk to around 15 F-4F ICE (Improved Capability Effectiveness) Phantoms, reflecting the drawdown in Luftwaffe Phantom operations. The unit will complete its last course for the Luftwaffe in 2004 when the mission of the 'Silver Lobos' will come to an end.

VX-30 – End of the Navy Phantom

The last US Navy fleet Phantoms were the F-4Ss of VFA-202 'Superheats' at NAS Dallas, Texas. This unit made the last ever Phantom trap, aboard USS *America* (CV 66) on 21 October 1986, and the last of the squadron's 'Rhinos' was retired to AMARC in May 1987. The US Marine Corps continued tactical Phantom operations with VMFA-321 'Hell's Angels' at NAF Andrews, Maryland, and VMFA-112 'Cowboys' at NAS Dallas. The 'Cowboys' became the last US Sea Service unit before retiring its F-4Ss on 18 January 1992 as it converted to the F/A-18A Hornet.

Over on the west coast of the US, at Naval Base Ventura County Point Mugu, test unit VX-30 'Bloodhounds' is the last US naval squadron to be associated with the popular F-4. The fact that the Phantom has been retained by the 'Bloodhounds' is testament to the fact that the type remains uniquely suited to the flight test requirements of the squadron. Like the USAF QF-4s, these Phantoms are also converted to QF-4 drone configuration, but are still heavily relied upon to fulfil manned test work. Unlike the USAF QF-4s, the Navy examples were converted at the Naval Air Depot (NADEP) at MCAS Cherry Point, North Carolina, under a separate programme. The Navy's QF-4s were originally to have been produced alongside the Air Force examples by Tracor, but differences in inter-service priorities eventually led to separate programmes.

The Navy QF-4s are converted to Full Scale Aerial Target (FSAT) configuration and can be flown manned or in 'NOLO' (No Onboard Live Operator) configuration, the latter missions being conducted from San Nicolas Island off the southern California coast during live missile tests. The VX-30 QF-4s have also been heavily involved in F/A-22 weapons trials over the nearby Pacific Missile Test Range.

The 'Bloodhounds' at Point Mugu have existed under a number of guises with the Naval Air Warfare Center (NAWC) – being known as Naval Weapons Test Squadron Point Mugu (NWTSPM) from 1995, and re-designated VX-30 (Air Test and Evaluation Squadron Three Zero) 'Bloodhounds' in 2002. The squadron enjoys a unique status within the Navy, flying a wide range of test missions over the vast Pacific Missile Test Range complex, conducting vital weapons research, development and evaluation roles – as well as training and support as tasked by the Naval Air Systems Command (NAVAIR).

Commanded by Commander Wade 'Torch' Knudson, the squadron retains a few non-airworthy QF-4Ns, as well as the all-white YF-4J ejection seat test-bed, but the majority of the airworthy Phantoms are QF-4S+ aircraft. They no longer feature the Dayglo orange control surfaces of their predecessor QF-4Ns, but are turned out in an immaculate two-tone overall grey scheme. The QF-4s are configured for many tasks, and manned missions vary from simulating offensive anti-shipping missiles to high-speed weapons testing. The Phantoms regularly support surface vessels for training exercises and employ the MA-31 high-speed aerial target (based on the Zvezda Kh-31 missile) to enhance threat awareness. The Phantoms are also called upon to support Tomahawk cruise missile testing, with the F-4s chasing the missiles. Control of the Tomahawk, and testing of particular systems, can be performed manually from the back seat of the Phantom. Tomahawks are usually escorted in 30-ft (10-m) separation visual formations on a restricted test route up to the Electronic Combat Range at China Lake, California.

Sadly, a decision in 2002 signalled the end for Naval QF-4 Phantom operations. Despite the type providing a valuable test platform for the Navy, rationalisation means that F/A-18 Hornets will take over the manned test duties, with unmanned requirements passing to the USAF's QF-4s. Only three naval aviators remain current on the Phantom at VX-30, and these skilled pilots are set to become the last Navy Phantom drivers. When the decision was made to draw down US Navy QF-4 operations, the NADEP at Cherry Point set about converting the last of the QF-4S Phantoms. On 3 March 2003, VX-30 test pilot Rich 'Mink' Bryant delivered the very last example (BuNo. 155847) to VX-30. The departure marked the end of an era for the depot and also signalled the beginning of the end for the Navy F-4. On 30 June 2004, the 'Bloodhounds' will cease Phantom operations completely, and after almost 44 years of service the US Navy will finally say goodbye to the ubiquitous McDonnell Douglas F-4.

VX - 30
126
VX-30
NAVY
3832
26

US Military King Air variants Part 2: C-12

VC-6A

Only two examples of the B90 King Air were taken into service and both were purchased in 1966. The US Army took delivery of a single VC-6A that was originally designated NU-8F. Serial 66-15361 was delivered in July 1966 and initially assigned to the Redstone Arsenal in Alabama to transport engineers (including Dr Werner von Braun) between Redstone and the US Army White Sands Missile Range in New Mexico. The aircraft was fitted with stores for mounting test devices when not operating as a personnel transport. It was eventually assigned to the Dugway Proving Ground in Utah and to the White Sands Missile Range as a test aircraft. It was retired in April 1994 and acquired by the White Sands Missile Range Museum in June of that year.

The second VC-6A was equipped with a VIP interior and was operated by the 89th Military Airlift Wing at Andrews AFB, Maryland, as a transport for dignitaries and high-level government personnel. Serial 66-7943 entered service in early 1966 and during the initial part of its career was used to transport President Lyndon B. Johnson and members of the 'first family' between Bergstrom AFB, Texas, near Austin, and the President's ranch near Johnson City. At the conclusion of Johnson's term the aircraft continued to serve with the 89th in the executive transport role until it was retired to the US Air Force Museum in 1985. The aircraft featured a redesigned wing of 50 ft 3 in (16.23 m) span and an area of 293.9 sq ft (27.30 m²).

Both examples were powered by PT6A-20 engines rated at 550 shp (410 kW) and driving three-bladed, non-reversing, constant-speed Hartzell propellers. The aircraft was rated for a maximum take-off weight of 9,650 lb (4377 kg) and could land at a maximum weight of 9,168 lb (4159 kg). The VC-6A carried a maximum of 384 US gal (1454 litres) or 2,496 lb (1132 kg) of fuel in nacelle and wing tanks.

The USAF's VC-6A wore a restrained version of the scheme applied to most of the 89th MAW fleet during the late 1960s, and was devoid of any titles except its serial number on the tail fillet. Today it is displayed at the USAF Museum, Wright-Patterson AFB, Ohio.

The US Army's VC-6A is seen at the White Sands Missile Range Museum, where it still resides today. Earlier it carried US Army Missile Command titles.

Serial	No.	Build No.
66-15361	1	LJ-153
66-7943	1	LJ-320

T-44A Pegasus

Similar in configuration to the Beech B90, the US Navy's T-44A Pegasus was assigned the Beech model number H90. It first flew in January 1977 and entered service in April that year as a replacement for the Grumman TS-2A and US-2A Tracker multi-engined trainers. Procured as a commercial-derivative aircraft, a total of 61 T-44As was built and 55 are currently operational with a single training squadron at NAS Corpus Christi, Texas, in support of the joint advanced multi-engine training curriculum. The Pegasus is used to train USAF pilots as well as those due to fly multi-engined navy types during their careers. As a transport, the aircraft is capable of carrying a crew of two and five passengers, although it normally carries an instructor in the right seat and a student in the left.

The Pegasus fuselage is 35 ft 6 in (10.82 m) long, its wingspan is 50 ft 3 in (15.32 m) and it is 14 ft 2½ in (4.33 m) tall. Two 550-shp (410-kW) PT6A-34B turboprop engines each drive a three-bladed, reversible, constant-speed Hartzell propeller. The trainer has a maximum take-off weight of 9,650 lb (4377 kg), and is rated for a maximum speed of 249 kt (462 km/h) and a maximum altitude of 31,000 ft (9449 m). Although capable of carrying up to 800 lb (363 kg) of cargo, it is not equipped with a cargo door.

Serial	No.	Build No.
160839/160856	18	LL-01/18
160967/160986	20	LL-19/38
161057/161079	23	LL-39/61

The T-44A was a minimum change version of the Beech Model B90. It replaced the Tracker in the multi-engined training role and continues to serve in that role, a secondary transport capability being seldom used. With the concentration of twin-engined training for all of the services within the Navy, the T-44A fleet was later augmented by a number of TC-12Bs.

King Air A90

The US Army operated at least two examples of the Model 65-A90 King Air acquired under the Confiscated/Excessed Aircraft Program. They were operated as non-standard aircraft using their commercial model numbers while in service with the Army. Capable of carrying two crew members and up to eight passengers, the A90 series is powered by two 500-shp (373-kW) Pratt & Whitney PT6A-20 turboprop engines, each driving a three-bladed, constant-speed Hartzell propeller. It can operate at a maximum take-off weight of 9,300 lb (4218 kg), and is rated for a maximum speed of 208 kt (385 km/h) and a maximum altitude of 30,000 ft (9144 m). 86-0092 was sold on as N50525 while 86-1683 became N7202L. No information is known about 86-0090.

Serial	No.	Build No.
86-0090	1	–
86-0092	1	LJ-159
86-1683	1	LJ-129

King Air C90/F90

Five King Air C90s and two F90s are operated by the Federal Aviation Administration. Based on the airframe of the model B90 (VC-6A), the C90 is powered by 550-shp (410-kW) PT6A-21 engines equipped with reversible-pitch Hartzell HC-B3TN-3/T10173 propellers. The F90 combines the fuselage and wings of the later Model E90 with the empennage and 'T'-tail of the model 200. It is powered by a pair of 750-shp (560-kW) PT6A-135 engines and reversible-pitch Hartzell HC-B4TN-3/T10173 propellers. Like the C90, it features the longer-span wing first used on the VC-6A but is capable of take-offs and landings at a maximum weight of 10,950 lb (4967 kg), and can operate at a maximum altitude of 31,000 ft (9449 m).

Type	Reg	No.	Build No.
F90	N15, N18	2	LA-138, LA-145
C90	N16, N17, N19/N21	5	LJ-893, LJ-896, LJ-909, LJ-912, LJ-902

C-12A Huron

Based on the Model A200, the C-12A was the second variant of the Super King Air to enter service, following the earlier RU-21J. It was powered by two 750-shp (559-kW) PT6A-38 engines driving three-bladed Hartzell propellers. Beech delivered 60 C-12As to the US Army and 30 to the US Air Force between 1974 and 1978. A single example was also delivered to the Greek Army.

The C-12A had an overall length of 43 ft 9 in (13.34 m), a wingspan of 54 ft 6 in (16.61 m) and a height of 15 ft (4.57 m). It was fully pressurised and capable of a maximum speed of 270 kt

76-22563 of USASC was typical of most early US Army C-12As in that it was finished in the chocolate brown and white scheme. The first 20 Army C-12As were originally ordered as U-25As.

Some US Air Force C-12As were operated by MAAGs in support of overseas embassies, and this example, 73-1207, served with the Armish MAAG in Iran prior to 1979.

(500 km/h) and a maximum altitude of 31,000 ft (9449 m). Although similar to the earlier RU-21J, the C-12A featured a redesigned cockpit and was equipped with weather radar and an autopilot. It had a useful load of 4,716 lb (2139 kg), carried a maximum of 549 US gal (2078 litres) or 3,569 lb (1619 kg) of fuel, and was rated for a maximum take-off weight of 12,500 lb (5670 kg). The commercial model is rated to carry a maximum of 13 passengers and two crew members, but the C-12A was most often configured with seating for eight passengers in the main cabin.

Serial	Number	Build No.
73-1205/1218	14	BD-01/14
73-22250/22269	20	BC-01/20
76-0158/0173	16	BD-15/30*
76-22545/22564	20	BC-21/33, 35-41**
77-22931/22950	20	BC-42/61

* C-12A 76-0167 was reserialled as 76-3239 **BC-34 to Greece

UC-12B

Assigned the model number A200C, the UC-12B is powered by two 850-shp (634-kW) PT6A-41 engines driving three-bladed, constant-speed Hartzell HC-B3TN-3G/T10178B series propellers. Although similar in configuration and overall performance to the earlier C-12A, the UC-12B featured a port-side cargo door and high-flotation landing gear. The Navy took delivery of 66 examples beginning in 1979. During the mid-1990s a consolidation of the joint-services OSA fleet resulted in the retirement of 23 UC-12Bs from this role. Although several were placed in storage at the Aerospace Maintenance and Regeneration Center at Davis-Monthan AFB, Arizona, 20 were converted for use as trainers as TC-12Bs to supplement the T-44A Pegasus fleet.

The Navy recently undertook a Flight Safety Upgrade that will equip its UC-12Bs with a flight management system, ground position warning system, traffic collision avoidance system, upgraded weather radar, wind shear detection equipment, an upgraded radar altimeter, and a cockpit voice recorder. The three-phase programme will be completed in 2010.

Serial	No.	Build No.
161185/161206	22	BJ-1/22
161306/161327	22	BJ-23/44
161497/161518	22	BJ-45/66

The UC-12B was the first of the line of naval utility UC-12s that continue to serve today. BuNo. 161196 is seen sitting outside a hangar at NAF El Centro, California.

UC-12B

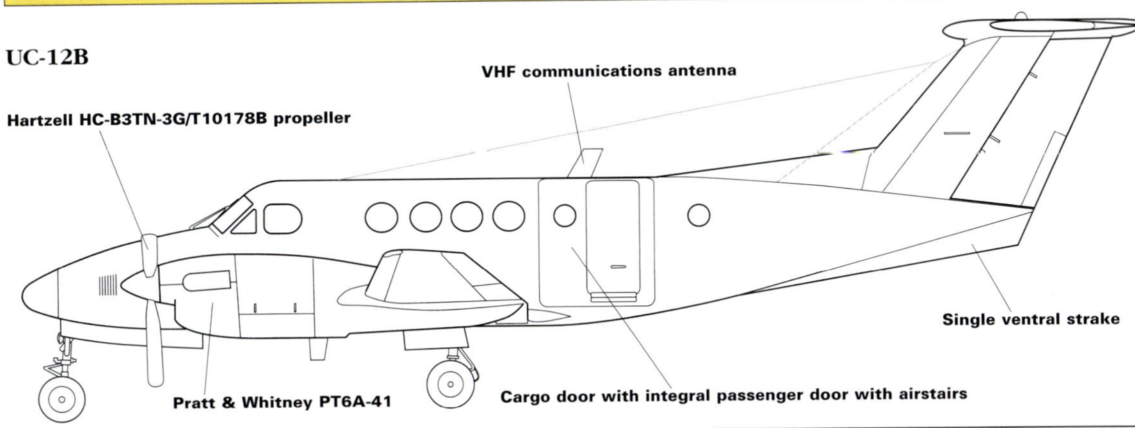

NC-12B

A single UC-12B (BuNo. 161311) was modified by Aeromet Inc. in Tulsa, Oklahoma, as an airborne testbed for sonobuoys under the designation NC-12B. The modification includes four sonobuoy launch tubes, a launch control system, an aerodynamic fairing, and nine sonobuoy storage racks. The aircraft is equipped to launch sonobuoys and 36 can be carried onboard the aircraft.

The NC-12B features a large fairing under the fuselage within which racks for up to 36 sonobuoys have been fitted. The aircraft replaced a P-3 Orion in the sonobuoy test role, as a more economical solution to the larger aircraft.

TC-12B

Following the removal of 23 UC-12Bs from service in the OSA role, 20 examples were converted as trainers and redesignated TC-12Bs. The aircraft are assigned to the Chief of Naval Air Training and operate in support of the joint advanced multi-engine training curriculum at NAS Corpus Christi, Texas. The conversions include BuNos 161188, 161193, 161194, 161197, 161199/161201, 161205, 161306, 161310, 161314/161316, 161498, 161505, 161508/161510, 161514 and 161518.

The main external distinguishing feature of the TC-12B is the red and white training scheme, also worn by the T-44A.

C-12C Huron

In 1978 the Army ordered 14 examples of the C-12C from Beech. Based on the Model A200, the aircraft was externally identical to the earlier C-12A and mirrored its performance. It was, however, powered by two 850-shp (634-kW) PT6A-41 engines driving three-bladed Hartzell propellers and equipped with a landing gear de-icing system. All surviving C-12As were eventually brought up to the later standard via a modification programme.

Serial	No.	Build No.
78-23126/23139	14	BC-62/BC-75

The engines of the C-12C were more powerful than those fitted to the C-12A. US Army CECOM operated 73-22262.

JC-12C

A number of C-12Cs have supported test programmes, but serial 76-22545 is the only example to be assigned the temporary test designation JC-12C. The aircraft was originally delivered as a C-12A.

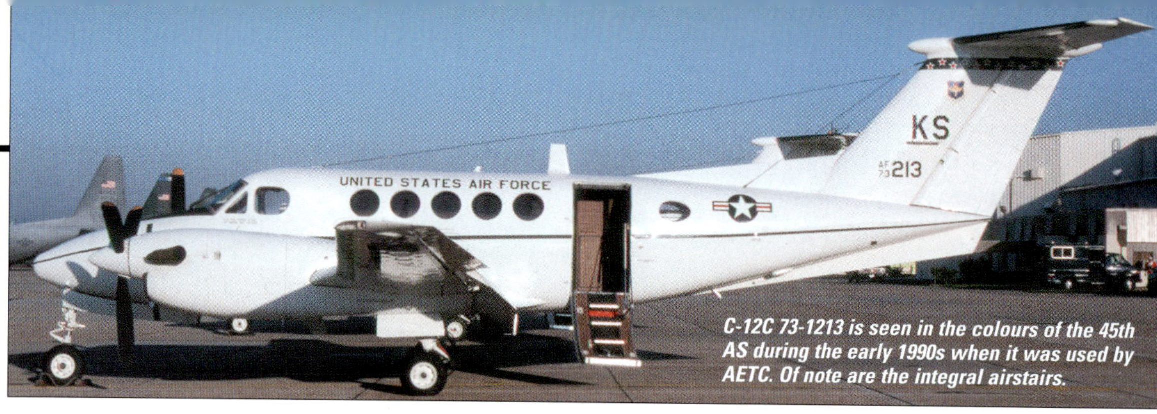

C-12C 73-1213 is seen in the colours of the 45th AS during the early 1990s when it was used by AETC. Of note are the integral airstairs.

C-12D Huron

Based on the commercial Model A200CT, the C-12D retained the 850-shp (634-kW) PT6A-41 engines and the three-bladed, constant-speed Hartzell HC-B3TN-3/T10178 series propellers that equipped the C-12C. It differed from the earlier model in being equipped with a port-side cargo door and high-flotation landing gear. The aircraft had provision for a pair of 52 US-gal (197-litre) or 338-lb (152-kg) auxiliary fuel tanks on the wingtips. The C-12D is rated for a maximum take-off and landing weight of 12,500 (5670 kg), a maximum speed of 260 kt (482 km/h) and a maximum altitude of 31,000 ft (9449 m). The Army took delivery of 40 examples and the Air Force accepted six. The final six aircraft delivered to the Army were equipped with 850-shp (634-kW) PT6A-42 engines and hydraulically-operated landing gear, and were known as C-12D2s. The earlier examples became known as C-12D1s.

Another 16 C-12Ds were modified as SEMA platforms and equipped with the Improved Guardrail V or Crazyhorse elint systems. Once modified, these variants were assigned different build numbers by Beech Aircraft. A single RC-12D (serial 80-23378) later had this equipment removed and reverted to the C-12D configuration.

***Above:* 83-0498 of the Andrews-based 89th AW was one of six USAF C-12Ds.**

***Left:* This C-12D1 was one of two aeromedical aircraft used by the USAAD.**

Serial	No.	Build No.
78-23140/23145	6	BP-01/06
80-23371/23380	10	BP-12/21
81-23541/23546	6	BP-22/27
82-23780/23785	6	BP-28/33
83-24145/24150	6	BP-34/39
83-0494/0499	6	BP-40/45
84-24375/24380	6	BP-46/51

RC-12D Improved Guardrail V

Based on the Model A200CT/C-12D, the RC-12D underwent a number of modifications that included the installation of wingtip pods and multiple blade antennas associated with the AN/URD-9A Improved Guardrail V sigint system. Beech modified 13 C-12D aircraft to RC-12D IGRV configuration. Originally rated for take-offs and landings at a maximum weight of 12,500 lb (5670 kg), the maximum take-off weight for the RC-12D was increased to 14,200 lb (6441 kg) and the landing weight jumped to 13,500 lb (6123 kg). Installation of wingtip antennas associated with IGRV increased the wingspan to 58 ft 4.6 in (17.80 m). The RC-12D was equipped with the same engine and propeller as the C-12D, but the modifications associated with IGRV caused the maximum operating speed to be reduced to 247 kt (457 km/h). Its fuel capacity was also reduced slightly, to 542 US gal (2051 litres) or 3,523 lb (1598 kg). Unlike the utility models, the RC-12D was equipped with aircraft survivability equipment that comprised the M130 chaff/flare dispensing system, the AN/APR-39 radar signal detecting set and the AN/APR-44 radar warning receiver.

RC-12D 80-23373 has had most of its mission equipment removed and is used by CECOM, possibly for foliage penetrating radar tests. It has a large pannier under the fuselage and aerials on the trailing edge of the wing.

The RC-12D was the first of the SEMA C-12s. VHF DF and radar signal detecting antennas were fitted in the wingtip pods, while a wide-band data link antenna pod was slung under the nose and dipoles attached to the wings.

Serial	No.	Build No.
78-23141/23145	5	GR-06/10
80-23371	1	GR-02
80-23373	1	GR-04
80-23374	1	GR-12
80-23375	1	GR-05
80-23376	1	GR-11
80-23377	1	GR-03
80-23378	1	GR-13
81-23542	1	GR-01

JC-12D

C-12D serial 78-23140 was assigned to the US Army Aviation Technical Test Center at Fort Rucker, Alabama, and was utilised for a variety of test programmes. It was assigned the temporary test designation JC-12D.

C-12E

The designation C-12E was reserved for an upgrade of 29 of the US Air Force's C-12As and Cs with the 850-shp (634-kW) Pratt & Whitney PT6A-42 powerplant, as used on the C-12F. The programme did not proceed and no C-12Es were ever produced.

C-12F3 84-0148 (seen during its service with the 3rd Wing) was one of the original F-models leased by the US Air Force from May 1984 before being purchased outright. They were the first C-12s for the US military with four-bladed propellers.

C-12F

The US Air Force selected the C-12F as an operational support aircraft in September 1983 and initially leased 40 examples, which were assigned to Military Airlift Command. Six more examples were delivered to the Air National Guard and served as support aircraft. The aircraft is powered by a pair of 850-shp (634-kW) PT6A-42 engines driving four-bladed, constant-speed McCauley 4HFR34C754/94LA-0 series propellers. The aircraft was certified to operate at the same weights as the earlier C-12D but was capable of a maximum altitude of 35,000 ft (10668 m).

85-51268 is a C-12F1, featuring the three-bladed propellers used on previous Army variants combined with the airframe of the commercial A200CT. The C-12F2 and F3 were based on the B200C airframe. Most of the surviving C-12Fs were fitted with Collins avionics, and were redesignated as C-12Ts or C-12Us.

Beginning in 1985 the Army purchased the first of 12 C-12Fs, which were based on the A200CT airframe but equipped with the same powerplant as the US Air Force C-12F. They retained the C-12D's three-bladed, constant-speed Hartzell HC-B3TN-3/T10178 series propeller and were capable of operating at a maximum altitude of 31,000 ft (9449 m). Another eight Army examples were based on the same B200C airframes as the US Air Force C-12F, but differed from earlier Army aircraft in being equipped with three-bladed, constant-speed McCauley 3GFR34C702/100LA-2 series propellers and being rated for operating at a maximum altitude of 35,000 ft (10668 m). All three sub-models of the C-12F were equipped with a cargo door and normally operated with a two-man crew. In order to differentiate between the three variants, the Army aircraft were eventually referred to as the C-12F1 and C-12F2 while the US Air Force examples became known as the C-12F3.

Type	Serial	No.	Build No.
C-12F3	84-0143/0182	40	BL-81/112
C-12F3	84-0484/0489	6	BL-118/123
C-12F1	85-51261/51272	12	BP-52/63
C-12F2	86-60084/60089	6	BP-64/69
C-12F2	87-70160/70161	2	BP-70/71

UC-12F

Based on the Model B200C, the US Navy's UC-12F is powered by 850-shp (634-kW) PT6A-42 engines but has three-bladed, constant-speed Hartzell HC-B3TN-3/T10178 series propellers common to the UC-12B and the US Army C-12F1. The service purchased 12 examples and the first was delivered in 1986. It is rated for a maximum take-off and landing weight of 12,500 lb (5670 kg), a maximum speed of 260 kt (482 km/h) and a maximum altitude of 35,000 ft (10668 m). The aircraft's configuration is similar to that of the earlier UC-12B but there are subtle differences within the avionics systems. It also features hydraulically-actuated landing gear.

Serial	No.	Build No.
163553/163564	12	BU-1/12

Unlike the three-bladed C-12F1s, the US Navy's UC-12Fs were based on the B200C airframe. BuNo. 163557 is seen in the colours of VRC-30, the FRS for UC-12s.

RC-12F

Two UC-12Fs (BuNos 163563/163564) of the 12 ordered by the US Navy were equipped with the Litton AN/APS-140/504 sea search radar fitted under the centre fuselage and modified as range support aircraft. They were assigned to the Pacific Missile Range Facility at Naval Station Barking Sands, Kauai, Hawaii, and provided range surveillance and missile range clearance capabilities (making sure the range was clear of sea traffic during missile test firings), and served in the logistic support roles. Initially they had wingtip pods, but these were removed in service. The King Airs were replaced in 2001 by similarly modified Fairchild C-26D Metroliners and transferred to VRC-30 at NAS North Island, California. VRC-30 acted as the Fleet Readiness Squadron for the US Navy UC-12 fleet. The radar and associated systems were eventually removed and the two aircraft now serve both in the Operational Support Airlift (OSA) role and as trainers.

An RC-12F range support aircraft is seen during its service with the Pacific Missile Range Facility at Barking Sands, Hawaii. The AN/APS-140/504 sea search radar under the fuselage is the major feature of the version, which early in its career also featured wingtip pods.

RC-12F

RC-12G Crazyhorse

The three RC-12Gs are based on the C-12D, model A200CT airframe. The aircraft's external configuration, internal systems and performance are similar to that of the RC-12D, but rather than carrying the Improved Guardrail V equipment the type is equipped with the AN/URR-75 Crazyhorse airborne radio direction finding equipment produced by Sanders. The system was developed specifically to meet the requirements of the US Army South and US Southern Command. Differences in its configuration resulted in the wingspan being reduced to 57 ft 9 in (17.60 m) and its maximum take-off and landing weights increased to 15,000 lb (6804 kg). Although its ASE fit was similar to that of the IGRV, Crazyhorse was capable of remote or airborne operation and the RC-12G featured two operator consoles in the main cabin.

The wingtip pods of the RC-12G contained HF/VHF antennas, while VHF intercept dipoles were fitted on the wings. An HF/VHF monopole antenna was attached to the top of the tail. The nose-mounted data link pod was not carried.

Serial	No.	Build No.
80-23379/23380	2	FC-1/2
80-23372	1	FC-3

JRC-12G

After being retired from SEMA duties, the Crazyhorse ARDF equipment was removed from RC-12G serial 80-23380, which was transferred to the US Army Aviation Technical Test Center at Fort Rucker, Alabama and assigned the temporary test designation JRC-12G.

The JRC-12G was shorn of all its wingtip pods and other mission-related antennas, leaving four on the fuselage (from back to front: TACAN, VHF air-to-air, VHF/AM/FM and VHF/UHF). It was also fitted with the twin strakes under the aft fuselage. As of early 2002 the aircraft was still in service with the Aviation Technology Test Center (ATTC).

RC-12H Guardrail/Common Sensor System 3 (Minus)

Based on the C-12D, the RC-12H retains the Model A200CT's PT6A-41 engines, propellers and performance. However, the RC-12H is equipped with the AN/USD-9B Guardrail/Common Sensor signals intelligence system. Differences between the RC-12H and the earlier RC-12D include a revised antenna configuration and wingtip pods. When fielded to Korea by the 3rd MI Bn (AE) in October 1988, the six RC-12Hs that were part of GR/CS System 3 (Minus) initially lacked the Advanced Quicklook and Communications High Accuracy Airborne Location System (CHAALS) systems, but this equipment was retrofitted in 1996. The RC-12H differs from the standard C-12D in being rated for maximum take-off and landing weights of 15,000 lb (6804 kg) and a maximum speed of 243 kt (450 km/h). Its new wingtip pods increased the overall wingspan to 58 ft 6 in (17.83 m).

Serial	No.	Build No.
84-24313/24318	6	GR-14/19

The RC-12H's antennas differed significantly from those of the RC-12D. The upper and lower 'barbed' antennas are low-band vertical bent blades while the black circle on the rear fuselage is a 'P' band antenna. The rear boom rotates.

C-12J

Based on the Beech 1900C Airliner, six C-12Js were purchased for the Air National Guard in 1986 for use as support aircraft. The C-12J is 53 ft 1½ in (16.19 m) long and provides seating for 19 passengers. It retains the King Air's wing, but is powered by a pair of 1,100-shp (820-kW) PT6A-65B turboprops that drive four-bladed Hartzell HC-B4MP-3A/M10877 series propellers. Fuel capacity is increased to 667 US gal (2525 litres) or 4,356 lb (1976 kg) and the aircraft can carry 5,300 lb (2404 kg) of cargo up to 1,570 miles (2527 km). The C-12J is certified for maximum take-off and landing weights of 16,600 lb (7530 kg) and 16,100 lb (7203 kg), respectively. It is rated for a maximum speed of 247 kt (457 km/h) and a maximum altitude of 25,000 ft (7620 m). During 1997 two of the C-12Js were transferred to the Army, after being declared excess by the US Air Force, and were assigned to headquarters units in Japan and Germany. Remaining US Air Force C-12Js are assigned to Pacific Air Forces and the Air Force Materiel Command.

A single 1900D was delivered to the Army in 1996. The aircraft is powered by two 1,279-shp (954-kW) PT6A-67D turboprops and equipped with four-bladed Hartzell HC-E4A-3/E10950 series propellers. The fuselage is 57 ft 10 in (17.63 m) long and its wingspan is 57 ft 11 in (17.65 m). The wing differs from that of the 1900C in being equipped with winglets. The 1900D's maximum take-off and landing weights are increased to 17,100 lb (7756 kg) and 16,765 lb (7604 kg), respectively, and its rated maximum speed is 248 kt (459 km/h).

The C-12J is a different aircraft to the rest of the C-12 family, but retains some features of the Beech 200. Only six were delivered to the US military, followed by a similar Beech 1900D.

Serial	No.	Build No.
86-0078/0083	6	UD-01/06
96-0112	1	UE-256*
*1900D		

C-12J

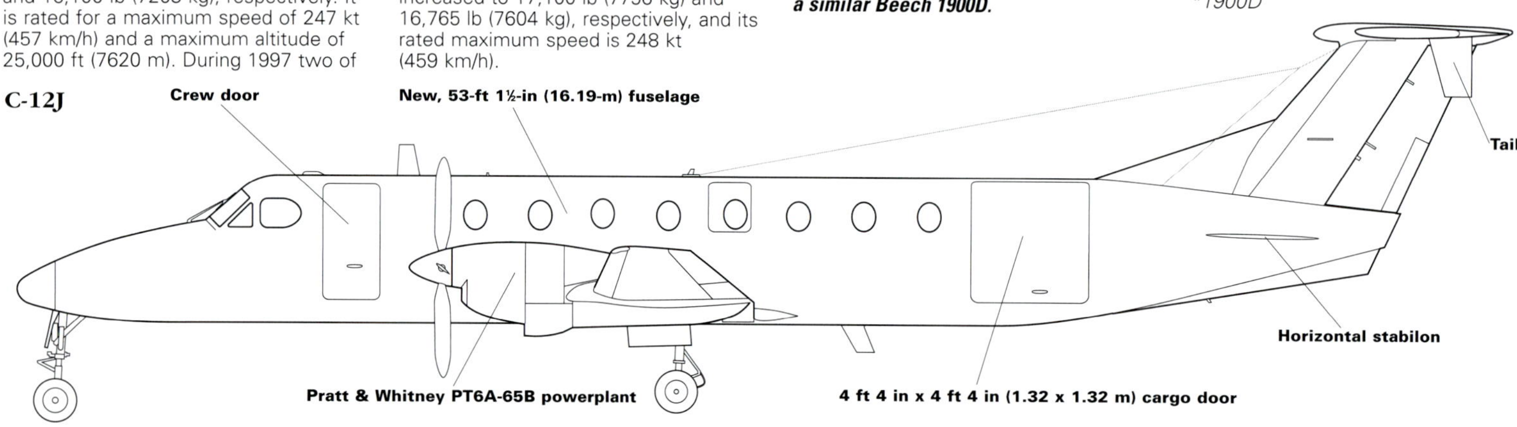

RC-12K Guardrail/Common Sensor System 4

Based on the Model A200CT, the RC-12K is powered by 1,100-shp (820-kW) PT6A-67 engines driving four-bladed McCauley model 4JFR34C758/106LNA propellers. It is the airborne component of the AN/URD-9D Guardrail/Common Sensor System 4 and is certified for maximum take-off and landing weights of 16,000 lb (7257 kg) and 15,200 lb (6895 kg), respectively. The aircraft carries 540 US gal (2044 litres) or 3,510 lb (1592 kg) of fuel, and is equipped with structural modifications that include vertical 'tailets' located beneath the horizontal stabiliser that are designed to increase the aircraft's directional stability. The RC-12K is rated for a maximum operating speed of 246 kt (456 km/h) and an altitude of 35,000 ft (10668 m).

Serial	No.	Build No.
85-50147/50155	9	FE-01/09

Similar to the RC-12H, the RC-12K has an extra antenna (the Lowband High Frequency Extension, hidden by the wingtip pod in this view of 85-50149) on the top of the fuselage, a CHAALS 'towel-rail' antenna on the rear fuselage side and more powerful engines to counter the drag. The RC-12 in a Guardrail system is not a stand-alone item, but the information collector of an intelligence-gathering and processing system which, in the case of the RC-12K, is the AN/URD-9D.

C-12L

Following the removal of the Cefly Lancer equipment from the RU-21Js, they were used in the OSA role as C-12Ls. All three were later sold on the civil market as N24SP, N200LN and N200KE (ex 71-21058/21060).

C-12M

The unofficial designation C-12M is used by the Department of Homeland Security's Bureau of Immigration and Customs Enforcement to identify six C-12Cs and King Air 200s that are equipped with maritime search radar.

UC-12M

Based on the Model B200C, the UC-12M differs slightly from the Navy's earlier UC-12F. It retains the PT6A-42 engines but this model features the same three-bladed McCauley 3GFR34C702/100LA-2 series propellers that equip the Army C-12F2. Although avionics differ between the UC-12M and UC-12F, aircraft performance is similar for the two models. The Navy received the first of 12 examples of the UC-12M in 1987, the last two becoming RC-12Ms.

Serial	No.	Build No.
163836/163847	12	BV-1/12

The UC-12M featured different propellers and cockpit avionics to the previous navy utility versions. Most UC-12Ms were deployed to support naval activities in Europe and the Middle East, but this example served at NAS Norfolk, Virginia.

RC-12M

The final two UC-12Ms (BuNos 163846/163847) were modified as range support aircraft. The aircraft were equipped with the same Litton AN/APS-140/504 sea search radar as the RC-12F. Initially assigned to the Pacific Missile Test Center at NAS Point Mugu, California, the aircraft were later transferred to the Air Operations Department at NS Roosevelt Roads, Puerto Rico. The aircraft were primarily tasked with range clearance duties over the Atlantic Fleet Weapons Training Facility.

As did the two RC-12Fs, the RC-12Ms carry AN/APS-140/504 radars (above) under the centre fuselage. BuNo. 163486 is seen at NS Roosevelt Roads, Puerto Rico.

RC-12N Guardrail/Common Sensor System 1

Based on the Model A200CT/C-12F airframe, the RC-12N is powered by 1,200-shp (895-kW) PT6A-67 engines driving four-bladed McCauley 4JFR34C758/106LNA-1 series propellers. The RC-12N is equipped with an avionics control system that includes multifunction displays/electronic flight instrumentation system and integrated aircraft survivability equipment. This equipment comprises the AN/APR-39 radar warning receiver, AN/APR-44 radar warning system, AN/ALQ-136, AN/ALQ-156 and AN/ALQ-162 countermeasure sets, and chaff/flare dispensers. The RC-12N is rated for maximum take-off and landing weights of 16,200 lb (7348 kg) and 15,400 lb (6985 kg), respectively, a maximum operating speed of 248 kt (459 km/h) and a maximum altitude of 35,000 ft (10668 m). The aircraft is the airborne component of the AN/USD-9C Guardrail/Common Sensor System 1. A total of 15 C-12Fs was originally modified to RC-12N configuration.

The AN/USD-9C Guardrail/Common Sensor System 1 uses the RC-12N collection aircraft. Like all of the later Guardrail aircraft, it uses the airframe of the C-12F, with the main difference being the exact layout of the antennas.

Serial	No.	Build No.
88-0325/0327	3	FE-10/12
89-0267/0276	10	FE-13/22
91-0516/0517	2	FE-23/24

RC-12P Guardrail/Common Sensor System 2

Similar in configuration to the RC-12N, the RC-12P is also based on the Model A200CT and is equipped with the same powerplant and propellers as the earlier variant. Nine RC-12Ps are components of the AN/USD-9E Guardrail/Common Sensor System 2. The RC-12P shares identical avionics with the RC-12N, but its mission equipment differs and the aircraft features fibre-optic cabling, smaller, lighter wing pods and a revised antenna configuration. These airframe modifications allow the RC-12P to operate at maximum take-off and landing weights of 16,500 lb (7484 kg) and 15,675 lb (7110 kg), respectively. Its remaining performance mirrors that of the RC-12N. In addition, the Comint/DF components are replaced by digital equipment developed by TRW for the US Air Force's U-2 Senior Smart programme, and the Communications High Accuracy Airborne Location System has been replaced by the more capable CHAALS – Exploitable (CHAALS-X). The aircraft also have increased onboard processing capability. A total of nine has been acquired for use by the US Army.

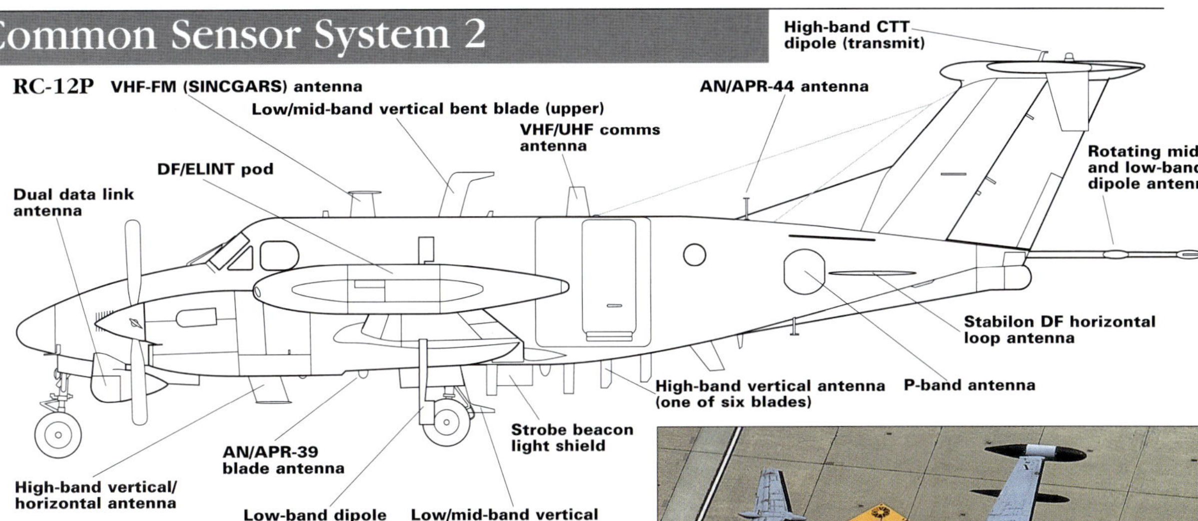

Serial	No.	Build No.
91-0518	1	FE-25
92-13120/13125	6	FE-26/31
93-0698, 0700	2	FE-33, 35

This view of an RC-12P show the dipoles on the wings just outboard of the engine nacelles, and the stabilon DF horizontal loop antennas.

RC-12Q Guardrail/Common Sensor System 2

The Army's three RC-12Qs are components of the AN/USD-9E Guardrail/Common Sensor System 2. With the exception of a large radome mounted on the upper fuselage, aft of the cockpit, the RC-12Q is externally identical to the RC-12P. Housed within this radome is the antenna for the AN/ARW-87 direct air-to-satellite relay system. Although the RC-12Q is also capable of performing the same Sigint missions as the RC-12P, the DASR allows it to operate as an airborne relay facility. The RC-12Q airframe, systems and performance are identical to those of the RC-12P.

Serial	No.	Build No.
93-0697, 0699, 0701	3	FE-32, 34, 36

Below: The RC-12Q is flown alongside the RC-12P as a component of the Guardrail Common Sensor System 2, and both RC-12s have a similar aerial array.

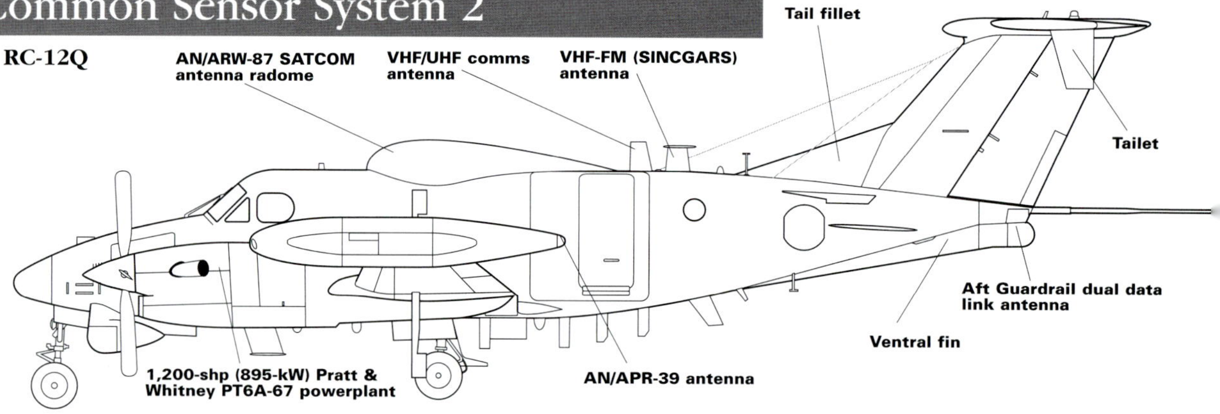

Right: The location of the AN/ARW-87 SATCOM radar antenna on the top of the fuselage of the RC-12Q direct air-to-satellite aircraft resulted in some of the communications aerials being moved aft. The rest of the aerials are similar to those on the RC-12P.

C-12R Huron

First flown in 1988, the C-12R is based on the Model B200C and is powered by 850-shp (634-kW) PT6A-42 engines driving four-bladed, constant-speed McCauley 4HFR34C7771/94LA-0 series propellers. While similar in overall configuration to the earlier C-12F3, the aircraft features updated avionics that include multifunction display/electronic flight instrumentation system, a digital autopilot, and a global positioning system. Its performance and weights mirror those of the C-12F3 series. The Army eventually purchased 29 C-12Rs and the majority were assigned to US Army Reserve Command theatre aviation companies. Those aircraft that are equipped with global air traffic management modifications have been redesignated C-12R1.

The R-model was the ultimate Operational Support Airlift version of the Huron purchased by the Army. This C-12R (92-3327 of the 2-228th AVN) has the original ventral fin.

Serial	No.	Build No.
92-3327/3329	3	BW-01/03
94-0315/0326	12	BW-04/15
95-0088/0101	14	BW-15/29

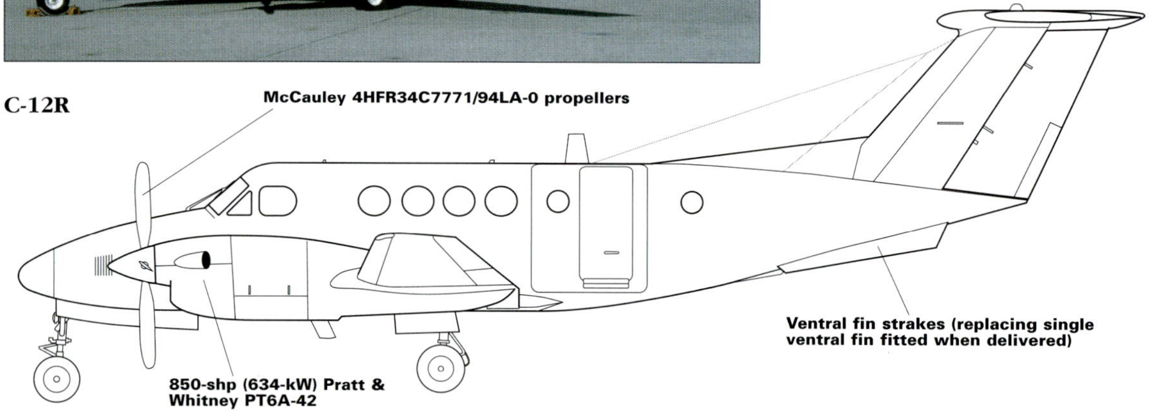

C-12S

In October 1998 Raytheon Aircraft was awarded a $US4.5 million contract for a single King Air 350 for the US Drug Enforcement Agency. Based on the Model B300, the aircraft can carry eight to 15 passengers (depending on configuration) and is assigned the MDS designation C-12S. It is powered by a pair of 1,050-shp (783-kW) Pratt & Whitney Canada PT6A-60A turboprops that drive four-bladed, constant-speed Hartzell HC-B4MP-3C/M10476 series propellers. The C-12S is 2 ft 11 in (0.89 m) longer than earlier models, and its wing is 3 ft 5 in (1.04 m) longer and features winglets. It is rated for a maximum take-off and landing weight of 15,000 lb (6804 kg), a maximum operating speed of 262 kt (486 km/h), and a maximum altitude of 35,000 ft (10668 m).

C-12T Huron

A total of 40 C-12F1, 2, and 3 aircraft assigned to units of the US Army National Guard's Operational Support Airlift Command was equipped with Collins cockpit upgrades. The modified aircraft were redesignated C-12T1, T2 and T3, respectively. The 'glass' cockpit is compatible with night vision goggles and comprises an FDS-255 electronic flight instrumentation system, TCAS II traffic collision avoidance system, RDR 2000 colour weather radar, FMS 800 flight management system, and AN/ARC-210 radios. Initially referred to by the non-standard designations C-12F1E, 2E and 3E, the aircraft were formally redesignated on 1 November 1999.

During March 2003 DynCorp International was awarded a $US6 million contract to equip 39 C-12Ts with a global air traffic management upgrade that will result in these aircraft being redesignated C-12Us, joining the other 23 examples of this variant.

C-12U Huron

The Army is currently modifying 23 C-12F3 aircraft (including 21 operated by active component US Army units and two remaining with the US Air Force) with an integrated flight management system comprising the Collins Proline 4 integrated flight management system, FDS-255 electronic flight instrumentation system, enhanced ground proximity warning system, TCAS II traffic collision avoidance system, WXR-850 colour weather radar, a cockpit voice recorder/flight data recorder and AN/ARC-210 radios. Once modified, the aircraft are redesignated C-12U. The initial examples were assigned to the 204th Military Intelligence Battalion at Fort Bliss, Texas, during 2002.

The upgrade from the three C-12F standards to C-12T is largely concerned with the cockpit avionics. C-12T3 84-0174 (left) of the Missouri Army National Guard started its service with the US Air Force as a C-12F3 before transfer to the Army and upgrade.

Super King Air 200/B200/300

In addition to the militarised C-12 series, a number of Super King Airs are in service with US military, government operators and government contractors. The National Aeronautics and Space Administration operates four civil King Air B200s as operational support aircraft from the Dryden Flight Research Center in California and from the Langley Research Center and Wallops Flight Facility in Virginia. Although the NASA aircraft are similar in configuration to the UC-12B/F and C-12F1, at least two have been upgraded through the installation of Pratt & Whitney PT6A-60A engines.

Registration	No.	Build No.
N7NA	1	BB-997
N8NA	1	BB-950
N529NA	1	BB-1091*
N801NA	1	BB-1164*

**Powered by PT6A-60A engines*

Right: The Beech 300 has a longer fuselage and wings than the Model 200.

Below: N701NA was based at the Ames Research Center before being transferred to the Dryden Flight Research Facility as N801NA in the late 1990s.

A component of the Department of Transportation, the Federal Aviation Administration operates a single Model 200 and 19 Model 300s. The former is similar to the UC-12B/F and C-12F1, while the latter is similar to the C-12S. The FAA's Model 300s differ from the standard civil variants in that they are not certified for single pilot operation. The Model 300's Pratt & Whitney Canada PT6A-60A turboprops drive four-bladed, constant-speed Hartzell HC-B4MP-3B/M10476 series propellers and the aircraft is rated for take-offs and landings at 14,000 lb (6350 kg). They are used primarily to inspect the operation of navigational aids and are equipped with an automated flight inspection system. The Model 300s were ordered in 1986 and initial deliveries were made in 1988.

Type	Registration	No.	Build. No.
200	N35	1	BB-88
300	N66/N84	19	FF-1/19

During 1988 the US Air Force acquired three examples of the B200C that are similar in configuration to the UC-12B/F of the US Navy and C-12F1 of the US Army. The aircraft wear civil registrations and are operated by government contractor EG&G in support of the Department of Energy. All are based at McCarran International Airport in Las Vegas, Nevada.

Serial	No.	Build No.
N654BA	1	BL-54
N661BA	1	BL-61
N662BA	1	BL-62

Beech 1900C/D Airliner

In addition to the six C-12J versions of the King Air originally purchased for the Air National Guard, the US Air Force acquired two examples of the 1900C airliner during 1988. These aircraft, which are based at McCarran International Airport in Las Vegas, Nevada, wear civil registrations. Although owned by the US Air Force they are operated by government contractor EG&G in support of the Department of Energy, alongside three B200Cs.

Serial	No.	Build No.
N20RA	1	UB-42
N27RA	1	UB-37

Raytheon Corporation's Range Systems Engineering Company operates two 1900D airliners in support of the US Navy Atlantic Underwater Test and Evaluation Center (AUTEC) on Andros Island, Bahamas. The aircraft carry civil registrations and are owned by Raytheon.

Serial	No.	Build No.
N45AR	1	UE-12
N46AR	1	UE-27

Raytheon's Range Systems Engineering Company use two Beech 1900Ds (this is N45AR) to transport personnel to Andros Island in the Bahamas.

Mirage III in RAAF service

The 'French Lady' Down Under

In November 1960, Dassault's Mirage III was selected to be the first all-weather supersonic interceptor for the Royal Australian Air Force, and a fond association began that lasted until its final flight in February 1989. Over the ensuing years, the aircraft earned such affectionate nicknames as 'The French Lady', 'La Belle de la France' or, more simply, the 'Miracle'.

The story of how Australia came to purchase the sleek, delta-winged fighter has its origin, in part, in the RAAF's involvement in the Korean War: unable to procure North American F-86 Sabres, No. 77 Squadron pilots were forced to face Communist MiG-15s in the outclassed Gloster Meteor Mk 8, and were placed at a significant disadvantage. This performance mismatch remained in the RAAF's 'corporate memory' and had a great influence on future planning.

As early as 1950, the Australian government had identified the Hawker P.1081 as its new fighter type, but development problems associated with that aircraft meant that an indigenous version of the North American F-86F Sabre was acquired as an alternative. In the meantime, of course, events in Korea had dictated the stop-gap purchase of the Meteor in lieu of off-the-shelf Sabres, and the rapid advance of airframe and engine design throughout the 1950s saw the performance of RAAF aircraft lag behind

their US, NATO and Communist bloc contemporaries.

Although the Australian Sabre, manufactured by the Commonwealth Aircraft Corporation (CAC) and fitted with a Rolls-Royce Avon engine, was (arguably) the best performer of the F-86 family, it was not an all-weather fighter. The increasing maximum speed of aircraft such as the F-100, F-102, MiG-17 and MiG-19, and even more advanced aircraft such as the F-104, F-106 and MiG-21 becoming established in squadron service by the end of

Throughout its long career the Mirage IIIO was moderately updated, mainly by the adoption of more modern weapons and, in the case of IIIO(F) aircraft like this No. 75 Sqn machine, the addition of ground attack capability. The original Martin-Baker Mk 84 seat was replaced by the 'zero-zero' Mk 4 in the early 1980s.

***Left:* No. 75 Squadron put up this Mirage IIIO six-ship toward's the end of the type's career. The squadron maintained some of its aircraft in camouflage right to the end, arguing that the old scheme was better suited for operations over the jungles of northern Australia.**

The Australian Mirage (this is the first IIIO) was based on the IIIE, and consequently had the longer forward fuselage. Compared to the IIIC, a 11¾-in (30-cm) plug was added aft of the cockpit to provide additional space for avionics.

the decade, meant that an all-weather aircraft with sustained supersonic performance was required.

Airframe selection

Lockheed had made strong representations to Australia on behalf of the F-104 Starfighter in the mid-1950s, and although several senior RAAF officers openly favoured the aircraft, RAAF pilots visited France, Belgium and Sweden at the end of the decade to fly the Mirage IIIC, F-104G and Draken. This activity culminated in a team being despatched to the USA and Europe in May 1960 to study in greater depth some of the aircraft, as well as other alternatives. The team, made up of RAAF and industry representatives, was tasked with recommending a Mach 2-capable, all-weather interceptor with a combat radius of 700 miles (1126 km), and an air-to-air missile 'kill' probability of no less than 85 per cent against a single opponent.

The field was eventually narrowed down to the Lockheed F-104G, Northrop N156 (then a private venture, but later to become better known as the F-5A Freedom Fighter) and Dassault Mirage III. Shortly before the team's departure from Australia, however, the English Electric Lightning and Republic F-105 Thunderchief were added to the list at the behest of the then-Minister for Air.

The two political additions were soon eliminated from the race: the F-105 was deemed too large and complex (as well as suffering significant performance shortfalls) and, as a point-defence fighter, the Lightning did not have sufficient range to meet Australia's needs. The Northrop candidate also was soon eliminated, as not only did it fall short of the Mach 2.0 yardstick but there was not yet any guarantee of it entering production.

This left the Starfighter and Mirage as the only real candidates for the order. Much discussion ensued, underpinned by the need to ensure that, whichever type was selected, it would not have become obsolete by the time it reached squadron service – no easy task, given the rapid development of military aviation at the time.

Another consideration was the general standard of airfields in the remote parts of Australia, as forward deployment to airfields other than established RAAF bases (only those at Darwin and Townsville were then extant in the northern part of the country) would be very much limited by the new type's needs. There were real concerns that the Starfighter, in particular, would have difficulty operating from the austere airfields throughout the more remote areas of the country.

In the end, a version of the proposed Mirage IIIE emerged as the clear winner, due in part to the airfield requirements of the Lockheed product, but also because the French fighter was seen as more of a multi-role platform. An initial order for 30 Mirage IIIO aircraft (the 'O' suffix reputedly denoted 'Orstralia', as the IIIA appellation had already been used) was announced

The Avon Mirage

Only one Mirage was built with a Rolls-Royce Avon turbojet. The aircraft carried the name City of Hobart **on the nose, along with the legend 'Mirage IIIO no. I'. Unlike the production aircraft which followed, it had a dorsal fin fillet. The Avon-powered Mirage first flew at Melun-Villaroche on 13 February 1961, and flight trials were generally satisfactory. However, the considerably greater thrust of the Avon compared to the Atar 9C did not bestow a sufficiently large enough increase in performance to warrant the cost of the changes necessary for production, and the British engine was quietly dropped. These two views show the Avon Mirage clean (left), and with a load of fuel tanks and 1,000-lb (454-kg) bombs (above).**

***Above:** The first Mirage IIIO, A3-1, is seen at its roll-out from the Dassault factory at Merignac. The first two aircraft were built and test-flown in France before delivery to Australia.*

A3-1 arrived in Australia inside the hold of an RAAF C-130A, packed with several boxes including one labelled 'Beans in Tomato Sauce'. C-130s later had their forward cargo doors sealed shut.

***Below:** One of the early aircraft, almost certainly A3-3, is prepared for an engine test at Avalon. It has an 'iron' nose, the radar being replaced by ballast for aerodynamic trials.*

by the Minister for Defence, Athol Townley, on 15 December 1960, and the contract was duly signed in Paris on 31 March 1961. The Mirage broke new ground as the first continental European aircraft for the RAAF, and therefore the first built with metric measurements.

The Mirage IIIO purchase was only one component of a significant capability upgrade, which had begun with a major increase in defence spending allocated to the RAAF from 1954. It was ultimately responsible for the acquisition of the C-130A/E Hercules, Macchi MB.326H, UH-1B Iroquois and (after much delay) F-111C.

Engine selection

Given that the CAC-built Sabres had been fitted with a Rolls-Royce Avon, Dassault believed that a Mirage III with the same engine, albeit in the shape of the afterburning Avon RA-67, would be more palatable to the RAAF. Accordingly, the French proposals were for an Avon-powered aircraft, and a modified Mirage III began flight trials from Villaroche on 13 February 1961. The Avon engine could offer 12,000 lb static thrust (53.67 kN) dry (16,000 lb/71.15 kN with reheat) as opposed to the Atar's 9,480 lb (42.16 kN) (13,000 lb/57.81 kN with afterburner) and it was felt that this, coupled with the fact that the Avon was a British engine, would sway the Australians away from the competition.

When the RAAF Mirage Project Team arrived in Paris in the last few days of 1960, it was suggested to the French that the SNECMA Atar 9C, as proposed for the Mirage IIIE, would be of interest, and Dassault obligingly rescheduled the testing phase of that engine/airframe combination to provide data that was directly comparable to the Avon trials. Shortly thereafter, SNECMA submitted a proposal to the Australians for the more powerful Atar 9K as fitted to the proposed Mirage IV nuclear bomber (and, later, to the Mirage F1 series), as a counter to the more powerful Avon, and this engine was accordingly added to the list of options. The Atar 9K ('K' supposedly denoting 'Kangaroo') was more in the Avon thrust class, promising 10,300 lb st (45.81 kN) dry, and 14,700 lb (65.38 kN) with afterburner.

Flight trials with the Avon-powered aircraft (named *City of Hobart*) were completed in late April 1961, and it was evident from the data gathered that the expected performance gains were not as high as had been hoped. The Avon engine had been expected to give the Mirage increased range, greater rate of climb and reduced take-off roll, but the small gains that were actually achieved were deemed not to be worth the engineering risk of adapting the airframe to suit the larger engine.

CAC and the RAAF had been faced with major structural changes to the Sabre when fitting the Avon, and a complete set of ground and flight trials was demanded as a result. These problems, and costs, were fresh in the minds of the RAAF engineering branch, and the choice of the lighter, cheaper Atar engine was made on 17 May 1961. At the time of selection, a choice between the Atar 9C and 9K had not been made; in the event, the latter engine was not available in the established timeframe, and was eliminated from the race in the following September. Thus, the Mirage IIIO was to be comparable to the Mirage IIIE of the French Air Force, but with numerous small equipment and engineering changes to suit Australia's needs.

Into production

Australia had built up its aircraft manufacturing capability during World War II, and by the end of the 1950s both CAC's factory at Fishermans Bend near the city of Melbourne, and the neighbouring plant of the Government Aircraft Factories (GAF), could boast world-class facilities. As an aircraft design and manufacturing company that had built a range of fighter and trainer aircraft, including the aforementioned Sabre, CAC, in particular, saw itself as the natural choice if the government were to decide to build the Mirage in Australia. On the other hand, GAF had been formed during the war as the Department of Aircraft Production, a wartime 'shadow' facility to produce other companies' designs (Beaufort and Beaufighter and, post-war, the Avro Lincoln). Most recently, GAF had completed the production of 48 licence-built Canberra bombers for the RAAF, and the indigenous Jindivik remotely-piloted vehicle programme was in full swing.

CAC's confidence in gaining the Mirage production contract was understandable, as Dassault had approached the company for assistance with the preparation of production data for its original bid, and a CAC representative had accompanied the study team that toured Europe and the USA in 1960 in the lead-

Although it was built by Dassault, A3-3 was the first aircraft to be assembled from kit form by GAF, and thus claims to be the first Australian aircraft. It is seen here landing at Avalon, where final assembly of the IIIOs took place.

The early Mirage test programme was undertaken from the final assembly site at Avalon. Above is the first Australian-assembled aircraft, in clean form, while the second (right) is shown carrying subsonic RP30 374-Imp gal (1700-litre) drop tanks. A3-3 survives today at Williamtown, while A3-4 was lost at Avalon in 1972.

up to the Mirage selection. It therefore came as a bitter blow to the company when, at the announcement of the Mirage selection, the Defence Minister announced that GAF would be the prime contractor for the Mirage airframe, and that CAC would build the Atar engine. The reason given for the allocation of Mirage production to GAF was that it was deemed to have a "more stable workload" with the Jindivik, whereas CAC's Sabre production was winding down. Naturally, CAC management was furious at what it saw as the political decision of awarding work to the government-owned GAF. After much remonstration, CAC was eventually, in April 1961, also awarded the contract to produce the wings, fin and tailcone, as consolation. CAC later built the Macchi MB.326H under licence for the RAAF, and assembled the Bell 206B-1 Kiowa for the Army and RAN, but the loss of the Mirage contract when the factory was facing drawdown at the conclusion of the Sabre project was a serious setback.

The prospect of producing a modern, complex aircraft required both GAF and CAC to invest heavily in upgrading their factories to cope with the demands that the Mirage would make, and many obstacles had to be overcome. The Mirage was not only the first metric aircraft in the RAAF, it was also the first type to be purchased from a manufacturer whose first language was not English. Much work was required to convert French-language technical publications into English and, because production drawings were not translated, to ensure that the workforce had the necessary information to hand. Interestingly, CAC converted its Atar engine drawings to Imperial measurements and tolerances, but GAF retained the metric system for the airframe.

Follow-on orders

Following the initial order for 30 Mirage IIIO(F) aircraft (optimised for air-to-air roles and equipped with the CSF Cyrano IIA radar), an additional 30 were ordered in October 1962, and the total was further increased to 100 the following May. Of these additional aircraft, 51 were to be of the multi-role IIIO(A) variant equipped with the Cyrano IIB radar, intended to replace the Sabre in the ground-attack role. Later still, a decision to procure the two-seat Mirage IIID for conversion and continuation training was taken, and 10 aircraft were ordered in October 1963. A final order for an additional six of this latter variant was placed in November 1970, bringing total RAAF Mirage production to 116. Over the project life, CAC also built no fewer than 140 of the 192 Atar 9C engines acquired by the RAAF.

As noted earlier, both GAF and CAC had factories at Fishermans Bend, a short distance downstream from Melbourne on the banks of the Yarra River. Airframe sub-sections and engines were shipped by road to Avalon for final assembly and flight testing. Avalon, about 34 miles (55 km) from Melbourne on the western coast of Port Phillip Bay, was a purpose-built facility commissioned by the Department of Supply in 1952, and administered by GAF. Fishermans Bend had its own runway, which had been used by CAC and GAF until the early 1950s, when the advent of jet aircraft (and proximity to the industries and suburbs of Port Melbourne) dictated a longer runway. Once assembled, Mirages were flight tested from Avalon by RAAF pilots from the Aircraft Research and Development Unit (ARDU) based at nearby Laverton. GAF did not employ civilian test pilots for Mirage testing, and ARDU pilots test flew the aircraft for both the contractor and the customer.

The first aircraft were built in France, then dismantled and shipped to Australia for reassembly. A3-1 flew for the first time at Bordeaux on 14 March 1963 and arrived at Avalon aboard an RAAF C-130A in January 1964. A3-3 previously had been shipped to Australia in component form and is regarded as the first Australian aircraft, making its first flight from Avalon on 16 November 1963. A3-2 was retained in France for flight trials with Dassault, and was not delivered to Australia until August 1965. This particular aircraft remained on test and trials duties for its entire career with the RAAF, being based first at Avalon, then Laverton, and finally Edinburgh with ARDU. It underwent many modifications throughout its testing career, including the installation of a non-standard 'radome' with cameras fitted. Its

Right: Early Mirage IIIO(F)s were completed with the red trim that also adorned French and Israeli aircraft.

Below: A3-4 and A3-11 fly over the suburbs of Melbourne, near the mouth of the Yarra and close to the GAF and CAC factories at Fishermans Bend. A3-11 was the first aircraft to be almost entirely built in Australia.

Resplendent in its silver lacquer finish, this No. 75 Squadron IIIO(F) has the unit's coat of arms applied to the fin. No. 75 was the first operational unit, and additionally had a primary air defence tasking.

non-standard nature meant it was the only surviving IIIO(F) aircraft not converted to IIIO(FA) standard. As Mirage assembly continued, Australian content increased, until, by 1966, 80 per cent of the airframe and 85 per cent of the Atar engine were manufactured in the country. Due to problems with its workforce around this time, however, GAF acquired 10 additional French-built fuselages to ensure that the production line flowed smoothly, so odd-numbered aircraft between A3-61 and A3-79 (inclusive) had less Australian content than their sisters.

Although based on the Mirage IIIE, Australia's aircraft differed in several respects, including the fitment of a Sperry twin gyro platform and the deletion of the ventral SEPR 844 rocket pack in lieu of an additional 550-litre (121-Imp gal) fuel tank. The later IIIO(A) was fitted with a radio altimeter, 'wet' wing leading edges (in later aircraft), and a Marconi Doppler radar navigation system with a flush underfuselage antenna.

Service entry

Prior to the first aircraft's arrival in Australia, the RAAF had set up a project office in Paris in order to oversee contract development, equipment selection and training requirements, and the first group of four pilots to undertake training with the French Air Force left Australia in May 1963. On their return it was intended that they would form the nucleus of the Mirage Conversion Unit within No. 75 Sqn, the first unit slated to fly the aircraft. In the event, No. 2 Operational Conversion Unit received the first aircraft, and this group formed the Mirage Training Flight within that unit.

Following a crash French language course at the RAAF School of Languages at Laverton, ground training began with the French Air Force in Strasbourg, and progressed to Colmar and Dijon. Flying was undertaken with the Centre d'Expérimentations Aériennes Militaires (CEAM) from Mont de Marsan on the Mirage IIIC, as the French had yet to introduce their Mirage IIIEs to service. It was during this time that the RAAF pilots experienced backseat rides in the Mirage IIIB, and their recommendations were instrumental in the later acquisition of a dual-seat Mirage for the RAAF, in the form of the Mirage IIID. Upon their return to Australia, these first Mirage-qualified pilots became heavily involved in the development of the flying training programme, including the writing of flying training and tactical manuals.

In order to cope with Mirage operations, most RAAF bases around Australia required improvements to runways, taxiways and servicing infrastructure, but none more so than Williamtown. RAAF Williamtown, north of Sydney, was (and still is) the traditional home of the RAAF fighter force, and was the natural choice as the hub of Mirage operations within Australia. Purpose-built facilities were constructed to cope with the maintenance requirements of such a complex aircraft; radar and modern avionics servicing areas, for example, simply did not exist in the Sabre era, and a large sum of money was spent on their construction. Airfield facilities also needed a great deal of work to bring them up to the standard required, with equipment such as a Ground Controlled Interception (GCI) facility and a (Swedish) runway barrier system purchased and installed before the first aircraft could be delivered.

Because of this work, early aircraft were retained at Avalon for a short period and the Training Flight crews were required to commute between Williamtown and Avalon to maintain their currency. The first aircraft (A3-3) was eventually delivered to Williamtown on 26 February 1964, followed three weeks later by the second, and enough aircraft had been delivered in time for the first students' course to commence in October 1964.

Introducing the RAAF's first all-weather and supersonic aircraft was a demanding task. The handling idiosyncrasies of the delta wing, the use of afterburner for the very first time (with the consequentially horrendous increase in fuel burn), introduction of tactical air navigation to the fighter force, and GCI procedures all combined to make the early courses a steep learning curve. The first Mirage loss occurred during December 1964 when A3-1 crashed near Avalon, after the pilot (an RAF exchange officer with ARDU) was forced to eject at extremely high speed, after the aircraft had entered an uncontrollable spin during flight trials. By the middle of 1965, however, the Mirage training system had trained enough ground and aircrew, and the factory had produced enough aircraft, to allow the first operational squadron to form at Williamtown.

No. 75 Squadron began receiving its first aircraft in January 1965, and finally stood up as a Mirage unit on the first day of August. As the RAAF's first truly supersonic squadron, No. 75 wasted no time in displaying its new craft to an appreciative Australian public, and deployed across the Tasman Sea to New Zealand for the very first time in January 1966. This deployment was followed shortly thereafter by the first live firings of the MATRA R530K missile by an operational RAAF squadron.

The Mirage IIIO(F) was optimised for the air-to-air role, so initial weapons were the afore-

Mirage production

All the RAAF's Mirages were assembled to some degree at the GAF airfield at Avalon, to the west of Melbourne, including the first two 'pattern' aircraft, the first IIIO(A) and the first IIID, which were wholly built in France before partial disassembly for delivery. Of the 116 aircraft, 79 were built almost entirely in Australia by GAF (fuselage and final assembly) and CAC (engine, wings, tail). Twelve were built by Dassault, and the remaining 25 had French-built fuselages, including 10 IIIO(A)s supplied to bridge a gap in production caused by strikes and 'go-slows' at GAF.

Above: Three Mirage IIIO(F)s from No. 2 OCU are led by a IIID over Sydney Harbour, during the period of the unit's brightest colour scheme.

Right: No. 76 Squadron was the second fighter unit to form. These two natural metal jets are carrying supersonic tanks and dummy AIM-9B Sidewinders.

mentioned MATRA R530K (universally carried on the centreline pylon), AIM-9B Sidewinder and two DEFA 30-mm (1.18-in) cannon. The RAAF had previously introduced the AIM-9B into service in 1959-60, when it was integrated into the Sabre's weapons system; a pair of these weapons could be carried by the Mirage, one each on an Aero 3B launcher on the outer wing pylons. Regarded as a fine weapon, the cannon later became well respected as an air-to-ground option once the Mirage IIIO(A)s began to reach squadron service. Unfortunately, the R530 earned itself a reputation for being unreliable, and of limited use in all intercepts but those where the target was behaving in a predictable manner, yet it remained in service until almost the very end of Mirage flying. The weapons fit can be regarded as the Achilles' heel of the Mirage force. Although adequate when the aircraft was first selected, the AIM-9B was very much an early generation version, which was only effective in a stern-chase interception. Like the MATRA missile, the AIM-9B served the Mirage fleet well beyond the 1960s, and the RAAF was not to regain parity with other air forces until the introduction of the MATRA R550 Magic missile in the mid-1980s.

No. 76 Squadron reformed on the Mirage in September 1966, as the second, and last, interceptor (IIIO[F]) unit. Future deliveries were of the IIIO(A) version and, for this purpose, A3-26 was built (and initially retained) in France for IIIO(A) prototype trials. The introduction of this model then allowed No. 3 Squadron to begin conversion from the Sabre in July 1967, followed in July 1969 by No. 77. As delivered, IIIO(F) aircraft were in a natural metal scheme (later painted overall with an aluminium lacquer), but the IIIO(A) aircraft left the factory in a green/grey camouflage scheme that befitted their 'mud-moving' role. With this later version coming off the production line, a decision to convert the earlier IIIO(F)s to the multi-role configuration was made and, after modification, these aircraft became known as IIIO(FA)s.

Thus, by the end of the decade Australia could boast four operational Mirage squadrons, two each in the air interception (Nos 75 and 76 Sqns) and ground attack (Nos 3 and 77 Sqns) roles. The fifth unit was, of course, No. 2 OCU, and the remaining handful of aircraft not allocated to these squadrons were on the strength of ARDU.

Butterworth deployment

RAAF Sabre squadrons had been based at RAF Butterworth in Malaya (now Malaysia) since 1958 as a component of Australia's contribution to the Commonwealth Far East Strategic Reserve. This force was created in response to the Communist insurgency in Malaya, and the air arms of Great Britain, Australia and New

Suitably attired for tropical duty, a marshaller beckons in a No. 3 Squadron Mirage IIIO(A) to its spot on the Butterworth ramp. The IIIO(A)s were delivered in a 'wrap-round' scheme with dark undersides, as befitted their primary ground attack tasking.

Zealand conducted operations against Communist forces for the best part of the 1950s, as part of Operation Firedog. As a result, the RAAF had reformed No. 78 Wing at Butterworth, to oversee its contribution to the air defence of the Malay Peninsula and Singapore.

Two of the four Mirage squadrons had been earmarked for forward deployment to Butterworth from an early stage (one air defence and one ground attack unit), and No. 75 was chosen as the first to deploy in 1967. The Sabres of Nos 3 and 77 Sqns were based at Butterworth at the dawn of the Mirage era, but the former unit was slated to withdraw to Australia for Mirage conversion, to be replaced by No. 75 Sqn.

In preparation for its relocation, No. 75 Sqn had carried out a series of long-range deployments, including the maintenance of a rotational forward deployment to Darwin. In earlier years, the Sabre squadrons had departed Australia from Darwin, through New Guinea and the Philippines to Malaysia, but the range constraints of the Mirage meant that a new route would have to be found.

A transit through Indonesian airspace had previously been ruled out, due to the lingering ill-feeling between the two countries over the 'confrontation' of 1962-66, but a diplomatic breakthrough came in time to allow a single-stopover transit of Indonesia for No. 75 Sqn. Accordingly, 23 aircraft (comprising 20 squadron aircraft, and three air-spares from 2 OCU) blasted off from Williamtown on 15 May 1967 for Butterworth as part of Operation Fast Caravan. The initial destination, however, was Darwin, for a two-day stopover. After routine maintenance, No. 75 Sqn departed Darwin on 18 May (with 19 of its own aircraft and an OCU replacement for one of its ailing jets), for a brief turn-around at the Indonesian Navy (TNI-AL) airfield at Djuanda, near Surabaya. After refuelling was complete and diplomatic protocols attended to, all aircraft departed for Butterworth to join the Sabres of No. 77 Sqn. No. 75 Squadron was to remain at Butterworth until withdrawn to Darwin in 1983.

Thoroughfare: Mirage changeover

Having travelled in the opposite direction in February 1967, No. 3 Squadron handed its Sabres over to the OCU at Williamtown and set about conversion to the new Mirage IIIO(A). After reforming on the Mirage, No. 3 Sqn began preparations to return to Butterworth in order to allow No. 77 Squadron to withdraw to Australia and begin its Mirage conversion process. The ferry to Malaysia occurred in February 1969 and the route taken was the same as that used almost two years previously by No. 75 Sqn. Known as Operation Thoroughfare, the Mirages of No. 3 Sqn actually crossed over in Darwin with the final batch of No. 77 Sqn Sabres returning home. During the stopover at Djuanda, one Mirage was found to have damaged a ferry tank and was flown 'clean' on the final leg, though prudence dictated a diversion to Changi instead of Butterworth.

After the arrival of No. 3 Squadron, the composition at Butterworth was settled, and did not change until No. 75 Sqn returned home in 1983. Following the withdrawal of British forces from the Far East, Butterworth became an RAAF station, and the Commonwealth Far East Strategic Reserve gave way to the Five Power Defence Arrangements (FPDA) in 1971, at which time control of Butterworth was passed back to the Malaysians. The Five Power Defence Arrangements was formulated in London in April 1971 by the governments of the United Kingdom, Australia, New Zealand, Malaysia and Singapore, and is still in force today, providing for the defence of Malaysia and Singapore.

Before the British withdrawal, the two RAAF Mirage squadrons participated in several joint air defence exercises with RAF Canberras, Hunters and Lightnings, and a healthy rivalry soon developed between the two forces. The exercises with RAF Lightnings from No. 74 Squadron at Tengah were a particular highlight, and became known as Tiger Rag exercises. The last Tiger Rag with Lightnings took place in April 1971, and the final one altogether (with RAF Hunters) followed in October that year. The Five Power Defence Arrangements provided for an Integrated Air Defence System to be set up at Butterworth, commanded by an RAAF officer and based around an RAAF Control and Reporting Unit.

Regular Mirage detachments to Singapore had been performed from the very beginning, but greater responsibilities would have to be shouldered once the RAF finally departed for home. As a result, the regular detachments became a permanent presence at Tengah (and in the final years at Paya Lebar), and the two Australian squadrons took turn-about to provide a flight of Mirages for a three-month

Mirage IIIO cockpit

These three views show the left, centre and right consoles of the Mirage's cockpit. The design was typical of its era: many of the system switches were located to the sides.

Left: The centre console was dominated by the weapons sight and the hooded screen for the radar – either Cyrano IIA in the IIIO(F) or IIB in the IIIO(A). The main instrument display – showing essential flight, fuel and engine data – was relatively uncluttered and easy to read.

Above: Non-standard colours on the Mirage IIIO were rare. Seen high over Williamtown, this aircraft was specially marked in the 1980s with the base badge and red/white 'dart'.

period. Nominally consisting of eight aircraft, the first No. 3 Squadron flight arrived at Tengah in January 1972 to commence its 12-week stint.

Butterworth was a popular posting within the RAAF, though the Mirage crews found the tropical environment operationally demanding – heat and humidity took their toll on both man and machine. Designed for a northern European climate, the Mirage suffered from inadequate equipment cooling and water ingress into the electrical system during the frequent downpours of the wet season. Despite this, morale remained high throughout the Mirage's tenure in Malaysia, and many pilots and ground crews enjoyed two or three postings to Butterworth.

One of the great advantages of being a pilot in a Butterworth Mirage squadron was the amount of dissimilar air combat training (DACT) available, first from the RAF squadrons and later from the Royal Malaysian Air Force (TUDM) with ex-RAAF Sabres and (later) A-4s and F-5s. The detachment on Singapore had similar opportunities with the Hunters, Skyhawks and F-5s of the Republic of Singapore Air Force. Several deployments to Clark Air Force Base in the Philippines were undertaken during the Mirage years for purposes of training against USAF assets, and RAAF Mirages participated in at least two Cope Thunder exercises.

Australia is a long way from anywhere, and deployments by the USAF or RAF were comparatively rare in the late 1960s due to the contraction of British foreign policy and the war in Vietnam, and the opportunities for DACT were extremely limited back at Williamtown. This situation had improved somewhat by the early 1970s, with RAAF Phantoms, and RAN and RNZAF Skyhawks to train against, but the opportunities were still not as plentiful as those available to the Butterworth squadrons.

Below: Two No. 3 Sqn Mirages IIIO(A)s take off from Butterworth. The squadron was resident in Malaysia from 1969 to 1986, initially manning the 'attack' half of the detachment.

The first decade

The first IIIO(A) had been delivered to No. 2 OCU in November 1966, prior to the adoption of the type by No. 3 Squadron. A very short time later this type was followed into the OCU (on 11 November) by the first Mirage IIID (A3-101), and it is worth noting here that (contrary to the information published in many magazines and journals on the subject of the Australian Mirage) RAAF dual-seat aircraft were designated IIID and not IIIO(D). The final single-seat aircraft (A3-100) was delivered in November 1968, and the final dual-seater (A3-116) in January 1974.

When the Mirage began entering service in the middle of the 1960s, the RAAF Fighter Training scheme consisted of the student commencing his initial flying training on the

Two-seaters

The first Mirage IIID (A3-101) for the RAAF was wholly built by Dassault and flown in France (right). The remaining 15 aircraft had Dassault-built fuselages, which were transported to Australia for assembly with locally-built components. The type was issued to No. 2 OCU for type conversion, and in small numbers to the front line (below, No. 75 Sqn at Darwin) for continuation training.

From January 1972 the RAAF maintained a detachment at Tengah to provide air defence for Singapore. Around eight aircraft were usually deployed from either No. 3 (illustrated) or No. 75 Squadron at Butterworth.

CAC Winjeel, followed by conversion to the two-seat Vampire, and it was felt that a posting to a Mirage squadron would be a big step. Accordingly, the Sabre was given a stay of execution to provide a stepping-stone to the Mirage, first with No. 2 OCU and then with the dedicated Lead-In Fighter unit, No. 5 OTU. This stepping-stone process is repeated in the modern RAAF, with the Hawk Mk 127 filling the void between PC-9 and Hornet.

In the event, the 1967 introduction into service of the CAC-built Macchi MB.326H, and the acquisition of the dual-seat Mirage IIID, rendered the Sabre step obsolete, and No. 5 OTU (and its aircraft) were consigned to history in 1971. Initial Mirage conversion within No. 2 OCU took about three months, with the student flying in the region of 100 hours on the Mirage (both dual- and single-seat models) before being posted to an operational squadron.

With the IIIO(A) entering service, new weapons had to be acquired and new tactics developed in order to operate the aircraft in the ground attack role. This burden fell to ARDU and No. 2 OCU, prior to No. 3 Squadron's formation. Initially, 500-lb (227-kg) and 1,000-lb (454-kg) low-drag bombs were utilised, the first live weapons being dropped in April 1968. These bombs were carried on an Alkan centre-line bomb beam, and later replaced with the Mk 82 and Mk 84 Snakeye series of high-drag weapons introduced in the 1970s. General Electric AN/ALQ-72 noise jamming pods were also introduced into service on the Mirage from 1972. In a chapter of the book *Mirage Story* (by Wing Commander M.R. Susans), Wing Commander J.A. Smith recalls a formation of four No. 3 Sqn aircraft dropping BLU-27/B 750-lb (340-kg) fire bombs (napalm) on the Singleton range in March 1968 – although no other use of this weapon could be uncovered.

Multi-role conversions

In June 1969, a decision was taken to rework the surviving 'fighter' IIIO(F) aircraft to the multi-role IIIO(FA) standard, and these aircraft were modified at Avalon by GAF as they fell due for servicing.

Mirages for reconnaissance

One secondary role envisaged for the Mirage IIIO fleet was photo-reconnaissance. A number of aircraft were converted for the role, housing a KA-56B camera in a modified radome. The camera was mounted on a cradle which swung down so that the film magazine could be removed (left). The lens peered through a two-pane window (below). Behind the camera window is the oval panel for the Doppler navigation system, which in most Mirage IIIE derivatives was housed in a bulged fairing. Nos 3 and 77 Squadrons operated the majority of the recce-capable Mirages, although this aircraft was on the strength of No. 2 OCU.

At this time, the fourth Mirage squadron, No. 77, was in the process of re-equipping with the delta-winged fighter. As recorded elsewhere, No. 77 Sqn had been withdrawn from Butterworth early in 1969, passing on its Sabres to No. 2 OCU at Williamtown. Conversion to the Mirage had commenced in March, and the squadron was declared fully operational on the last day of June 1969.

This saw the completion of the fighter re-equipment plan, and the RAAF could now boast four operational Mirage squadrons: Nos 3 and 75 Squadrons at Butterworth, and Nos 76 and 77 Squadrons at home in Australia, at Williamtown. At the commencement of Mirage operations, No. 78 Wing at Butterworth and No. 81 Wing at Williamtown were the parent units of the flying squadrons, but they were disbanded in 1967 and 1966, respectively. As mentioned in an earlier section, these squadrons were role-specialised and each devoted approximately 70 per cent of its time to its 'core' role. The decision to convert the IIIO(F) to multi-role IIIO(FA) standard saw the entire force become multi-role.

This was not the case at the end of the decade, however, and some role swapping occurred as aircraft were released back to GAF for conversion. In September 1969, No. 76 Squadron was given the task (along with five pilots from the newly-formed No. 77 Sqn) of ferrying IIIO(A) aircraft to Butterworth. This allowed No. 75 Squadron to re-equip with the multi-role aircraft, and the surplus IIIO(F)s were returned to Williamtown for No. 77 Squadron use, until they could be incorporated into the factory modification process. Therefore, at the same time that No. 75 Squadron at Butterworth was learning ground attack skills, the ground attack-optimised No. 77 Squadron at Williamtown found itself re-tasked as an air defence unit. This situation lasted until July 1970, when sufficient Mirage IIIO(FA)s had returned from Avalon.

The introduction of the Mirage IIIO(A) provided the opportunity for a battlefield reconnaissance capability to be added to the two ground attack units. Several aircraft were sent back to GAF for modification, commencing with A3-83 from No. 3 Sqn in July 1968. A single KA-56B camera was installed in a modified radome, together with the necessary wiring changes to enable a cockpit switch panel to be fitted in lieu of the radar scope and controls. After modification, these aircraft were not dedicated reconnaissance aircraft, and the camera nose could be exchanged for a standard unit, containing the Cyrano IIB radar, as required. In the mid-1970s a pair of F95 split-image vertical cameras was added, installed in a modified cannon bay tank. Both Nos 3 and 77 Squadrons trained a cadre of pilots in battlefield reconnaissance operations (TAC PR), and this capability remained the RAAF's sole tactical reconnaissance asset until the introduction of the RF-111C into service in the mid-1980s.

The RAAF celebrated its Golden Jubilee in 1971, and it was decided that a Mirage formation aerobatic team would be an appropriate platform to showcase the skills of the fighter force to the public. The task of forming such a

Left: From the early 1980s GBU-12 Paveway II laser-guided bombs could be carried on the centreline and on RPK-10 fuel tank bomb carriers. Here the combination is tested on ARDU's A3-76. Designation was provided by ground-based sources.

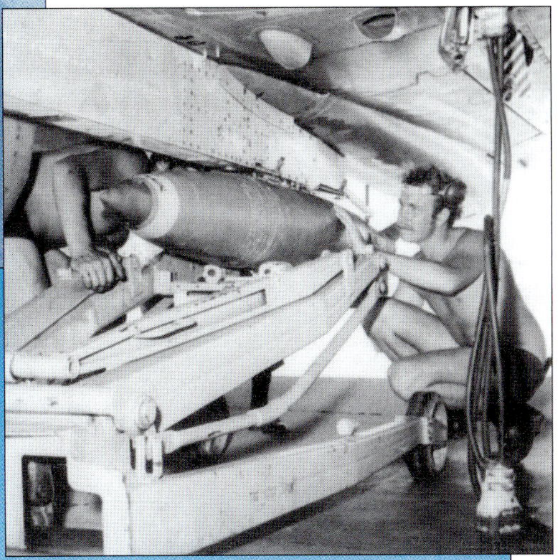

Attack weapons

When it entered service the Mirage IIIO(A) employed UK-style bombs, but from the 1970s the US Mk 80 series was used. ARDU's long serving testbed A3-2 releases six Mk 82s from the trials AMSER rack (right), the aircraft also carrying underwing camera pods. At far right armourers load a Mk 82 on to a Mirage's centreline.

team fell to No. 77 Squadron at Williamtown in November 1970, which lost no time in selecting pilots and working up a routine. Initial flying was based around a five-aircraft formation demonstration, but the large turn radius of the Mirage meant that there were several moments of 'empty sky' during each display, and a synchronised pair was added to fill the gaps. A series of air displays had been planned at RAAF bases around the country to celebrate the Jubilee, and the 77 Sqn team, by now known as 'The Deltas', was the centrepiece of each show. The team eventually disbanded after the last in the series of Jubilee displays in April 1971.

In the 1950s and 60s, several aerobatic teams were formed by RAAF fighter squadrons,

Below: General-purpose bombs and the twin DEFA 552 cannon were the primary weapons in the ground attack role, although 68-mm SNEB rocket pods were also available. Here A3-2 fires a full salvo.

Fighter weapons

Below: The standard radar-guided weapon was a single MATRA R530K on the centreline. The missile was limited in its engagement envelope and had an effective range of about 9 miles (15 km).

Below: A No. 79 Squadron Mirage III fires off a Magic missile. This unit, which was only briefly in existence, undertook the last Magic firings from the Mirage IIIO in November 1987.

Above: Until the early 1980s the AIM-9B Sidewinder was the IR-guided weapon, here represented by an inert training round on a No. 75 Squadron Mirage. This early 'Winder was essentially a tail-chase weapon.

Displayed on a No. 3 Sqn aircraft (note the Southern Cross rudder marking) at Butterworth is the standard 1980s armament of two Magics and a single R530. The latter had interchangeable IR and radar seeker heads.

mostly flying the Sabre, but the cost and complexity of the Mirage meant that this practice largely died out with the retirement of the older type. No. 77 Squadron revived the tradition briefly in 1976, when the 'Miracles' were formed for a series of displays, and again in 1981 when a synchronised pair was created to celebrate the RAAF's Diamond Jubilee. This pair, although not named, was unusual in that, for the very first time, the aircraft were given a display colour scheme – painted white overall, with red and blue trim. Both the 'Deltas' and the 'Miracles' had used aircraft that carried the standard camouflage pattern and markings of the day.

Almost from the beginning of operations, there was concern in RAAF engineering circles that the Mirage fleet was consuming fatigue life at a rate far greater than their French counterparts. This was particularly exacerbated by the switch to multi-role operations, where a greater part of each aircraft's working life would be spent in the more dense and turbulent air at low level. This reason was used (in part) for the announcement in 1973 that No. 76 Squadron was to disband; its aircraft spread through the remaining squadrons in order to dilute fleet fatigue life.

There were other factors involved, such as budget constraints and soaring fuel costs, but the disbandment of one of only two fighter squadrons on Australian soil came as a surprise. Despite lobbying to the contrary, No. 76 eventually disbanded on 24 August 1973, and No. 77 Squadron was left to defend Australia alone, until No. 75 Squadron was withdrawn to Darwin in 1983.

The loss of No. 76 Squadron did improve matters initially, as more aircraft, ground and aircrews, and spares became available to the other units, but the change to multi-role operations resulted in a certain dilution of skills. Instead of spending 70 per cent of his time maintaining proficiency in air defence (for example), a Mirage pilot was now able to devote only 50 per cent of his efforts to that role.

Fade to grey

By the middle of the 1970s fatigue was becoming a real concern, and a study of wing fatigue life began at the Aeronautical Research Laboratories in Melbourne. As a result of the findings, a boron patch was developed that could be applied to damaged areas in order to restore structural integrity. This was a breakthrough in aeronautical engineering, and was used most successfully on a number of Mirage wings. Today, the same technology is in use in the F-111 wing fatigue programme. The Mirage was originally due to be withdrawn from service in 1979, but it soon became clear that a successor would not enter service before then, and some form of Life of Type Extension (LOTEX) would have to be performed. An additional 30 sets of wings were purchased and progressively fitted to the fleet. One-third of these new wings were manufactured by CAC at Fishermans Bend, and the remainder were sourced from France. The wing LOTEX ran between 1980 and 1984 – and it came none too soon, as several aircraft had been restricted to +5g manoeuvring because of wing fatigue issues by 1982.

At the same time that the wing purchase was announced, the Mirage force was informed that it would get a new heat-seeking WVR (within visual range) missile to replace the obsolete AIM-9B. The MATRA R550 Magic was selected for this purpose, and began integration into the Mirage fleet in 1982. The last live AIM-9B firing occurred in June 1984, by No. 3 Sqn, which also had the honour of firing the first live Magic in October 1984. Around the time that the Magic was selected, the Paveway series of laser-guided bombs also became available, and was first used by No. 77 Squadron during Exercise Western Reward in October 1980. The laser-guided bombs were initially carried on the centreline bomb beam, but after the introduction of the RPK-10 bomb carrier/fuel tank in the early 1980s, one weapon could be carried on each underwing tank. If standard Mk 82 iron bombs were used instead, two could be carried on each tank. Late in the Mirage's service life, an ARDU aircraft was fitted on the centreline

For high conspicuity during weapons and systems trials the ARDU at Laverton, then Edinburgh, adopted this unusual scheme, which also incorporated photo-calibration marks. This aircraft has a modified radome housing an aft-facing camera to record weapon drops. An even more visible orange/white scheme was later applied to the unit's permanent test aircraft.

with an Australian Multiple Store Ejector Rack (AMSER), then under development for the F/A-18 Hornet. This rack was capable of carrying six Mk 82 bombs, but appeared too late to serve operationally on the Mirage.

Throughout the 1970s the Mirage replacement issue waxed and waned at the mercy of Australian politics. Many new types were offered, including refurbished Mirage IIIOs, but a successor did not appear on the horizon until the end of the decade, when it was announced that the competition had narrowed to a choice between the General Dynamics F-16 and McDonnell Douglas F/A-18. The Hornet was declared the winner in October 1981, with service entry to occur in 1985; this timing meant that the Mirage would need to remain in service until almost the end of the 1980s, a full decade later than the initial retirement date.

As the replacement issue dragged out, the lay media began to sensationalise each and every accident. Newspaper headlines echoed some of the more radical 'widow-maker' references found in the West German press of the time when referring to the loss rate of Luftwaffe and Bundesmarine F-104Gs. The Mirage loss rate was high, but it should be kept in perspective: the aircraft was far more advanced than anything that the RAAF had previously operated, and the performance leap from the Sabre to the Mirage was substantial. The fact that it was a single-engine design also meant that an engine failure was usually catastrophic (though at least two 'dead-stick' landings were successfully performed in RAAF service). In the 21 years that the Mirage was operational with the RAAF, no fewer than 43 aircraft were lost (at the time of writing, the RAAF has lost only four Hornets over a similar period of operation). These losses resulted in 25 ejections and, sadly, 14 fatalities. Of the 43 losses, 16 were engine-related, six occurred due to undercarriage malfunction (where ejection is mandatory), and two were caused by tyre failure on take-off.

By the early 1980s, the Malaysian and Singapore Air Forces were far more capable than they had been a decade earlier, and a gradual RAAF drawdown at Butterworth was planned. The permanent detachment of aircraft to Tengah gave way to a series of deployments to Paya Lebar in October 1983, and by then No. 75 Squadron had withdrawn to Darwin. For the first time in a decade, Australia had more than one fighter squadron based in the country.

Mirages in the north

Darwin was never intended to be a permanent base, but No. 75 Squadron relocated from Butterworth in August 1983 and remained there until it stood down as a Mirage unit in September 1988. The 'bare' base at Tindal, near Katherine, had been selected as Australia's 'forward-deployed' fighter base, but a massive construction effort had to take place before it could be used by one of the new Hornet squadrons. In the meantime, No. 75 made Darwin its home, and was actively engaged in exercises and deployments (including at least one back to Paya Lebar) to the very end.

With the Hornet introduction fast approaching, significant changes and restructuring to the Mirage force began to occur. The first, and most visible, was the gradual adoption of a new camouflage scheme, intended to trial colours and patterns that might be utilised on the new Hornets.

A series of low-visibility camouflage schemes was tested, mostly by No. 3 at Butterworth (but also by No. 75 Squadron before it left Malaysia), with most consisting of a two- or

Above: Aircraft from four of the six RAAF units to operate the Mirage formate over Penang island during a sortie from Butterworth in 1977. In the lead is a No. 75 Squadron aircraft, followed by aircraft from Nos 77 and 3 Squadrons, and No. 2 OCU in the foreground.

Left: Due to Australia's geographic isolation, the opportunity for exercising with and against aircraft from other air forces is limited, and even more so since the permanent Butterworth deployment came to an end. Consequently, the RAAF grabs every chance when foreign aircraft visit. Here two Mirages from the OCU formate with two F-15s visiting from the 18th TFW at Kadena. Eagles have been relatively regular visitors to Australian shores for either airshow appearances or exercise participation.

Mirage IIIO operators

No. 78 Wing

No. 78 Wing was formed at Townsville in November 1943, and relocated to Nadzab as a component of No. 10 Operational Group the following year. At the end of World War II, it was withdrawn to RAAF Williamtown and reduced to cadre status. In 1948, No. 78 Wing was reactivated with Nos 75 and 76 Squadrons, and moved to Malta to reinforce NATO in the Mediterranean. Operating Vampire FB.Mk 9s on loan from the RAF, the wing remained on Malta until it was disbanded in 1954.

The wing was formed once again at Williamtown in 1955 with the Meteor and (from 1957) Sabres. In 1958, it forward deployed to Butterworth as part of the Commonwealth Far East Strategic Reserve. It controlled RAAF Mirage operations in southeast Asia. Disbandment came with a restructure of the RAAF in 1967. In its final form, No. 78 Wing stood up at Williamtown in February 2002 to oversee RAAF fast-jet aircrew training as a component of the Air Combat Group.

No. 81 Wing

Formed at Ballarat on 12 April 1944, No. 81 Wing initially comprised Nos 76, 77 and 82 Squadrons. A move to Townsville and thence to Los Negros followed, for wartime operations, and later bases included Noemfoor, Morotai and Labuan. After the war No. 81 Wing moved to Bofu (Japan) in 1946, as a component of the British Commonwealth Occupation Forces (BCOF), and relocated to Iwakuni in February 1948. Following the withdrawal of all its component squadrons, with the sole exception of No. 77 Sqn, the wing disbanded later in 1948.

No. 81 Wing reformed at Williamtown on 3 January 1961, overseeing Nos 75 and 76 Squadrons, and No. 2 OCU, but disbanded again, shortly after the first Mirages began arriving, in December 1966. With a return to the 'wing' structure within the RAAF, No. 81 Wing formed once again at Williamtown on 2 February 1987 as part of the Tactical Fighter Group. Today it is still based at Williamtown, as a component of the Air Combat Group, and controls the RAAF's Hornet fighter squadrons.

No. 2 Operational Conversion Unit

Formed at Port Pirie, in South Australia, as No. 2 Operational Training Unit (OTU) on 6 April 1942, the unit was initially equipped with the CAC Wirraway. After relocating to Mildura in May 1942, No. 2 OTU began to receive a range of service types to fulfil the training duties. In 1943 the Spitfire Flight within the unit moved to RAAF Williamtown, and began an association that remains to the present day. After the war, the dramatically reduced training requirements of the peacetime RAAF saw many training units disband, and No. 2 OTU was no exception, being stood down in March 1947.

RAAF operational training requirements increased dramatically due to No. 77 Sqn's involvement in the Korean War, and 2 OTU was reformed as a result, on 1 March 1952. Based at Williamtown and initially operating Wirraways, Mustangs and Vampires, the unit has an unbroken history to the present day. Sabres were introduced in November 1954, and re-identification as No. 2 Operational Conversion Unit occurred on 1 September 1958. The Mirage arrived at Williamtown for the Mirage Trials Flight within the unit in March 1964, and was followed by the Macchi in October 1969. Except for a short period between April 1970 and July 1971, when the Macchi Flight passed to the control of No. 5 OTU, the composition of No. 2 OCU did not change until the introduction of the F/A-18 Hornet in 1985.

The unit passed its Mirage and Macchi aircraft to No. 77 Squadron in January 1985, and began preparing for the arrival of the first Hornet. After conversion to the F/A-18, the Macchis were handed back to the OCU in July 1987, and remained within the unit until Macchi operational conversion duties were passed to No. 76 Sqn in 1989. Today No. 2 OCU is the RAAF Hornet training unit within No. 78 Wing, and remains based at Williamtown.

No. 2 OCU gave up the Mirage training commitment in 1985, passing the role to No. 77 Squadron.

Markings

The first markings that No. 2 OCU applied to its Mirages were of a low-key nature, consisting of a fin band (that stopped short of the rudder) near the top of the fin, with black and yellow diagonal stripes. This soon gave way to the more outrageous 'sunburst' pattern that saw yellow and black lines radiating out from a point at the forward lower edge of the fin, practically covering the entire tail.

In the early 1970s a couple of standard patterns were developed for use by Mirage squadrons, and No. 2 OCU adopted what is probably the better known of the two. Consisting of three triangular 'darts' that taper to a point at the forward lower point of the fin, with the squadron markings applied to a white disk superimposed over them. In the case of No. 2 OCU, these darts were black (centre) and yellow with a Tiger's head in the disk. Though smaller than the initial 'sunburst' design, these markings nevertheless covered a large portion of the fin.

No. 3 Squadron

One of the RAAF's most historic units, No. 3 Squadron AFC was formed at Point Cook on 25 October 1916, by the expedient renumbering of No. 2 Sqn AFC. It became the first Australian flying squadron to arrive in France when it relocated to Cambrai in September 1917. No. 3 Sqn spent the remainder of World War I in France and was eventually withdrawn to Australia and disbanded in June 1919.

Reformed at Point Cook on 1 July 1925, No. 3 Sqn RAAF was initially equipped with DH.9, DH.9A and SE.5a aircraft and relocated to Richmond. The outbreak of World War II found the unit operating the Hawker Demon, but it soon handed its aircraft in and departed Australia by sea to Ismalia in May 1940. No. 3 Sqn fought in the Western Desert flying Gladiators, Tomahawks and Kittyhawks and deployed to Italy following the invasion by Allied forces in July 1943. At the end of the war, the unit was based at Udine, operating Mustangs, but was shortly thereafter withdrawn to Australia and disbanded.

Selected as one of the peacetime RAAF units, No. 3 Sqn formed once again in Canberra in 1948, with CAC-built Mustangs, and moved to Williamtown in 1956 to re-equip with the CAC Sabre. In 1958 it was forward-deployed to Malaya, with No. 77 Sqn RAAF, and was based at Butterworth. No. 3 Sqn withdrew to Williamtown in 1967 to re-equip with the Mirage, before returning to Butterworth in 1969. Finally withdrawn to Australia again in 1986, No. 3 Sqn re-equipped with the F/A-18 Hornet at RAAF Williamtown.

Markings

As the first operational IIIO(A) squadron, No. 3 Sqn only ever operated camouflaged aircraft. Its initial markings were of the 'dart' type, utilising orange and black, with a frilled-neck lizard over the numeral '3' in the disk. These markings did not find favour with higher authority, and were soon changed for a brief period, with the lizard replaced by No. 3 Sqn's official emblem (a Flaming Grenade and Fleur de Lys). This was itself replaced by the second of the two standard designs: this time with a horizontal orange stripe (edged in blue) dissecting the fin at about mid-height and the 3 Sqn emblem superimposed in a white disk in the centre. Later, the famous light blue rudder with a white 'Southern Cross' was added. This is a No. 3 Sqn tradition that was first applied to Kittyhawks in Italy. With the advent of the grey experimental schemes, the standard unit markings were reapplied in a 'low-viz' form of greys, and occasionally without the band.

Part of No. 3 Squadron's task in southeast Asia was to provide a detachment to Singapore, first to Tengah and latterly to Paya Lebar. Note the Southern Cross rudder marking.

No. 75 Squadron

No. 75 Squadron formed at Townsville on 4 March 1942 with Curtiss Kittyhawks, and deployed to Port Moresby just over a fortnight later, in a desperate attempt by the Allies to stem the Japanese advance on Australia. After seeing immediate action, the squadron returned to Australia for re-equipment, and returned to the fray in July 1942. Based at Milne Bay, No. 75 Squadron (alongside No. 76 Sqn) was instrumental in repulsing a Japanese invasion on the eastern end of New Guinea, and wrote a chapter in Australian folklore in the process. As Japanese forces began to be pushed back, the squadron was to move many times throughout the South-West Pacific area in pursuit, and were operating from Tarakan when World War II ended. After VJ Day, No. 75 Squadron was withdrawn to Australia and re-equipped with the P-51D Mustang, before disbandment in 1948.

Reformed the following year with the Australian-built de Havilland Vampire, No. 75 Sqn was based at Williamtown until being re-assigned to Malta as a component of No. 78 Wing in 1951. After returning home to Australia in January 1955, the squadron re-equipped with the Gloster Meteor, and converted to the Avon Sabre in April 1955.

After further conversion to the Mirage in 1967, the squadron relocated to Butterworth that year, and remained there, as a component of the Commonwealth Far East Reserve (later Five Power Defence Arrangements), until being withdrawn to Darwin in August 1983. In 1988, No. 75 Sqn transferred to Williamtown, to begin conversion to the F/A-18 in 1988. Today the unit is Australia's forward deployed fighter squadron, based at RAAF Tindal in the Northern Territory.

This aircraft was unusual in reverting from the 'low-viz' grey scheme back to green/grey camouflage.

Markings

No. 75 Squadron's colours are black and white, and this trim was soon applied to the first Mirages to enter operational service. Initially consisting of a large squadron Coat of Arms, markings were soon changed to a black and white 'dart' with the famous Magpie in the disk. As with No. 3 Sqn, these markings were toned down somewhat, to black and grey, with the introduction of the light grey aircraft, and this version was also applied to the green/grey camouflaged aircraft.

No. 76 Squadron

Initially equipped with the Curtiss P-40E Kittyhawk at Archerfield on 14 March 1942, No. 76 Squadron relocated to Townsville and was declared operational the following June. The month after, the squadron deployed to Port Moresby and thence to Milne Bay, and successfully drove back repeated Japanese attacks with the assistance of No. 75 Sqn. In September 1942, No. 76 Sqn withdrew to Darwin, and remained based in the Northern Territory and North-West Australia, engaged in the defence of the homeland, until April 1943 when it was brought back to the Sydney area for conversion to the P-40M. Following conversion, the unit deployed to Goodenough Island and steadily moved forward, as Japanese forces retreated, until the end of the war found them at Labuan.

Following the war, No. 76 Squadron converted to the P-51D Mustang and deployed to Bofu in February 1946, and this was followed by a further move, to Iwakuni, in February 1948. With the gradual wind-down of British Commonwealth Occupational Forces in Japan, the squadron disbanded at Iwakuni in October 1948. Reformed at Williamtown on 24 January 1949 with CAC-built Mustangs, No. 76 Squadron converted to the Vampire in late 1951, and shipped to Malta as part of No. 78 Wing in July 1952 (flying British-built Vampires). A return to Australia occurred in 1955, with the unit being disbanded in March of that year.

No. 76 Squadron formed once again at Williamtown on 11 January 1960 with two-seat Vampires, and converted to the CAC Sabre in May 1961, before being re-equipped with the Mirage in 1968. As a result of defence budget cuts, No. 76 Squadron disbanded again in August 1973.

In its latest form, No. 76 Sqn was reformed at Williamtown on 1 January 1989 with Macchi MB.326H and CAC Winjeels as a Lead-In Fighter (and Forward Air Control Training) unit, and is today equipped with the Hawk Mk 127.

The black panther badge of No. 76 Squadron adorned the Mirage IIIO(F) for just five years. The badge and fin dart were particularly striking on the silver lacquer finish which most of the unit's aircraft wore.

Markings

Although relatively short-lived on the Mirage, No. 76 Squadron's colours are arguably the most attractive of all: utilising the 'dart' form, the markings of No. 76 Sqn differed from those of other units in that all three of the 'dart' segments were red, trimmed with black, superimposed with a white disk containing a black panther's head. By the time of disbandment in 1973 not all of the unit's aircraft had received the green/grey camouflage paint, and the squadron is best remembered in the heyday of the late 1960s, with the overall-silver aircraft providing the perfect canvas for the 'dart' style of marking.

No. 77 Squadron

Formed at Pearce in Western Australia on 16th March 1942, No. 77 Squadron was initially equipped with the P-40E Kittyhawk, moving to Darwin four months later. Initially charged with defending Darwin from air attack, the unit deployed to Milne Bay in November 1943, and remained operational for the rest of World War II at various locations throughout the South West Pacific theatre. It was chosen as one of the RAAF units to be based in Japan at the war's end, initially based at Bofu until moving to Iwakuni in February 1948. At the beginning of the Korean War, No. 77 Sqn was the only RAAF fighter unit remaining in Japan (although only just, as it was officially on 'stand-down' prior to withdrawal to Australia), and was immediately committed to action. Equipped with the P-51D Mustang, No. 77 Sqn began offensive operations from Iwakuni on 2 July 1950, before deploying to the first of several bases on the Korean Peninsula in October 1950. In April 1951, the squadron withdrew back to Iwakuni in order to begin conversion to the Gloster Meteor. After training was completed, No. 77 returned to Korea in July 1951, based at Kimpo. The first Meteor sortie was flown on 29 July, but the unsuitability of the Meteor to air combat led to a switch to ground attack operations from January 1952. Even so, No. 77 Sqn gave a good account of itself, and by the time it returned home to Australia in 1954 the pilots could claim at least three MiG-15s destroyed.

Based at Williamtown on its return, No. 77 Sqn stood down in August 1956 for conversion to the Avon Sabre. This process was completed in November 1956, with the squadron commencing Sabre operations, and was forward deployed to Malaya with No. 3 Sqn in 1958. In 1969 the unit withdrew to Australia once again, to convert to the Mirage at Williamtown, and has been based within Australia ever since. Today No. 77 Squadron flies the GAF-built Hornet from Williamtown as a component of No. 81 Wing, Air Combat Group.

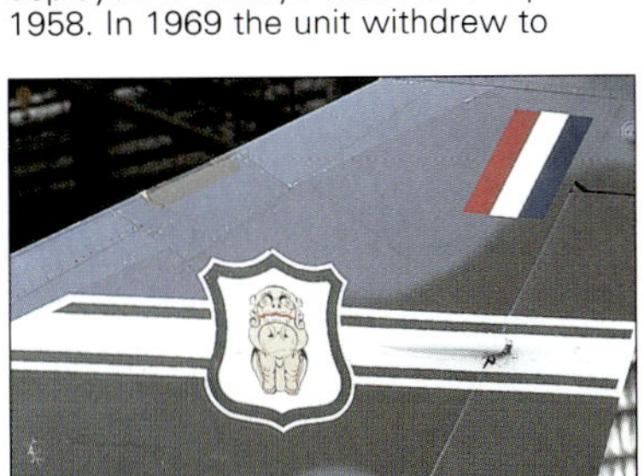

Markings

This unit is perhaps the best-known Mirage squadron to the Australian public, as it was based in Australia for its entire time with the Mirage. After early experiments with the 'dart' design, trimmed in the green and white squadron colours, No. 77 Squadron changed to a white horizontal strip (edged with green) and superimposed with a shield carrying the numerals 77. This in turn gave way to the definitive version, where the squadron number gave way to a gold 'Peace Lion' – an emblem adopted by the squadron in Korea, and irreverently known as the 'Grumpy Monkey'. On the later grey scheme the white was removed from the markings of some aircraft, but the green/grey machines kept their colours until the end.

No. 77's 'Peace Lion' markings are seen in both high- and low-visibility versions.

No. 79 Squadron

As one of the RAAF's Spitfire squadrons, initially equipped with the Spitfire Mk VC, No. 79 Squadron formed at Laverton in Victoria on 26 April 1943 and deployed to Goodenough Island eight weeks later. As with most RAAF fighter units, No. 79 Sqn was in action continuously until the very end of the war, when VJ Day found it based on Morotai, operating the Spitfire Mk VIII.

Following disbandment in November 1945, the squadron was reformed with Avon Sabres at Ubon in Thailand on 1 June 1962 as a component of Australia's SEATO commitment to the defence of Thailand during the Vietnam War. This arrangement lasted until July 1968, when the unit's aircraft were flown to Butterworth and the squadron disbanded.

When No. 3 Squadron was withdrawn to Australia from Butterworth in March 1988 to commence its conversion to the F/A-18 Hornet, No. 79 Squadron was reformed to maintain a Mirage presence in Malaysia. In this particular existence, No. 79 Sqn flew the Mirage for only a relatively brief period before returning its aircraft to Australia (and storage) in 1988, disbanding at Butterworth on 30 June of that year.

Reformed a decade later at Pearce, No. 79 Squadron took over the Lead In Fighter role of No. 25 (Reserve) Squadron. Initially flying the CAC-built Macchi MB.326H, the unit currently operates the BAE Systems Hawk as a component of No. 78 Wing.

Markings

During No. 79 Squadron's brief existence as a Mirage unit, it operated the aircraft left behind at Butterworth by the departing No. 3 Squadron. Squadron colours are maroon and gold, and its emblem is a phoenix. These made a most attractive scheme when applied in full colour to the few camouflaged aircraft remaining. As applied, the markings were of a maroon horizontal fin stripe, trimmed with gold and superimposed by a white disk containing the Phoenix emblem. On the grey aircraft, the squadron markings were rendered in various shades of grey (above).

Aircraft Research and Development Unit (ARDU)

Formed as No. 1 Air Performance Unit at Laverton on 1 December 1943 as the RAAF's Testing and Trials unit, the Aircraft Research and Development Unit has operated many and varied types of aircraft over the years. ARDU moved from Laverton to Edinburgh in February 1977, and continues to support all three services of the ADF. ARDU's involvement with the Mirage began even before deliveries to Australia commenced, and all aircraft produced locally were test-flown by ARDU pilots. Many RAAF Mirages have passed through ARDU, on loan from their parent unit in the support of various trials and tests, but the unit always maintained a 'core' fleet of specially instrumented aircraft: these included A3-1 (w/o 7/12/64), -2, -16 (w/o 24/10/74), -76 (w/o 3/5/84) and -112. It was an ARDU Mirage IIID that flew the last ever RAAF Mirage flight on 8 February 1989.

Markings

Initially, Mirages allocated to ARDU wore no unit markings, but with the advent of camouflage a green and yellow fin strip was applied, superimposed with the ARDU logo. This logo was designed especially to reflect Mirage operations, and consisted of the letters ARDU shaped like a delta-winged aircraft in planform view. This arrangement was carried forward on to the green and yellow and, later, orange and white schemes.

A3-53 was the third IIIO(A) to be built in Australia. During its assignment to ARDU it retained standard camouflage but added the unit's Mirage-shaped badge and green/yellow fin stripe.

RAAF Mirage serials

Serial	Type	Built by	Fate
A3-1	IIIO(F)	Dassault	w/o Avalon 7/12/64
-2	IIIO(F)	Dassault	to Pakistan 10/90
-3	IIIO(F)/IIIO(FA)	Dassault	pres. Williamtown
-4	IIIO(F)/IIIO(FA)	Dassault	w/o Avalon 3/8/72
-5	IIIO(F)/IIIO(FA)	Dassault	to Pakistan 10/90
-6	IIIO(F)/IIIO(FA)	Dassault	to Pakistan 10/90
-7	IIIO(F)/IIIO(FA)	Dassault	scr. Darwin 1987 (fuselage to Darwin Museum)
-8	IIIO(F)/IIIO(FA)	Dassault	w/o East Sale 17/2/78
-9	IIIO(F)/IIIO(FA)	GAF assembled	to Pakistan 10/90
-10	IIIO(F)/IIIO(FA)	GAF assembled	to Pakistan 10/90
-11	IIIO(F)/IIIO(FA)	GAF	to Pakistan 10/90
-12	IIIO(F)/IIIO(FA)	GAF	to Pakistan 10/90
-13	IIIO(F)/IIIO(FA)	GAF	to Pakistan 10/90
-14	IIIO(F)/IIIO(FA)	GAF	w/o Williamtown 16/3/76
-15	IIIO(F)/IIIO(FA)	GAF	to Pakistan 10/90
-16	IIIO(F)/IIIO(FA)	GAF	w/o Tullamarine 24/10/74 (rebuilt and pres. Adelaide)
-17	IIIO(F)/IIIO(FA)	GAF	to Pakistan 10/90
-18	IIIO(F)/IIIO(FA)	GAF	w/o Butterworth 1/4/74
-19	IIIO(F)/IIIO(FA)	GAF	to Pakistan 10/90
-20	IIIO(F)/IIIO(FA)	GAF	Inst. airframe 1 CAMD 11/86
-21	IIIO(F)/IIIO(FA)	GAF	to Pakistan 10/90
-22	IIIO(F)/IIIO(FA)	GAF	to Pakistan 10/90
-23	IIIO(F)/IIIO(FA)	GAF	to Pakistan 10/90
-24	IIIO(F)/IIIO(FA)	GAF	to Pakistan 10/90
-25	IIIO(F)/IIIO(FA)	GAF	to Pakistan 10/90
-26	IIIO(A)	Dassault	w/o Butterworth 6/7/76
-27	IIIO(F)/IIIO(FA)	GAF	to Pakistan 10/90
-28	IIIO(F)	GAF	w/o Queanbeyan 29/7/66
-29	IIIO(F)/IIIO(FA)	GAF	w/o Bluff Downs Homestead 9/4/84
-30	IIIO(F)/IIIO(FA)	GAF	w/o Bluff Downs Homestead 9/4/84
-31	IIIO(F)/IIIO(FA)	GAF	to Pakistan 10/90
-32	IIIO(F)/IIIO(FA)	GAF	w/o Butterworth 29/10/81
-33	IIIO(F)/IIIO(FA)	GAF	to Pakistan 10/90
-34	IIIO(F)/IIIO(FA)	GAF	to Pakistan 10/90
-35	IIIO(F)/IIIO(FA)	GAF	to Pakistan 10/90
-36	IIIO(F)/IIIO(FA)	GAF	w/o Darwin 27/5/85 (under restoration Tindal)
-37	IIIO(F)	GAF	w/o Tengah 18/3/69
-38	IIIO(F)/IIIO(FA)	GAF	to Pakistan 10/90
-39	IIIO(F)/IIIO(FA)	GAF	to Pakistan 10/90
-40	IIIO(F)/IIIO(FA)	GAF	w/o east of Williamtown 2/5/86
-41	IIIO(F)/IIIO(FA)	GAF	gate guardian Wagga
-42	IIIO(F)/IIIO(FA)	GAF	pres. Wangaratta
-43	IIIO(F)	GAF	w/o Williamtown 1/9/67
-44	IIIO(F)/IIIO(FA)	GAF	pres. Bankstown
-45	IIIO(F)/IIIO(FA)	GAF	pres. Moorabbin
-46	IIIO(F)/IIIO(FA)	GAF	w/o near Darwin 3/4/67
-47	IIIO(F)/IIIO(FA)	GAF	w/o Williamtown 7/8/79
-48	IIIO(F)/IIIO(FA)	GAF	scr. Butterworth 5/87
-49	IIIO(F)/IIIO(FA)	GAF	to Pakistan 10/90
-50	IIIO(F)/IIIO(FA)	GAF	w/o Tea Gardens 2/2/77
-51	IIIO(A)	GAF	pres. Point Cook
-52	IIIO(A)	GAF	w/o Gloucester Tops 25/9/67
-53	IIIO(A)	GAF	to Pakistan 10/90
-54	IIIO(A)	GAF	to Pakistan 10/90
-55	IIIO(A)	GAF	pres. Amberley
-56	IIIO(A)	GAF	to Pakistan 10/90
-57	IIIO(A)	GAF	scr. 10/86
-58	IIIO(A)	GAF	w/o Williamtown 2/5/80
-59	IIIO(A)	GAF	pres. Butterworth
-60	IIIO(A)	GAF	to Pakistan 10/90
-61	IIIO(A)	Dassault fus.	w/o Williamtown 24/6/76
-62	IIIO(A)	GAF	to Pakistan 10/90
-63	IIIO(A)	Dassault fus.	w/o Butterworth 17/7/72
-64	IIIO(A)	GAF	w/o Butterworth 6/7/76
-65	IIIO(A)	Dassault fus.	to Pakistan 10/90
-66	IIIO(A)	GAF	to Pakistan 10/90
-67	IIIO(A)	Dassault fus.	w/o Butterworth 8/6/76
-68	IIIO(A)	GAF	to Pakistan 10/90
-69	IIIO(A)	Dassault fus.	w/o Singapore 30/3/83
-70	IIIO(A)	GAF	w/o Limeburners Creek 30/10/68
-71	IIIO(A)	Dassault fus.	to Pakistan 10/90
-72	IIIO(A)	GAF	pres. Point Cook
-73	IIIO(A)	Dassault fus.	to Pakistan 10/90
-74	IIIO(A)	GAF	w/o Darwin 27/4/72
-75	IIIO(A)	Dassault fus.	w/o East Sale 18/2/80
-76	IIIO(A)	GAF	w/o St Vincent's Gulf 3/5/84
-77	IIIO(A)	Dassault fus.	w/o Newcastle 11/5/67
-78	IIIO(A)	GAF	scr. Williamtown 11/87 (fuselage to Pakistan)
-79	IIIO(A)	Dassault fus.	w/o Gloucester 3/4/73
-80	IIIO(A)	GAF	w/o Williamtown 9/9/81
-81	IIIO(A)	GAF	to Pakistan 10/90
-82	IIIO(A)	GAF	w/o Bega 2/8/83
-83	IIIO(A)	GAF	to Pakistan 10/90
-84	IIIO(A)	GAF	to Pakistan 10/90
-85	IIIO(A)	GAF	w/o Mt Gunong Tahan 4/5/72
-86	IIIO(A)	GAF	to Pakistan 10/90
-87	IIIO(A)	GAF	to Pakistan 10/90
-88	IIIO(A)	GAF	to Pakistan 10/90
-89	IIIO(A)	GAF	w/o 100 km west of Darwin 21/6/85
-90	IIIO(A)	GAF	scr. Williamtown 1986
-91	IIIO(A)	GAF	scr. Williamtown 11/86
-92	IIIO(A)	GAF	pres. Point Cook
-93	IIIO(A)	GAF	to Pakistan 10/90
-94	IIIO(A)	GAF	w/o Butterworth 5/12/77
-95	IIIO(A)	GAF	w/o Williamtown 16/3/87
-96	IIIO(A)	GAF	to Pakistan 10/90
-97	IIIO(A)	GAF	w/o Butterworth 6/3/84
-98	IIIO(A)	GAF	w/o Jahore 6/7/72
-99	IIIO(A)	GAF	to Pakistan 10/90
-100	IIIO(A)	GAF	pres. Darwin
-101	IIID	Dassault	to Pakistan 10/90
-102	IIID	Dassault fus.	pres. Williamtown
-103	IIID	Dassault fus.	to Pakistan 10/90
-104	IIID	Dassault fus.	to Pakistan 10/90
-105	IIID	Dassault fus.	w/o Darwin 26/4/84
-106	IIID	Dassault fus.	scr. Williamtown 1985
-107	IIID	Dassault fus.	scr. Butterworth 6/81
-108	IIID	Dassault fus.	to Pakistan 10/90
-109	IIID	Dassault fus.	w/o Williamtown 6/10/71
-110	IIID	Dassault fus.	to Pakistan 10/90
-111	IIID	Dassault fus.	to Pakistan 10/90
-112	IIID	Dassault fus.	to Pakistan 10/90
-113	IIID	Dassault fus.	to Pakistan 10/90
-114	IIID	Dassault fus.	w/o Williamtown 10/8/76
-115	IIID	Dassault fus.	pres. Edinburgh
-116	IIID	Dassault fus.	scr. Williamtown 1987 (fuselage to Pakistan)

w/o = written off
pres. = preserved
scr. = scrapped
fus. = fuselage

RAAF Mirage III colour schemes

Initial Mirage deliveries to the RAAF were in a polished natural metal colour scheme with red trim around the intakes, similar to the French colour scheme of the day. As such, the aircraft became a canvas for colourful unit markings, and the first two operational squadrons to convert to the aircraft – Nos 75 (below) and 76 – wasted little time in applying their colours.

Tropical trials in Darwin highlighted the fact that the bare metal did not provide enough protection against the elements, and this resulted in an overall coat of aluminium lacquer being applied to the fleet. If anything, this uniform coat of silver paint made the Mirage look even sleeker, especially as the red trim and colourful unit markings remained. Below is a factory-fresh IIID at Avalon in 1968.

With the introduction of the ground-attack Mirage IIIO(A), and taking in to account American experiences in the Vietnam War, it was felt that a more suitable camouflage pattern paint scheme was needed. Accordingly, IIIO(A) aircraft were delivered in a matt green/grey scheme, consisting of an overall coat of Extra Dark Sea Grey (BS 381C.640) with a disruptive pattern of Olive Drab (BS 381C.298) applied to the upper surfaces and fuselage sides. This scheme was progressively applied to all the surviving silver aircraft also. After a short period of relative anonymity, the colourful squadron markings began to resurface, and units began experimenting with colours and designs that would soon became familiar to all Mirage enthusiasts. Note the Sabres behind this 'wrap-round' Mirage below.

After a little while, it became apparent that the dark grey undersides, while effective for low-level operations, were not ideal for concealment against the sky. A new scheme was introduced, consisting of Dark Green (BS 381C.641) and Extra Dark Sea Grey (BS 381C.640) disruptive camouflage, with Light Gull Grey (FS.26440) undersides. This scheme also utilised semi-matt paints, as the original matt scheme was difficult to maintain. Once again, full colour unit markings were applied, and this scheme is probably the best remembered, as it was worn by the fleet for the majority of their service lives. Below is a No. 77 Sqn aircraft.

With the introduction of the Hornet on the horizon, various experiments were conducted with the Mirage fleet to find a scheme that would suit the new aircraft in its most likely areas of operation, namely, northern Australia and southeast Asia. The two Butterworth squadrons at the time (Nos 3 and 75) experimented with several different schemes of grey, mostly in two-colour disruptive patterns, but also utilising three colours on at least one aircraft. Various colours and hues were tried before a scheme was selected that would become standard on RAAF Hornets. This scheme consisted of Dark Blue-Grey (FS.25237) upper surfaces and sides, with Light Ghost Grey (FS.26375) lower surfaces. Unit markings began losing some of their colour also, and though the light grey scheme suited the Mirage well, the loss of the colourful markings of the 1960s and 1970s was lamented. RAAF Mirages began to be repainted in the new scheme, but it was far from universally applied. No. 75 Squadron at least, by now back in Darwin, kept a handful of aircraft in the old green/grey camouflage as the unit felt it was better suited to low-level operations. In fact, the squadron repainted at least one aircraft (A3-100) back into the old scheme. Below is a No. 77 Sqn Mirage IIID – on some aircraft the serial was painted white.

Special schemes were a rarity on the Mirage and, apart from the odd high-visibility markings applied for exercises, the only major departure were three No. 77 aircraft that were given a special colour scheme for the RAAF Silver Jubilee in 1981. These aircraft were overall white, with red and blue speed trim and the No. 77 Sqn 'Korean Lion' badge on the rear fuselage (below). It has been asserted in the past that this trio did not carry their serial numbers externally, but they actually had them applied (in very small numerals) to the brake parachute housing, aft of the rudder. Likewise, not many RAAF Mirages wore names or nose art, with the most famous exception being No. 77 Squadron's Mirage IIID *Daphne* (later *Daphne de Dual*), which had an outline of a reclining lady on the nose. The tradition of naming the squadron two-seater *Daphne* was briefly carried on into the Hornet years.

Perhaps the most colourful RAAF Mirages were those on the strength of ARDU. Initially, this unit operated aircraft in the standard colours of the day but, by the late 1970s, had applied overall yellow and green high-visibility markings to its fleet.

This was followed in the mid-1980s by an overall orange and white scheme.

Lancaster

Backbone of Bomber Command

Together with the Spitfire and Hurricane, the Avro Lancaster is one of the few RAF aircraft to be widely known to the general public. The Lancaster was a legendary weapon of war in its own time, and has since entered the general history books.

Below: Even with the adoption of the wide-span tailplanes and large, twin fins being developed for the Lancaster, the Manchester remained a woeful aircraft. Chief complaints were poor handling, especially at high weights, and the unreliability of the Rolls-Royce Vulture engines. Seventy-six of the 200 production aircraft were lost. This Manchester Mk IA carries the 'EM' codes of No. 207 Squadron.

In a letter of thanks to A.V. Roe and Co. in December 1945, Sir Arthur Harris described the Lancaster as the greatest single factor in winning World War II. The Lancaster formed the backbone of Bomber Command, and spearheaded the Command's bloody war of attrition against German industry and its Trenchardian doctrine of strategic bombing for 'moral' (moral because Trenchard famously could not spell 'psychological') effect.

Built in smaller numbers than the Vickers Wellington, the Lancaster flew a staggering total of 156,192 operational sorties during the war (nearly four times as many as the rival Halifax), losing 3,836 aircraft in the process. The Lancasters' sortie total included 107,085 bombing raids against Germany, with the attendant loss of 2,508 aircraft. The sorties against German targets included 23,204 daylight sorties, during which 179 aircraft were lost. The scope and extent of Lancaster losses, and the enormity of the sacrifice made by so many Lancaster aircrew, added to the aircraft's legendary status. Lancaster aircrew won 10 of the 32 'air' Victoria Crosses awarded to aircrew during World War II.

If the story of the Lancaster had been just one of empty sacrifices and heavy losses, the aircraft would be remembered in the same light as the Fairey Battle or the Bristol Blenheim. The Avro Lancaster, however, was a legendary success, and was responsible for dropping almost two-thirds of the RAF's wartime bomb tonnage (608,612 tons of the 955,044-ton total). The type equipped more squadrons than any other RAF bomber type. There were 56 front-line Lancaster squadrons by the war's end, and Bomber Command (which was then well on the way to becoming an all-Lancaster force) could routinely call upon around 750 front-line Lancasters for its operations, with some 300 more serving with conversion and training units.

Conception

The Lancaster's legendary status may also be due in part to the fact that its story is one of great success being dramatically grabbed from the jaws of ignominious failure. Bomber Command's most important and successful bomber was a refined, four-engined derivative of the almost irredeemably hopeless Manchester – in effect, an aircraft that has come to be regarded as one of the world's greatest was actually developed from one of the very worst.

Left: In a scene repeated at many airfields across eastern England, No. 57 Squadron's Lancasters stand silently in the late afternoon sun. Briefing is over, pre-flights completed, and soon the tranquility will be shattered by the sound of Merlins as they crackle into life and the bombers taxi out for take-off. As the last one heads to the east, the calm returns, but a question hangs heavily in the air – how many will come back?

Main picture: A 'big-belly' Mk I, PP687 was photographed while on air test from Castle Bromwich in March 1945, prior to its delivery to No. 149 Squadron. Lancasters continued to pound the shrinking Reich until the last days of the war, flying their last bombing mission with a raid on Tonsberg on 25/26 April 1945, less than a fortnight before VE Day.

Two Air Ministry Specifications for new heavy bomber-transports were issued during 1936. Specification B.12/36 called for a four-engined heavy, carrying a bombload of 12,000 lb (5443 kg) (or 24 troops when used in the transport role) and a crew of six. Two prototypes each of the Supermarine Type 317/318 and of the aircraft that became the Short Stirling were ordered to meet this Specification.

Specification P.13/36 outlined a smaller aircraft, powered by twin Rolls-Royce Vulture engines, still with a six-man crew but carrying only an 8,000-lb (3629-kg) bombload or 12 troops. The Specification also called for the smaller aircraft to be capable of carrying a pair of 21-in (53-cm) torpedoes, meaning that the aircraft could not use the same highly-compartmentalised bomb bay as the Stirling, which could carry nothing larger than a 2,000-lb (907-kg) bomb. Of the designs submitted to meet the Specification, the Air Ministry ordered prototypes from Handley Page and Avro.

Handley Page expressed severe doubts about the Vulture engine and was ordered to redesign its bomber around four of the new Merlin engines, producing the HP.57. The addition of two more engines required an increase in wingspan, but this was restricted by the requirement that the new aircraft be able to use existing hangars, thereby limiting the HP.57's span to 98 ft 8 in (30.07 m) (4 in/10 cm less than the maximum allowable width). The aircraft was required to be able to operate from existing airfields with standard 1,500-ft (457-m) grass runways. Handley Page would have preferred to use radial engines, but the Air Ministry favoured the Merlin, which gave longer range at the expense of lower speed.

Avro pressed ahead with the Vulture-powered Type 679. Two Avro Type 679 prototypes had been ordered on 8 September 1936, six days after the final issue of P.13/36. The first of these, L7246, made its maiden flight on 25 July 1939. Avro began work on what would eventually become the Lancaster (the four-engined Type 683) even before the prototype Manchester made its maiden flight, and the company hoped that the Ministry of Aircraft Production would eventually switch to production of the larger bomber, as the Manchester B.Mk III. More and more hope was pinned on the Type 683 as it became increasingly clear that the Manchester itself was a deeply flawed aircraft.

The Vulture engines were less powerful than had been planned and they proved unreliable, and it was found necessary to increase the aircraft's wingspan from 80 ft 2 in (24.43 m) to 90 ft 1 in (27.46 m) and to add redesigned aileron hinges. The aircraft also received mass-balanced elevators and redesigned tailfins and rudders. A lack of directional stability was cured by an additional central fin, while severe airflow disturbances required refinement of the gun turret installations.

Four-engined bombers

The aircraft that became the Manchester was ordered in large numbers, despite its problems, with the aim of equipping 20 squadrons by March 1942; however, on 29 July 1940, even before service trials were underway, the orders for 800 aircraft were cut back to 200 (plus the two prototypes). Air Marshal Sir Charles Portal, AOC-in-C of Bomber Command, had made the far-reaching decision that Britain's strategic bomber arm would, from the earliest possible opportunity, be equipped solely with four-engined aircraft, and Avro was directed to switch to licence-production of the rival Halifax after the 200 Manchesters were complete.

Avro immediately proposed that the Manchester should instead be replaced by the Type 683, an aircraft that enjoyed 70 per cent component commonality with the Manchester, and which, the company pointed out, would therefore be quick, easy and economic to manufacture on

A trio of early production No. 207 Squadron Lancasters flies from the squadron's base at Bottesford in early 1942. The highly conspicuous national insignia and code letters were swiftly toned down to present enemy fighter pilots with less of a target.

Lancaster prototypes

The first Lancaster (BT308) was a Manchester fitted with long-span wings and Merlin X engines in the Beaufighter's 'power egg' installation. This had an unusual single exhaust pipe on the cowling. Initially the aircraft was completed with the Manchester's three-fin, short-span tail unit (above), but within six weeks had the definitive twin-fin arrangement fitted (right).

DG595, the second prototype, made its first flight on 13 May 1941. It was closer to the production standard, and featured a ventral turret and cowlings with individual exhaust stubs for the 12 cylinders. Tests with this aircraft led to an extension of the wing. Both it and the first prototype were finished with high-visibility national insignia and yellow undersides.

the two separate Manchester production lines which were then already in operation. Avro was authorised to produce two prototypes of the new four-engined Manchester B.Mk III, but was still expected to prepare for the possibility of manufacturing the Halifax when Manchester production ceased. As the Manchester's problems continued, there was less and less incentive for the Air Staff to order a four-engined version. The Manchester was, in short, a disaster.

When it finally reached the front line in February 1941, the Manchester was severely handicapped by the unreliability of its Vulture engines, inadequate crew facilities, armament shortcomings, and poor handling characteristics at high all-up weights. Despite these problems, Manchesters equipped seven front-line squadrons within No. 5 Group. The group's existing Hampdens had been under-armed and the cramped fuselage had made the type uncomfortable on long-range missions, but the aircraft had been fast and surprisingly agile. The average Hampden flew 27.25 operations before being lost. The Manchester offered increased payload/range and (theoretically) an even better maximum speed, but proved much more vulnerable, clocking up only 16.7 sorties before being lost.

Later aircraft were delivered to B.Mk IA standards with the increased-span tailplanes and increased-height fins and rudders designed for the Type 683, and with no central fin. Despite the improvements, the Manchester suffered a heavier loss rate than any other Bomber Command type, and flew only 1,269 operations. The type dropped 1,826 tons of bombs, but in doing so suffered 76 operational losses.

Avro's proposal to replace the twin-Vulture-powered Manchester with the four-engined Manchester Mk III was accepted by the Air Ministry in September 1940, before the full extent of the Manchester's weaknesses were apparent. When Shorts had designed the Stirling and Handley Page had drawn up the Halifax, both companies had been constrained by details in the peacetime specifications. In particular, the need for these aircraft to fit inside existing hangars imposed a strict limit on wingspan, and the peacetime requirement for multi-role capability dictated a particular fuselage cross-section. By contrast, by the time detail design of the Avro 683 began, the company was able to draw on real operational experience. This showed that multi-role capability was irrelevant, and that bombers would be kept and serviced in the open, making a wingspan limit quite unnecessary. Much greater emphasis could be placed on the type's ease and economy of production, and on providing a more optimised and undiluted bomber capability.

Prototype construction

Although a formal contract for the Manchester Mk III would not be signed until 15 July 1941, Avro pressed ahead with construction of a Type 683 prototype. This was based on an unfinished Manchester airframe (with the original-style short tailplane and triple fins) taken from the production line. To this Avro added a new extended-span wing and four Rolls-Royce Merlin engines in Beaufighter Mk II-type power-eggs. Stuart Davies, the head of the Avro Experimental Shop, was told firmly that the aircraft should be flying by 31 May 1941 – two months earlier than the Ministry of Aircraft Production had specified. This deadline, imposed by Avro's managing director, Roy Dobson, proved easy to beat, and the first Manchester Mk III prototype flew on 9 January 1941 under a clearance to fly certificate which already used the new Lancaster name (though this was not officially approved until 27 January).

Davies had originally aimed for a first flight by 31 December 1940, and only failed to meet his own dead-

Depicted in January 1942 on a misty Woodford day, shortly after the Lancaster had been released to service, this aircraft is from the first batch of 43 Merlin XX-powered production machines. Most aircraft were fitted with shrouds covering the exhaust stubs to hide the flames from night-fighter pilots. The rectangular panel on the rear fuselage was the crew entry door.

From March 1942 a fairing was added around the base of the Lancaster's mid-upper turret to smooth the airflow around it. The fairing was shaped so that the guns were restricted from firing into the aircraft itself. A feature of early Lancasters – inherited from the Manchester – was the row of small windows down each side. R5689 'N-Nuts' was issued to No. 50 Squadron in May 1942 and is depicted here in August. The aircraft had only a brief operational career, burning out after crashing at Thurlby on its return from a minelaying raid on 19 September.

line because of disruption caused by a German air raid on Ringway, and by the need to change large parts of the hydraulic system to avoid the problems that were then afflicting the basic Manchester.

As first flown, the Manchester Mk III prototype was powered by the same 1,145-hp (854-kW) Merlin X engines as the production Halifax, though they were housed in very low-drag, modern nacelles. The aircraft was sent to the Aeroplane and Armament Experimental Establishment at Boscombe Down on 27 January, but by the time a handling report was issued on 3 March 1941 BT308 was back with Avro, undergoing modifications.

Within six weeks BT308 had received the enlarged twin-fins and extended-span tailplanes associated with the Manchester B.Mk IA, and had lost its central fin. The aircraft was also re-engined with Merlin XX engines which produced 135 hp (100 kW) more power per engine than the Merlin Xs used by the Halifax. With these new engines fitted, speed trials at Boscombe Down were something of a revelation, because the aircraft proved to be about 10 per cent faster than Avro's own estimates.

A second Lancaster prototype, DG595, made its maiden flight on 13 May 1941 but a backlog of trials and evaluation work at Boscombe Down delayed delivery to the A&AEE. Chief designer Roy Chadwick then used the time to refine and improve the aircraft further.

When Avro finally received a formal contract for its two prototypes on 15 July 1941, the document also covered two more prototypes, to be fitted with Bristol Hercules radial engines, as an insurance policy. The first of these aircraft, DT810, made its maiden flight on 26 November 1941, but the second was cancelled. At virtually the same time, Avro received an order for 100 aircraft originally ordered as Manchesters to be completed as Lancasters: 43 by Avro and 57 by Metropolitan Vickers. On 6 June 1941 this order was increased to cover the production of 454 Lancasters.

The first production Lancaster, L7527, flew initially on 31 October and was the first aircraft flown in the dark green, dark earth and black colour scheme which came to characterise the type. As such, the machine looked more warlike than the 'yellow-bellied' prototypes, although a close examination would have revealed that the 'guns' installed in the tail turret were no more than painted broomsticks.

Service trials at Boscombe Down were generally successful, but showed the need for improved wingtips; the outer wing panels were extended, bringing the total span to 102 ft (31.01 m), and the number of wing fuel tanks was increased from four to six. A handful of early production aircraft had to be retrofitted with the new outer wing panels. Freed from all of the constraints imposed on the early Halifax, and with more powerful, more modern, lower-drag powerplants, the Lancaster finally demonstrated superb performance. Accordingly, the type was ordered into large-scale production by a group of factories, including Austin Motors, Armstrong Whitworth, Avro, Metropolitan Vickers and Vickers Armstrong. Sir Frederick Handley Page must have looked on in envy, quite probably wondering how his Halifax would have performed had he been allowed to give it an unrestricted wingspan and Merlin XX engines.

Into service

In mid-September 1941 the Lancaster prototype visited RAF Waddington for familiarisation and to ensure that aircraft and airfield were fully compatible. The aircraft received its formal release to service on Christmas Eve 1941, and on the same day, No. 44 Squadron received its first three Lancasters.

Bomber Command was in crisis as the first Lancasters arrived at Waddington. Still largely equipped with obsolete or unsuitable aircraft, and with insufficient navigation equipment to achieve the degree of bombing accuracy required, Bomber Command had limited and inadequate resources to mount the kind of campaign demanded by the politicians. The Butt report had minutely examined indi-

Early service

B.Mk I, No. 97 Squadron
This aircraft was one of the first delivered to the RAF's second Lancaster squadron, and is depicted how it appeared in March 1942. The grey codes were changed to red in May/June. Note the unfaired dorsal turret.

© Osprey Publishing Ltd

Lancaster Mk Is take shape in Armstrong Whitworth's factory at Baginton (Coventry). Avro made the most Lancasters, building 3,673 at its Manchester (Chadderton factory – Woodford assembly) plants and another at Yeadon in Yorkshire. Other suppliers were: Armstrong Whitworth (1,329 at Whitley/Baginton), Austin (330 at Longbridge), Metropolitan-Vickers (1,080 at Trafford Park – Woodford assembly), Vickers-Armstrong (300 at Castle Bromwich and 235 at Chester), and Victory (430 at Malton, Ontario).

Above right: Lancasters were built in sections before final assembly – here noses are completed.

No. 83 Squadron sets out to raid Bremen from Scampton on 25 June 1942. This was the third of Harris's Thousand Bomber Raids, and involved 96 Lancasters, of which only one was lost. The squadron had not quite completed its conversion to the Lancaster, and retained a few Manchesters. This group contains two of the older aircraft (third and fifth from left). This was to be the last major attack in which the Manchester took part.

vidual aircraft bombing photographs and found that the Command was achieving very much lower accuracy figures than had been assumed. Overall, only one in four crews claiming to have bombed a target had got within 5 miles (8 km) of it, and in cloudy conditions the figure dropped to one in 15 crews. Even on clear summer nights, fewer than half of the crews dropped their bombs within 5 miles of their target. Figures for targets in Germany were slightly worse. Only the best crews' aircraft were equipped with cameras at that stage of the war, and the Butt report studied only those photographs from crews who had claimed to have bombed successfully.

Even such modest results had been obtained at major cost. During the four-month period from 7/8 July to 10 November 1941, Bomber Command lost 526 aircraft, which was equivalent to its entire front-line strength. Aircraft were relatively easy to replace, but the loss of trained crews (including many of its most senior and highly skilled aircrew, who had trained before the war) was a much more difficult problem. The AOC-in-C, Sir Richard Peirse, was ordered to scale down operations while the future of the bomber offensive was examined and debated.

Accordingly, Peirse switched the focus of his offensive to easier, lower-priority targets that were less well protected by flak, searchlights, fighters, and the natural defence of mist and haze that characterised the Ruhr area of Germany. Even against these targets, it soon became clear that crews were becoming less willing to press home their attacks. Although Peirse had done his best with the resources at his disposal, he was forced to shoulder the blame for the Command's failure and was removed from his post on 8 January 1942, being temporarily replaced by the AOC No. 3 Group, Air Vice Marshal J.E.A. Baldwin.

Early in the war, Bomber Command had been forced to abandon day bombing and to seek the sanctuary of night, which had now itself been shown to be insufficient to protect the bombers. The idea that strategic bombing alone would bring Germany to its knees was largely discredited.

In the end, the decision to continue the strategic bombing campaign was made largely on pragmatic and political grounds. With the new Russian ally fighting for its life in the east, Churchill was eager to demonstrate that Britain was taking the fight to the enemy in the west, and Bomber Command was a highly visible way of showing this.

The War Cabinet and the Air Ministry therefore drew up a plan for a revolutionary new offensive, under which 43 leading German industrial cities, with a combined population of 15 million, would come under continuous attack. These attacks would be made against the cities themselves, and not just against factories within them; Churchill himself approved the switch from attempts at precision bombing to what was a deliberate area bombing strategy.

The 'bomber enthusiasts' calculated that a bomber force of 4,000 aircraft could produce complete collapse within six months, but Churchill rejected this plan, insisting that Bomber Command would carry out its strategic offensive at its existing strength.

Change of emphasis

On 14 February 1942, Bomber Command received a directive confirming the change of emphasis. "The Primary Objective of your operation should now be focused on the morale of the enemy civil population and in particular of the industrial workers." Portal, the Chief of the Air Staff, insisted that henceforth Bomber Command's "aiming points are to be built-up areas, and not, for instance, the dockyards or aircraft factories..." Bomber Command had little input into the new strategy, though it was fiercely approved and championed by its new AOC-in-C, Air Chief Marshal Sir Arthur Harris, who was appointed eight days later.

In some respects Harris was lucky, and was able to capitalise on events that had been set in train by his predecessors. The hopeless Manchester, the ageing Hampden and the prehistoric Whitley would all be gone by the end of 1942, replaced by Halifaxes, Stirlings and Lancasters. The Gee navigation system would allow bombers to find closer targets (including Bremen, Emden, the Rhineland, the Ruhr and Wilhelmshaven) with greater accuracy, and would help lost bombers recover to Britain.

No. 44 Squadron began conversion to the Lancaster in December 1941, and No. 97 Squadron followed in January 1942. No. 97 Squadron's first Lancaster was the first prototype, BT308, which arrived on 10 January before the unit started gaining production aircraft.

The Lancaster's first operation was to have been a mining operation mounted on 25 January against Aas Fjord, aimed at bottling up the *Tirpitz*. This had been frustrated

by the unserviceability of the planned forward operating base at Wick. Instead, the Lancaster flew its first operation on 3 March 1942, when No. 44 Squadron despatched four aircraft to 'garden' (mine) at 'Yams' and 'Rosemary' (the approach to Heligoland and an area farther north), each aircraft carrying four 'vegetables' (mines). A week later, on 10/11 March, No. 44 Squadron contributed two aircraft to a raid on Essen, and sent another against Köln three nights later. No. 97 Squadron began Lancaster operations on 20 March, sending six aircraft on a 'gardening' mission at 'Willows' (Swinemünde in the Baltic). Mining missions represented a relatively low-risk introduction to operations.

War against the cities

Harris was obsessed with bombing enemy cities and saw in the Lancaster the perfect tool for wreaking destruction, able to carry the largest blast weapons. Harris aimed to use blast bombs to blow off roofs and blow out windows, to block roads with rubble, and to hinder the fire brigades, while simultaneously dropping very large numbers of incendiaries that would cause huge fires which would take hold in the roofless, windowless buildings. "It's easier to burn a city down than it is to blow it up," he once famously remarked. Harris believed that the use of blast bombs was useful even when fires did not take hold. High explosive was terrifying, and the temptation to repair buildings whose windows had merely been blown out would cost the enemy dearly in resources and manpower. To maximise the destruction and psychological effect, Harris was single-minded in his effort to concentrate raids into the shortest possible timeframe, in order to overwhelm the defences and the emergency services.

Concentrating raids in this way also made certain that bombers would pass quickly through the minimum number of German night-fighter boxes, which would reduce the number of interceptions and help ensure that flak and searchlight defences at the target would be overwhelmed. German airspace was divided into a network of boxes, each controlled by a single GCI controller who could direct a maximum of only six intercepts per hour, regardless of the number of aircraft passing through his box.

Early Lancaster operations included a host of missions which Harris almost certainly regarded as unwelcome diversions from the serious business of winning the war through strategic bombing. The first of these was a daring low-level daylight raid on the MAN diesel works at Augsburg on 17 April 1942. After a series of low-level training flights, Nos 44 and 97 Squadrons expected to be ordered to attack a coastal target like Kiel, but instead were briefed for a long-range mission to Augsburg. The planned route crossed the Channel at Le Havre, running south to a point below Paris, where they turned east close to Sens. Skimming the Swiss border, the formation then flew towards the Ammer See and Munich, but then turned north to Augsburg and the MAN diesel engine factory. This was a long haul by any standards and meandered through the heart of enemy-held Europe, in daylight, at low level. Despite the elaborately-planned diversionary attacks by 30 Douglas Bostons and some 700 fighter sorties, there was expectation of heavy interference by Luftwaffe fighters and many people viewed the raid as being dangerously close to a suicide mission.

The raid was led by Squadron Leader John Dering Nettleton, the charismatic South African acting CO of No. 44 Squadron. By the time of the Augsburg raid, No. 44 was actually commanded by Roderick Learoyd, VC, but Learoyd was still converting to the Lancaster and left the leadership of this raid to his more experienced deputy. No. 44 Squadron at Waddington provided a first wave of six

While the USAAF persisted in day bombing throughout the conflict, the RAF had turned to nocturnal bombing following devastating losses during the Battle of the Heligoland Bight in December 1939. What was lost in accuracy was made up for in heavier bombloads and reduced fighter opposition. As the war progressed, electronic aids such as G-H, Oboe and H_2S allowed accuracy to improve beyond that achieved by the USAAF's day bombers.

W4114 was the prototype for the Lancaster Mk III, which differed from the Mk I by having Packard-built Merlins (initially Merlin 28s). The Packard engines had Stromberg carburettors in place of the British SU units. W4114 was on test at the A&AEE at Boscombe from October 1942 to November 1943.

Augsburg raider

B.Mk I, No. 44 Squadron

R5508 was the aircraft flown by Squadron Leader John Nettleton during the daring low-level raid on the MAN diesel works at Augsburg. Nettleton was awarded the Victoria Cross for his courage in leading this attack.

© Osprey Publishing Ltd

Fog often hampered Bomber Command's operations, leading to the development and installation of FIDO (Fog Investigation and Dispersal Operation) equipment. This consisted of pipelines and burners being laid either side of the main runway. Petrol was pumped through the pipes and ignited, the heat generated being sufficient to burn off fog. Although visibility was greatly improved, and large numbers of aircraft were recovered safely due to its use, FIDO did introduce the risks of heavy turbulence and of burning should the aircraft veer off the runway into the flames. Here a Lancaster lands at Graveley.

aircraft, each fully laden with 1,154 Imp gal (5246 litres) of fuel and four 1,000-lb (454-kg) bombs with 11-second delay fuses; six more similarly-armed aircraft from No. 97 Squadron at Woodhall Spa formed the second wave. The diversionary Boston raid was exceptionally badly timed, serving only to draw massive numbers of enemy fighters into the air directly in the path of the incoming Lancasters. A full *gruppe* of JG 26 Fw 190s and Bf 109s shot down four of the No. 44 Squadron aircraft, and another was lost as it pulled off target, leaving Nettleton to return to Waddington alone. Two No. 97 Squadron aircraft were shot down after releasing their bombs on target, but the remainder returned to base safely (though one was so badly damaged that it never flew again). Seventeen bombs were dropped on the target, of which 12 exploded, causing heavy damage to a small part of the factory and destroying 3 per cent of its machine tools. Nettleton received the VC for his "unflinching determination, leadership, and valour of the highest order". The raid was presented as demonstrating Bomber Command's ability to hit outlying areas of Germany, though the impact on production at the MAN plant was limited and other factories made up the shortfall in production of diesel engines, so that tank and U-boat output was barely affected.

Maritime operations

During April 1942 Nos 44 and 97 Squadrons mounted seven-aircraft detachments to RAF Lossiemouth, near Elgin, for operations against the German battleship *Tirpitz* as it lay at anchor at Trondheim. Six aircraft from each squadron were sent on 27 April, together with 31 Halifaxes, but the raid, and a repeat attempt the next night, was unsuccessful.

Both Peirse and Harris had steadfastly refused permanently to transfer modern long-range aircraft to Coastal Command. This made it inevitable that Bomber Command would regularly lose aircraft (including Lancasters) on temporary attachments to Coastal Command. In June 1942, Bomber Command temporarily lost half of its longest-serving and most experienced Lancaster units when five No. 44 Squadron Lancasters and their crews were detached to Nutts Corner in County Antrim, Northern Ireland, for anti-U-boat operations. The detachment attacked two submarines during its brief stint, the second being sunk by Flight Lieutenant T.P.C. Barlow, DFC.

No. 61 Squadron detached eight aircraft and 12 crews to RAF St Eval in Cornwall on 16 July, for a five-week period, scoring a U-boat kill during its first sortie from St Eval. The Lancasters began to show signs of unreliability, suffering several engine failures. Finally, on 19 August, three of the squadron's Lancasters failed to return from sorties against the *Corunna*, a German blockade runner, and one more was damaged by flak. No. 50 Squadron joined the cat and mouse game with the *Corunna* the next day, when No. 61 Squadron lost a fourth aircraft. The latter squadron had mounted 96 patrols totalling 878 hours, and had lost four aircraft and 28 aircrew, achieving the destruction of a single U-boat. Whether a month's normal operations would have achieved anything greater remains uncertain, though this was Harris's unspoken assumption when he used the episode to justify his opposition to the "diversion" of resources to "irrelevant sideshows".

Harris's opposition to the diversion of Lancasters from his strategic bombing campaign restricted and limited the success of Coastal Command's equally vital campaign to combat the U-boat menace, to safeguard Allied convoys, and to impose a blockade on Germany. It may even have lengthened the war. It has been suggested that a single four-engined bomber allocated to Coastal Command had 20 times more effect on the German economy than the same aircraft allocated to Bomber Command. Harris, however, always viewed Coastal Command as "entirely defensive" and even witheringly described it as being "merely an obstacle to victory".

Bomber Command had failed to destroy the *Tirpitz* in its Norwegian lair, despite making numerous attempts, and then failed to stop the battle-cruisers *Scharnhorst* and *Gneisenau*, and the heavy cruiser *Prinz Eugen*, as they

Opposite page, top: In its element, a Chadderton-built Lancaster flies on an air test from Woodford. At the time the factory was delivering Mk Is and IIIs side-by-side, the two differing only in the provenance of the engine.

Operations in support of Coastal Command required the Lancaster to undertake some daylight missions, including the hunt for U-boats. The Lancaster's brief wartime detachment to Coastal Command was notable mainly for the abortive Capital Ship Bomb raid on Gdynia. Thereafter, the Lancasters were almost exclusively targeted against Germany's heartland, where Harris felt they could wreak the most havoc.

Maritime special

B.Mk I, No. 106 Squadron
W4118, named *Admiral Prune*, was one of six aircraft modified with a bulged bomb bay and strengthened bomb-carrying beam to carry the Capital Ship Bomb. It was a regular mount of Wing Commander Guy Gibson, who flew it during the Gdynia raid.

© Osprey Publishing Ltd

escaped in the infamous 'Channel dash' on 12 February 1942. Part of the problem had been the lack of suitable weapons, and when a new capital ship bomb was devised the decision was taken to mount an attack on Gdynia, where the aircraft-carrier *Graf Zeppelin* was apparently nearing completion and the ships involved in the 'Channel dash' were believed to have taken refuge. The capital ship bomb was a 5,600-lb (2540-kg) monster with a hollow (shaped) charge warhead, and was designed to penetrate an armoured deck before exploding and blowing out the ship's bottom.

The new capital ship bombs were incredibly expensive and difficult to produce, and only four were available. One was used for ballistics trials, which showed that the bomb had an extremely irregular and unpredictable trajectory that would make accurate aiming impossible, except from very low level. In this context, even 1,000 ft (305 m) was regarded as being far too high. No. 106 Squadron, led by Guy Gibson, was detailed to carry out the mission, and six crews began training for the then-unspecified operation. Six aircraft were modified to carry the weapon.

The raid was doomed to failure when it was decided that the Lancasters would bomb from 6,000 ft (1829 m) – far too high to achieve the degree of accuracy required. Nevertheless, on the night of 26/27 July 1942, three No. 106 Squadron Lancasters captained by Gibson, Whamond and Hopgood staggered into the air at an unprecedented 67,000 lb (30391 kg) AUW (7,000 lb/3175 kg above the aircraft's authorised maximum take-off weight) and flew to Gdynia, which they found covered by mist. The aircraft remained in the target area for about one hour, despite heavy flak, and caught fleeting glimpses of two capital ships. After making about a dozen bombing runs, all three aircraft had dropped their weapons, but all missed – Gibson's bomb falling wide by a reported 400 yd (366 m).

The modification to carry the capital ship bomb was not entirely wasted, since it also allowed the aircraft to carry the new 8,000-lb (3629-kg) HC (high capacity) bomb. No. 106 Squadron began operations with the new weapon on 31 July/1 August, initially against Düsseldorf, and later carried the weapons to Nuremberg and over the Alps to Turin. The Halifaxes of No. 76 Squadron had already quietly introduced the weapon during a raid on Essen on 10/11 April 1942.

Successes and failures

Bomber Command did enjoy some modest successes during 1942, most notably during night raids on Lübeck and Rostock, and during another daylight raid, this time against the U-boat yards at Danzig. This was the most distant target yet attempted by Bomber Command. Some 44 Lancasters were despatched, flying out over the North Sea at low level and in formation before splitting up to fly to the target independently, taking advantage of cloud cover

Below: Photographed from a Lockheed Hudson, R5852 'OL-Y' was a Metrovick-built Mk I assigned to No. 83 Squadron in 1942, before its loss in September. The 'EM' codes of its initial operator – No. 207 Squadron – can be seen beneath the 'OL'. No. 83 was the first of the Lancaster Pathfinder squadrons, flying its first target-marking sortie on 18/19 August during a raid on Flensburg.

Following on from the two Merlin-engined prototypes, two prototypes of the Bristol Hercules VI-engined Mk II were ordered, but only the first (DT810, illustrated) was built as it performed so well that the second (DT812) was deemed unnecessary. DT810 first flew on 26 November 1941, complete with yellow undersides and the circled 'P' marking (for prototype).

over Denmark and the Baltic. Twenty-four Lancasters bombed the target from 'normal height', as dusk fell, then returned to England in darkness, having lost two aircraft in the target area.

For every success there were many failures, and during many operations the death toll on the ground was smaller than the number of bombers (let alone the number of Bomber Command aircrew) lost. Before Harris took over, the Bomber Command loss rate had been as low as 2.5 per cent, giving a typical crew a 28.2 per cent chance of surviving an initial tour of 30 operations (that would be followed by a 20-operation second tour after a rest). Under Harris's leadership, the casualty rate quickly reached 4.3 per cent, meaning that the average bomber aircrew had only an 11 per cent chance of surviving their full expected tours of duty. This was unsustainable for the force as a whole, threatening to leave insufficient experienced crews to provide the necessary continuity and leadership. Anything more than a 4 per cent loss rate was also a real threat to morale. This had been starkly demonstrated in the Halifax squadrons of No. 4 Group, which suffered a 6.2 per cent casualty rate between March and August 1942, leading to a one-month rest from operations. On some individual raids, the loss rate climbed far higher: during one raid against Düsseldorf, for example, Bomber Command suffered a 7.1 per cent loss rate, and on a raid against Essen the Command sustained 10.6 per cent losses.

Thousand-bomber raids

Bomber Command's future remained under threat. What was needed was a morale-boosting spectacular, which Harris provided in the form of the first Thousand Bomber Raid, against Köln on 30 May 1942. Harris could call on only about 400 aircraft and trained crews from Bomber Command's front-line units, and had to augment the numbers with aircraft and crews from his operational training units, including both instructors and students. Of 208 aircraft provided by No. 91 Group, 49 took off captained by student pilots, for example.

A common sight in eastern England was a motorcade carting away sectioned Lancasters for repair. Avro's primary repair depots were at Langar and Bracebridge Heath, aided by Brush Ltd at Loughborough and Northampton, Field Aircraft Services at Tollerton, and the LMS Railway workshops at Derby. No. 1660 Heavy Conversion Unit's R5845/'YW-T' is seen here departing Swinderby in October 1943 after an accident. It went on to survive the war, including service with No. 3 Lancaster Finishing School and No. 90 Squadron.

Harris had originally hoped to be able to use up to 250 bombers from Coastal Command and a handful from Flying Training Command. The Admiralty, unwilling to participate in what it perceived to be a public relations exercise and fearful of losing control of its handful of long-range aircraft, forbade Coastal Command participation altogether. Flying Training Command initially offered 50 Wellingtons, but most of them were inadequately equipped for night bombing and, in the end, only four of the Command's Wellingtons participated in the first Thousand Bomber Raid. Harris managed to launch 1,047 aircraft against the magnificent cathedral city of Köln, whose proximity to Bomber Command's East Anglian bases and position on the River Rhine (highly visible to H_2S bombing/navigation radar, and within Gee range) made it a relatively easy target.

Some 73 Lancasters were committed to the operation, all drawn from No. 5 Group. These included 15 aircraft from the recently formed No. 83 Squadron, 11 from the even-newer No. 106 Squadron, and one aircraft sent out by No. 50 Squadron. The Lancasters augmented 131 Halifaxes from No. 4 Group, 88 Stirlings from No. 3 Group, and 602 Wellingtons, 299 of them from Bomber Command's front-line groups. In addition to the bombers, 49 No. 2 Group Blenheims, plus 39 from Fighter Command and 15 from Army Co-operation Command, were despatched on intruder missions against German night-fighter aerodromes along the route to be followed by the bomber stream to Köln.

Some 868 aircraft bombed the primary target, dropping 1,455 tons of bombs (mainly incendiaries) as they funnelled through the target in just 90 minutes. They destroyed 3,330 buildings and damaged many others. The raid killed between 469 and 486 Germans, most of them civilians, and injured 5,027 more. Some 45,132 people were bombed out of their accommodation, and an estimated 150,000 fled the city. Forty-one bombers were lost, including a single Lancaster.

Raid on Essen

Harris kept his Köln strike force together for a follow-up attack against Essen, two nights later on 1/2 June 1942. This time Harris assembled 956 aircraft, including 74 Lancasters. Although losses were light – only 31 bombers failed to return, including four Lancasters – damage to the target was not significant. Most aircraft failed to bomb Essen, due to haze or thin cloud cover, and the town suffered the loss of only 11 houses, 15 civilian dead, and a burned-out PoW working camp. One hundred and fifty civilians were killed in Oberhausen, Duisberg and Mülheim, and bombs fell on 11 Ruhr towns. Such 'accuracy' was by no means atypical, and it was estimated that between one-half and three-quarters of all bombs dropped were not hitting the cities at which they were aimed.

Although no-one knew it at the time, the night of 4/5 June 1942 (four nights later) marked the exact mid-point of the war. Although half the war had passed, Bomber Command had barely got into its stride and its offensive was to step up considerably, with 82 per cent of

its sorties and 94 per cent of its bomb tonnage – and 77 per cent of its losses – still ahead. Put another way, delivering the first 6 per cent of its bombs took 18 per cent of Bomber Command's sorties and cost 23 per cent of its losses.

The third, final and largest Thousand Bomber Raid was made against Bremen by 960 Bomber Command aircraft (10 per cent of which were Lancasters, including 16 aircraft each from Nos 83, 97 and 207 Squadrons, 17 from No. 106, 14 from No. 61, and 12 from No. 44). The Bomber Command aircraft included the normally-diurnal Bostons and Mosquitoes of No. 2 Group, and were augmented by 102 Coastal Command Hudsons and Wellingtons and by five Army Co-operation Command Blenheims. This brought the total to 1,067 aircraft, which transited through the target area in only 65 minutes.

Bremen was within Gee range, so the leading aircraft were able to bomb despite cloud cover and achieved a degree of accuracy. A strengthening wind fanned the flames and helped the destruction, which totalled 572 houses (plus another 6,108 damaged) and killed 85 Bremeners. Most of the force area-bombed the city, but No. 5 Group was specifically tasked against the Focke-Wulf factory (which suffered significant damage) while Coastal Command hit the Deschimag shipyard. Although the RAF regarded the Bremen operation as a success, confidential German reports indicated that the success was not proportional to the effort.

The Germans believed that only 80 RAF aircraft had bombed the town and that subsequent BBC broadcasts claiming that 1,000 bombers had been despatched were propaganda aimed at making the losses more palatable to the British public; the extent of the damage fell far short of what might have been expected from so large an enemy force. German defences claimed to have shot down 52 enemy aircraft (they were fairly close to the mark: 44 aircraft were shot down in the target area and four more fell on their way back across the North Sea). In absolute numerical terms, this was Bomber Command's heaviest loss of the war so far, though in percentile terms the loss was relatively light, at 5 per cent.

A raid against Düsseldorf on 31 July/1 August marked the first time that more than 100 Lancasters were sent out against a single target.

Heavy losses for poor returns

Although the Lancaster fared better than any of the Command's other heavy bombers, losses were alarmingly high and accuracy was generally poor. In September, 33 Lancasters (equivalent to two squadrons) failed to return, and 18 more were lost during October followed by 16 in November. The lower loss rate in October and November still represented a loss of a squadron equivalent per month, and reflected the lower intensity of operations and poor

This evocative colour photograph depicts No. 50 Squadron's early Mk Is setting out on a raid. The aircraft nearest the camera-ship, R5687 'VN-D', was lost on the night of 27/28 July 1943 during the second night of the Hamburg raids, when large parts of the city were set ablaze in a devastating firestorm.

H_2S radar

Developed at the Telecommunications Research Establishment under the direction of Dr Lovell, H_2S was a navigation radar which worked (in its initial form) at a 9-cm wavelength. Scanning through 360°, the radar provided a map display on a cathode ray tube, and allowed the operator to distinguish between land and water, and built-up areas and open country. There were four available ranges: 10-, 30-, 50- and 100-mile (16-, 48-, 80.4- and 161-km) radii. Some sets incorporated the Fishpond equipment, which used a part of the unused radar signal to provide a tail warning function on a separate display.

Named* Bendall's Folly, *W4963 was a B.Mk I used for several trials by the A&AEE at Boscombe Down. It is seen here in June 1944 while engaged in trials of the H_2S radome, which occupied W4963 for several periods between 1943 and 1945.

DS604 was a Mk II assigned to No. 61 Squadron (in No. 5 Group), which began converting to the Hercules Lancaster at Syerston in October 1942. The squadron was unusual in that it flew both Merlin- and Hercules-powered aircraft for a few months. In January 1943 it took the Mk II into battle for the first time, although by March it had standardised on Mk Is and Mk IIIs. The four Hercules engines of the Mk II combined to increase empty weight by 1,615 lb (733 kg) compared to the early Merlin-powered aircraft, while gross weight remained the same at 60,000 lb (27216 kg). However, the combined power output of the Hercules VI powerplant was 1,080 hp (806 kW) more than the combined output of the Merlin XX. As Merlin 'Lanc' production continued, the power output of the engine rose for only a small increase in empty weight. Gross weight was increased first to 63,000 lb (28577 kg), and then to 65,000 lb (29484 kg). By 1945 an overload capability of 72,000 lb (32659 kg) was possible, of which 36,169 lb (16406 kg) was 'removable' (that is, fuel, oil, crew and bombload).

winter weather rather than any real underlying improvement. It was clear that the Command was continuing down the road to complete annihilation.

Despite the heavy losses, Bomber Command's Lancaster force built up quickly. Following the conversion of No. 44 Squadron from Hampdens, Nos 97, 207, 61, 83, 106, 50, and 49 Squadrons converted from the Manchester. They were followed by Nos IX and 57 Squadrons, former No. 3 Group Wellington units which transferred to No. 5 Group in order to convert to the Lancaster, bringing the group to a strength of 10 squadrons.

No. 61 Squadron was the first to introduce a new crew complement, in which the traditional 'second pilot' was replaced by a dedicated flight engineer. Eleven former second pilots were given 'captaincies' very quickly (perhaps too quickly), but five of them were killed within six weeks. The experiment did, however, reduce the demand on the flying training schools, and ensured that when a Lancaster went down, only one pilot went with it.

Bomber Command's long-term intention was for Nos 1, 3 and 5 Groups to operate the Lancaster, and for No. 4 Group to operate the Halifax. With No. 5 Group's re-equipment complete, the next priority was to convert No. 1 Group to the Lancaster. No. 3 Group was already well advanced in the process of conversion to the Stirling, while No. 1 Group was the only one of the night bomber groups not already equipping with four-engined heavies. Since the retirement of the appalling Manchester, No. 1 Group's elderly Wellingtons had been suffering the worst loss rate of Bomber Command's remaining night bombers. Nos 101, 103 and 460 Squadrons were therefore the next units to convert to the Lancaster.

Lancaster versus Halifax

While there were no immediate plans to re-equip No. 4 Group with the Lancaster, that was Harris's long-term aim. Whereas the Halifax and Stirling had been designed in peacetime for the anticipated needs of war, the Lancaster had been designed in the light of real operational experience. It had become clear that the multi-role capability which compromised the Halifax and Stirling was not needed, and that ease and economy of production were more important than crew comfort. Since it was clear that heavy bombers would not be housed within hangars, it was equally clear that the 100-ft (30.48-m) wingspan limit that so hampered the Halifax did not have to be applied to the new Lancaster. Thanks to its streamlined low-drag nacelles, longer wing and capacious bomb bay, the Lancaster could tote a heavier bombload farther than the rival Halifax, and it was much cheaper to produce. These factors were all-important to Bomber Command's AOC.

'Bomber' Harris himself had summarised the conventional wisdom when he baldly stated, "One Lancaster is to be preferred to four Halifaxes. The Halifax suffers about four times the casualties for a given bomb tonnage... Low ceiling and short range make it an embarrassment when planning attacks with Lancasters."

Judgements like this were based on the original Merlin-engined Halifax, and Harris never bothered to re-evaluate the aircraft after he had formed an impression of the initial, troubled Halifax I and II. Few senior officers at Group Headquarters, or at Bomber Command HQ at High Wycombe, had first-hand experience of the new Halifax versions. The Handley Page aircraft was judged on the record of the inferior Merlin-engined variants which many of them had flown and which had never been a match for the Lancaster, and the poor performance of the early Merlin-powered Halifax had sealed its fate. Though the Halifax did not have the performance or payload of the Lancaster, the aircraft had suffered a lower loss rate than the Lancaster during early Pathfinder operations (even in its Merlin-engined form), and in its later radial-engined form, during the latter part of the war, it enjoyed a considerably lower loss rate than the Lancaster (0.56 per cent compared to 0.74 per cent). Moreover, the late-mark Halifax gave its crews a much better chance of survival if they were hit by enemy fire: 29 per cent of Halifax crews who were shot down survived the experience, compared to only 11 per cent of Lancaster aircrew. However, live Halifax aircrew in the enemy Stalags or Oflags (prisoner of war camps) were of no more use to Harris than dead Lancaster crews, and so survival statistics were of no interest.

Peenemünde Pathfinder

B.Mk III, No. 7 Squadron

No. 7 Squadron was the unit 'donated' by No. 3 Group to establish No. 8 Group – the Pathfinder Force. The aircraft was used on a number of raids, including the heavy attack on Peenemünde on 17/18 August 1943.

© Osprey Publishing Ltd

The Lancaster could still drop its bombs from higher altitude and carry a heavier bombload farther into the Reich than even a late-mark radial-engined Halifax. There is some belief that this advantage was less significant than it first appeared, because the Lancaster crews – bombing from higher altitude, in greater discomfort, farther from base, over more heavily-defended heartland targets such as Berlin – may have been less accurate in delivering their bombs. In Arthur Harris's strategic bombing campaign, though, large numbers of inaccurate bombs dropped on area targets in the vicinity of Berlin were worth more than smaller tonnages of bombs dropped more accurately on pinpoint targets closer to home. Getting bombs precisely onto a pinpoint target was of less concern than saturating what were usually area targets, so another Halifax advantage was effectively negated.

The Lancaster was a superb pilot's aircraft and was preferred by most pilots over rival bomber types. This was a crucially important attribute in an air force that was almost entirely run by pilots and former pilots, and where virtually all senior officers wore pilot's wings on their chests. In an air force run by pilots, being a great pilot's aircraft was of almost incalculable value, and predisposed the Air Staff to view the Lancaster in a better light.

Pathfinders

When the Germans began jamming the Gee navigation system during the summer of 1942, the need for a dedicated force of target-finders and target-markers was highlighted. Wing Commander S.O. Bufton had pioneered the technique of marking targets with coloured flares while leading Nos 10 and 76 Squadrons, and as Director of Bomber Operations at the Air Ministry he argued for the formation of an elite Target Finding Force, taking the most experienced and highly skilled crews (who had proved consistently to be best able to find and bomb their targets) to form dedicated units. Harris agreed with the need for target-marking, but vehemently opposed the formation of any elite target-marking unit within Bomber Command, preferring to keep the best captains and crews within normal front-line units where they could encourage, lead and inspire through example.

Harris decided to assign one squadron within each group to target-finding duties each month, based on its

DS771 is seen on test prior to delivery to No. 467 Squadron, RAAF. The aircraft was lost on the Stuttgart raid of 15/16 March 1944. Later Mk IIs had the improved Hercules XVI, although power output remained the same as that of the Hercules VI – 1,650 hp (1231 kW). The Mk II was never popular with Bomber Command, and was withdrawn from the fray in September 1944.

B.Mk II - the radial 'Lanc'

Avro Lancaster B.Mk II

No. 408 (Goose) Squadron, RCAF Linton-on-Ouse, 1944

No. 408 (Goose) Squadron flew the Lancaster B.Mk II from Linton-on-Ouse from August 1943. LL725 was assigned to the unit until its loss over Hamburg on 29 July 1944. The aircraft featured an extended bomb bay, as tested by the second Lancaster prototype, and had a Frazer-Nash FN 64 ventral turret which could fire through an aft-facing 100° arc.

The radial-engined 'Lanc' climbed faster than the Merlin aircraft and was as fast, but its lower operational ceiling meant that it was generally overlooked. Furthermore, the ready supply of Packard-built Merlins meant that an alternative engine source never became a crucial requirement. Although 1,000 B.Mk IIs were ordered, only 300 were completed by Armstrong Whitworth.

Following on from ED765, ED817 was the second Lancaster converted to the so-called Provisioning standard for the Dams raid. It was used for trials of the Upkeep mine at Reculver, Kent, but owing to damage sustained during the trials it was not used on the actual raid, being replaced as 'AJ-C' by ED910.

The Lancaster made up only a small part of the three 'Thousand Bomber' raids launched by Harris in 1942, but increasingly shouldered the main burden of Main Force operations. By the end of May 1943 it was not unusual to launch over 300 Lancasters on a single raid, and the numbers kept on growing. The largest single raid by Lancasters was the infamous Dresden attack on 13/14 February 1945, when no fewer than 796 were dispatched. At least 80 Lancasters can be counted in this photograph showing the Main Force stream setting off for mainland Europe some time in 1944.

success in the previous month's 'bombing ladder'. He believed that this would spread the load more fairly and would encourage an overall improvement in accuracy through constant competition. Other people believed that dedicated, elite squadrons were required in order to allow them to develop equipment, techniques and tactics and to exchange ideas.

When the decision was finally taken to form a dedicated Pathfinder Force (which later became No. 8 Group), it was created by taking one squadron from each of the existing heavy bomber groups, selecting the squadrons by taking the units which were then at the top of their respective group bombing ladders, including one of No. 5 Group's Lancaster squadrons (No. 83). No. 156 Squadron traded its Wellingtons for Lancasters in January 1943, and No. 8 Group gained more Lancasters when No. 97 Squadron was transferred in from No. 5 Group in April 1943. The group's other squadrons re-equipped with Lancasters during the remainder of 1943 and 1944.

'Pansies' and 'Blobs'

Pathfinding was initially a fairly unsophisticated business, using makeshift target-markers made by filling 250- or 4,000-lb (113- or 1814-kg) bomb cases with a mix of incendiary materials, including benzole, rubber and phosphorus. The 250-lb markers were known as 'Pink Pansies', and the larger ones as 'Red Blob Fires'. These early target-markers proved easy for the Germans to simulate on the ground, thereby drawing bombers away from the real aiming point. New markers, developed in association with the pre-war fireworks industry, were introduced at the beginning of 1943 and were easier for bomber pilots to recognise while being harder for the Germans to imitate. The first of the new markers consisted of a 250-lb bomb containing 60 12-in (30-cm) candles, which were ejected and ignited at a pre-set height using a barometric fuse, making a distinctive 60-yd (55-m) pattern on the ground in red, green or yellow.

Initially, the Pathfinders marked the aiming point and then left, but techniques soon evolved. Dedicated 'backer-ups' were used to replenish the markers, while 're-centerers' could drop flares which effectively moved the aim point – often doing so to compensate for any creep-back if Main Force bomb-aimers released short of the aim point, or to shift an aim point within an area target. The Pathfinder Lancasters gained H_2S bombing/navigation radar after some Main Force squadrons, and after the Pathfinder Stirlings and Halifaxes, but were soon using the equipment to improve bombing accuracy, especially where a target had a particularly distinctive radar signature – perhaps an identifiable bend in a river. The introduction of VHF radio further enhanced the Pathfinders' usefulness, allowing a Master Bomber to broadcast corrections, draw attention to enemy decoy flares, and prevent creep-back. A Master Bomber was used for the first time during the attack on Turin on 7/8 August 1943.

In perfect conditions the Pathfinders used visual ground-marking (code-named Newhaven), and when the target was partly obscured by cloud they dropped the ground-markers using H_2S (code-named Parramatta). When the ground was totally obscured by cloud, the Pathfinders used 'sky-marking' (code-named Wanganui) and dropped parachute flares, with the Main Force bombing 'through' the flares.

The Pathfinder crews were highly trained and were expert in precise navigation, but were themselves guided by specially-equipped Mosquitoes which were able to use the radar-based Oboe navigation system to find their targets and mark them with unsurpassed accuracy. The code-prefix Musical (e.g., Musical Parramatta) was applied when Oboe had been used rather than H_2S.

More diversions

Despite the widespread introduction of new aircraft, Bomber Command's losses were continuing to mount, and accuracy remained at best patchy and at worst very poor. An October attack on Köln by 289 aircraft (18 of which failed to return) was carefully assessed and catalogued by the city's efficient bureaucrats, who counted one 'Luftmine' (4,000-lb/1814-kg bomb), three HE bombs and 210 incendiaries as having fallen on the city, injuring four residents and seriously damaging two houses. Bomber Command records showed that the bombers had carried 71 4,000-lb bombs, 231 smaller HE bombs and 68,590 incendiaries.

During the autumn of 1942 Bomber Command switched some of the focus of its offensive away from Germany and flew a number of missions against Italy, sending four major raids across the Alps to Genoa, and four more to Turin. On 28/29 November, Guy Gibson and another No. 106 Squadron crew each took an 8,000-lb (3629-kg) bomb to Turin. Losses against the Italian targets were very low, with three 'no loss' attacks.

Despite the heavy losses and inconclusive results achieved during the Augsburg raid, the mission acted as a blueprint for another, even larger, daylight low-level raid against another target deep inside enemy-held territory. This time the objective was the French Schneider armament factory at Le Creusot, where the risk of collateral damage to French civilians could not be ignored. Even with the new Pathfinders, no night attack would have been accurate enough to ensure the degree of precision required and Harris reluctantly accepted that the only option would be a set-piece daylight raid by low-flying Lancasters.

Nine squadrons offered 94 Lancasters for the raid and these set out in the early afternoon, so that the force would bomb in daylight (in order to ensure accuracy) but would then return to base under cover of darkness. The aircraft, led by Wing Commander Leonard Slee of No. 49 Squadron, made a rendezvous over Upper Heyford before setting out for France in a great gaggle, staying below 1,000 ft (305 m) in order to avoid detection by enemy radar. The Lancasters reduced height to 300 ft (91 m) and crossed the French coast at Nantes, before looping around Tours en route to the target.

Six aircraft (two each from Nos 106, 9 and 61 Squadrons, led by Wing Commander Guy Gibson) pushed on to bomb the Henri Paul transformer station at Montchanin, 6 miles (9.6 km) farther south. These aircraft attacked in line astern from a nominal 500 ft (152 m), though one aircraft was damaged by the blast from its own bombs at 150 ft (46 m) and another crashed in the target area, almost certainly due to the same cause. This was the only loss during Operation Robinson (a code-name that, incidentally, ignored the British convention of avoiding connection with the target or mission – "Robinson Creusot"). Another aircraft (from No. 57 Squadron) was damaged when it flushed out a covey of partridge on the return trip.

Nearing the target, the remaining aircraft (five aircraft having turned back early) fanned out and climbed to 4,000 ft (1219 m) for their bombing runs. The plant was put out of action completely for more than three weeks by the ensuing attacks, and repair work disrupted production for more than eight months.

Overseas theatres

Harris was never willing to allow heavy bombers to be sent to overseas theatres, which he considered to be less important. Although overseas commanders requested Lancasters, Harris was reluctant to send any heavy bombers, unless they were aircraft types which he felt were useless to Bomber Command, like the Liberator, which was discarded by Bomber Command because of its modest payload and altitude and was thus sent to the Middle East and India.

No. 617 Squadron was formed specifically for the Dams raid, drawing the best crews from around Bomber Command under the leadership of No. 106 Squadron's former CO – Wing Commander Guy Gibson.

Bomber Command's night offensive had cost it dearly during 1942, while causing little real damage to Germany. Again and again, large formations went out only to find the vast majority missing their (city-sized) targets. Aircrew casualties often outnumbered the deaths of German civilians. Meanwhile, No. 2 Group, derided as a sideshow by Harris, regularly despatched squadron-strength raids by Douglas Bostons, most of which suffered no losses even though they carried relatively tiny bombloads, could not range far into Germany itself and tended to hit semi-tactical targets. No. 2 Group carried out some extremely accurate bombing raids and every bomb carried tended to count, while most of the heavies were still scattering their bombs harmlessly over empty farmland. It is extremely doubtful whether the Lancaster and its like caused anything near as much damage and trouble to the Germans in 1942 as did the Boston, Mitchell, and even the Blenheim – and that is without counting the contribution of No. 2 Group's star performer, the de Havilland Mosquito.

The first few days of 1943 brought the emergence of two new bomber groups. The Canadian units gathered within No. 4 Group became operational as No. 6 (RCAF) Group on New Year's Day, and the Pathfinder Force became No. 8 (PFF) Group on 8 January.

The radial-engined Lancaster B.Mk II began operations with a flight of No. 61 Squadron at Syerston on 11 January 1943. The Hercules-powered Lancaster had been designed in case the supply of Merlin engines was disrupted by America's entry into the war. Some people expected that US industry would cease production of this 'foreign' engine, and that the pressure on Merlin production would become unbearable. In fact, the USA dramatically increased Merlin production, making US-built Packard Merlin engines

Provisioning Lancasters

Below are the 19 Lancasters which took part in the Dams raid, together with the pilot's name.

AJ-A*	ED887	Young
AJ-B*	ED864	Astell
AJ-C*	ED910	Ottley
AJ-E*	ED927	Barlow
AJ-F	ED918	Brown
AJ-G	ED932	Gibson
AJ-H	ED936	Rice
AJ-J	ED906	Maltby
AJ-K*	ED934	Byers
AJ-L	ED929	Shannon
AJ-M*	ED925	Hopgood
AJ-N	ED912	Knight
AJ-O	ED886	Townsend
AJ-P	ED909	Martin
AJ-S*	ED865	Burpee
AJ-T	ED825	McCarthy
AJ-W	ED921	Munro
AJ-Y	ED924	Anderson
AJ-Z*	ED937	Maudsley

** lost during raid*

The four Type 464 Provisioning aircraft not used were ED765, ED817, ED915 and ED933.

ED825 was a Chastise veteran, flown by Flt Lt McCarthy in the second wave which attacked the Sorpe dam. Clearly visible in this view are the mounting arms for the Upkeep mine, and the pulleys and belt drive which span it at 500 rpm before it was released, driven by a motor in the lower fuselage. The 'bouncing bomb' had to be accelerated and stabilised at its spin rate for 10 minutes before it could be released.

Dams raider –Type 464 Provisioning

B.Mk III, No. 617 Squadron

ED912 had participated in the famous Dams raid (Operation Chastise) as 'AJ-N', flown by Pilot Officer Knight and credited with a hit on the Eder dam. Its codes were later changed to 'AJ-S'. It retained the Upkeep mounting arms for some time after the raid.

© Osprey Publishing Ltd

Production of Lancasters was built up steadily though the war as new factories were added to the Production Group. From the mid-1942 level of around 54/55 aircraft per month, production reached a peak in September 1944, when eight factories delivered 281 aircraft. Thereafter the two Avro plants scaled back their output and monthly numbers were consequently reduced. Combined with a decreasing number of losses, the high rate of production allowed a fast-growing roster of squadrons to convert to the Lancaster. No. 218 Squadron, in No. 3 Group, converted to the Lancaster in July 1944, this aircraft being LM257 which survived the war.

available to a range of British-built aircraft types. It soon became clear that the shortage would be of Hercules engines, also needed for the Halifax, Stirling and Sunderland, and orders for 1,000 Lancaster B.Mk IIs from Armstrong Whitworth were cut back to just 300 aircraft. The Hercules-engined Lancaster B.Mk II climbed more quickly than the Merlin-powered aircraft up to around 18,000 ft (5486 m), and was about as fast in the cruise and probably faster 'flat out'. Its air-cooled radial engines were much less vulnerable to battle damage than the Merlin, and, as another bonus, the type enjoyed much better ditching characteristics. Unfortunately, the aircraft had a lower operational ceiling and carried a smaller bombload. It was particularly useful for squadrons whose previous equipment had been the Wellington or Stirling, whose engineers did not require time-consuming training on an all-new powerplant.

French port attacks

The year began with another diversion from Harris's strategic offensive, when, on 14 January, the War Cabinet ordered Bomber Command to mount area attacks against a number of French port cities which included U-boat operating bases. The attacks against Lorient and St Nazaire were quite specifically not intended to be pinpoint attacks against the U-boats themselves, nor even the docks in which they were based, but instead were aimed to devastate maintenance facilities and the cities' basic services such as power, water and light, and (though this was left unstated) the locally-recruited workers who serviced the submarines. The attacks made no significant damage to the U-boats and their associated support facilities, but inflicted terrible damage on the French cities and their civilian populations. Planned attacks against Brest and La Pallice were abandoned.

Bomber Command returned to Berlin (for the first time since 1941) on 16/17 January, despatching 190 Lancasters and 11 Halifaxes against the German capital. Some 145 Berliners and 53 Allied PoWs were killed in the scattered bombing, and the Deutschlandehalle (Europe's biggest covered hall) was burned out, after the 10,000-seat venue had been evacuated without casualties. Only one Lancaster failed to return. Berlin would never be attacked so cheaply again, and during subsequent attacks against the city losses were usually heavy. Twenty-two aircraft were lost over Berlin the very next evening, for example.

Attacking the 'happy valley'

Harris's main offensive began in the spring of 1943, opening with an operation that became known as the Battle of the Ruhr. The Ruhr was an obvious target for Bomber Command, since it was a densely-packed concentration of important (and not merely symbolic) industrial city targets. It was far more important to the German war economy than Berlin, and was closer and more accessible from Bomber Command's East Anglian bases. However, the cities in the Ruhr valley (the 'happy valley') were not an easy target, since they were heavily defended by belts of searchlights, AA guns and night-fighter control boxes. The defences were so heavy, in fact, that even at the height of the battle, Bomber Command had to keep hitting other targets, from Stettin and Pilsen to Turin and St Nazaire, to discourage the Germans from concentrating even more resources on the defence of the Ruhr.

The first raid of the Battle of the Ruhr was targeted against Essen on 5/6 March. Marking was accurate, despite a ground haze, and the 442 attacking aircraft (including 157 Lancasters, four of which failed to return) inflicted heavy damage during their 40-minute attack, killing between 457 and 482 people and destroying 3,018 houses. Two-thirds of the bombs dropped were incendiaries, and of the remainder one-third were fused for long delays.

Whereas maximum-effort raids during 1942 had seldom included more than 250 bombers, the number of aircraft committed to raids against the Ruhr was much higher, increasing steadily during the battle and occasionally including as many as 600 aircraft. An ever-higher proportion of the attacking aircraft were now four-engined bombers, and the numerical importance of the Lancaster grew as new squadrons joined the fray. The first raid that included more than 200 Lancasters was sent out against Essen on 3/4 April, and by 23/24 May as many as 343 'Lancs' were available to attack Dortmund. With much improved target-marking, enemy casualties also began to increase: 493 Germans (and 200 Allied PoWs) were killed in Dortmund on 4/5 May, and a deadly firestorm at Wuppertal on 29/30 May razed 1,000 acres (405 ha) to the ground and killed an estimated 3,400 people. Even there, 33 bombers failed to return, giving an exchange rate of about 100 civilians per bomber lost. Some 4,377 German civilians died in the last raid of June, against Köln, which also made 230,000 people homeless.

Bomber Command did not have it all its own way. Fine weather in June, particularly, allowed the German night-fighters to achieve considerable success, and bomber losses ranged from below 4 per cent to 8.4 per cent during a raid against Oberhausen. During June alone, 107 Lancasters failed to return.

Wartime trials

As might be expected, the Lancaster was involved in many trials during the war, mostly aimed at improving its operational capability and its weapons. Some notable tests were torpedo dropping by Mk II DS606, arrester hook trials by L7529 in 1942, the long-range saddle tank in 1944 (HK541 and SW944), glider-towing (JA903 and JA904), paddle-blade propellers (JB127), Bristol turret with two 20-mm cannon (JB456), radar-controlled tail guns (KB805, in USA), remote-controlled gun barbettes (LL780 and RF268), large bomb bay doors to enclose a Grand Slam (PP741), and H_2S Mk III radar (JB558, Telecommunications Flight at Defford). There were also a number of wartime engine testbeds, fitted with a variety of Merlins or Hercules.

Left: L7529 was the third production Lancaster and was used for various trials. Here it is seen testing a runway arrester hook in 1942.

Below: In anticipation of deployment to the Far East, two Lancasters were tested with a large saddle tank to increase range. The tank was found to adversely affect the aircraft's handling.

Canadian production – Victory 'Lancs'

Under the guidance of Alfred Sewart, a factory was established at Malton, in the outskirts of Toronto, to build Lancasters. Known as Victory Aircraft Ltd, the Malton factory built Mk X aircraft with Packard Merlins, initially the 1,300-hp (970-kW) Merlin 28, then the 1,390-hp (1037-kW) Merlin 38 and finally the 1,640-hp (1223-kW) Merlin 224. Otherwise, the Mk X was essentially similar to the Mk I/III, with interchangeable components. Mk Xs were built in two batches: the first 300 (KB700 to KB999) were delivered between September 1943 and March 1945. The second batch of 130 (FM100 to FM229, cut from 200) was delivered between March and May 1945. Subsequently the Malton works changed its name to Avro Canada and modified many Mk Xs for post-war duties.

***Above:** Mk Xs were the first Lancasters to be fitted with the low-drag Martin turret. KB783 was sent to the A&AEE in October 1944 for trials with the new turret, which was mounted further forward to maintain the centre of gravity.*

***Left:** KB700* **Ruhr Express** *was the first Mk X from the Canadian line, and was sent to Woodford in October 1943 for inspection by Avro engineers. It was allocated to No. 405 Squadron. Most Mk Xs had the bulged bay.*

***Below:** Fire and flak frame a Lancaster as it bombs Hamburg on the night of 30/31 January 1943. By this time the use of Pathfinders was making itself felt, with improved bombing accuracy being achieved. In July/ August Lancasters would return to Hamburg with devastating consequences, raising a massive firestorm.*

Harris would probably have preferred to strike against Berlin, because of its symbolic importance, but realised that a serious offensive against the German capital would have to wait for longer nights, improved bombing and target-marking techniques, and an answer to the menace posed by flak and fighters.

Dambusters

In the early summer of 1943 a number of 'special' missions took place against what Harris would have seen as 'panacea targets'. On the night of 16/17 May, 19 specially-modified Provisioning Lancaster B.Mk Is of No. 617 Squadron attacked the Mohne, Eder and Sorpe dams in three waves, using the Upkeep mine (better known as the 'bouncing bomb'). The first wave of nine aircraft lost one Lancaster to flak en route to the target and used five Upkeeps against the Mohne dam, losing one aircraft in the process. The three remaining Upkeeps were used against the Eder dam, which was again breached. Two more aircraft were shot down as they flew home.

The Dambusters' second wave of five aircraft was to have attacked the Sorpe dam, which was of very different construction, having a concrete core reinforced by concrete-capped earth banks. The aircraft ran into defences that had been awakened by the first wave: two were shot down and two more forced to turn back. The final aircraft attacked the dam with inconclusive results. The third wave took off from midnight as a mobile reserve, and drew even heavier opposition. Two aircraft were shot down en route to the target, but Flight Sergeant Brown attacked the Sorpe dam and Flight Sergeant Townsend attacked what he thought was the Enepe dam, navigation having been made difficult by the spreading waters left by the breached Mohne dam. The brave pilot then obeyed orders to re-visit the Mohne dam for a final visual reconnaissance, before flying home at maximum boost and minimum altitude, his aircraft zipping along at a creditable 240-270 mph (386-434 km/h), climbing only to avoid trees, houses and powerlines. Dawn began to break while Townsend was still over Germany, and by the time he ran out between the Dutch islands of Vlieland and Texel, enemy gunners could see him well enough to bounce their shells off the sea surface in their attempts to shoot him down. After 90 minutes of daylight, Townsend finally landed at 06.15, having shut down one engine before crossing the English coast. Townsend's flight was regarded by many pilots as having been the most courageous example of airmanship performed that night, but went largely ignored in subsequent accounts of the mission. The final aircraft in the third wave was forced to turn back, its turret out of action and its pilot and navigator uncertain of their position because of gathering mist.

Some 330 million tons of water were released when the two dams burst. Downstream from the Mohne dam 1,294 people were drowned, almost half of them Russian and

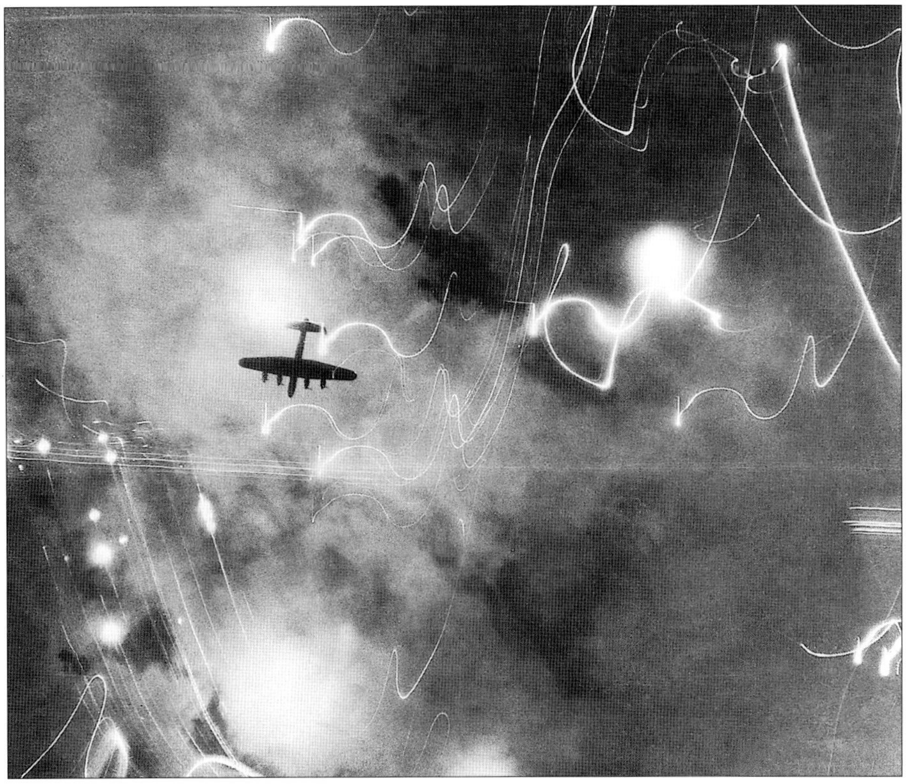

Bombing *Tirpitz* with Tallboys

B.Mk I, No. 9 Squadron
Getting Younger Every Day was one of No. 9 Squadron's 12,000-lb (5443-kg) Tallboy carriers, and it flew on two raids against the *Tirpitz*. Both of these were shuttle-bombing missions launched from the Russian base of Yagodnik.

© Osprey Publishing Ltd

The first of many civil-registered Lancasters and Lancastrians, G-AGJI began life as Mk I DV379, but was delivered, without armament, to British Overseas Airways Corporation in November 1943. BOAC used the aircraft to evaluate the type's usefulness in the airliner role, leading to the development of the Lancastrian – a dedicated airliner version with a streamlined nose and seating 9 to 12 passengers. The first Lancastrian flew on 17 January 1945, and it formed the basis of BOAC's post-war airliner fleet.

Proud ground crew pose with the King of the Air – otherwise known as Lancaster Mk III JB663, or No. 106 Squadron's 'A-Able'. The H2S-equipped Lancaster racked up a total of 111 operations, and flew over 1,000 hours by the time it was sent to No. 15 MU in March 1945.

Ukrainian PoWs. Three power stations and five pumping plants were destroyed, and 12 more were severely damaged. Much livestock was killed, and water supplies throughout the Ruhr were disrupted for a month. Domestic electricity supplies to Bochum, Dortmund, Hagen and Hamm were also constrained.

The breaching of the Eder dam caused fewer casualties but heavier damage: 25 road and rail bridges were swept away, and the railway station at Gifflitz, the bridge and the main line to Hamburg were all destroyed. The Fulda and Weser rivers needed dredging, and bridges and banks needed repairs, although these took only six weeks to complete.

Huge numbers of personnel were diverted from other duties for the repair effort, and may even have prevented the Germans from completing the Normandy defences before D-Day. The vital earth dam at Sorpe remained intact, however (and would probably have withstood the impact of all 10 Upkeeps carried by the second and third waves). Of 133 aircrew and 19 Lancasters despatched, 56 aircrew failed to return, only three of whom survived, and eight aircraft were shot down. These men were the best of Bomber Command and their loss represented a major blow. At the time, No. 617 Squadron seemed to have succeeded beyond all expectations and won great acclaim.

The squadron was to have been disbanded after Operation Chastise, but was reprieved and returned to action as a specialised heavy attack unit, using a range of weapons including the new 12,000-lb (5443-kg) HC bomb. No. 617's Lancasters were fitted with a new gyro-stabilised bombsight, the Stabilised Automatic Bomb Sight (SABS). This took account of the aircraft's speed and height, barometric pressure, wind speed and drift, and provided steering commands to the pilot. Bombing from heights of 20,000 ft (6096 m), the squadron started to achieve average errors of fewer than 100 yd (91 m). From September 1943 the squadron began attacking viaducts and targets along the Dortmund-Ems canal, which carried raw materials and manufactured goods (including prefabricated U-boat sections) between the Ruhr and Germany's northern ports.

Though it remains largely unknown, the next Lancaster special mission was probably more significant than the dams raid, and probably did more damage to the German war economy. It was certainly very much less costly. On 20/21 June some 60 No. 5 Group Lancasters attacked the Zeppelin works at Friedrichshafen, on the shores of Lake Constance (known as der Bodensee to the Germans). The attack was guided by a Master Bomber, Wing Commander G.L. Glomm of No. 467 Squadron. Heavy flak forced the Lancasters to bomb from 15,000 ft (4572 m) instead of the planned 5,000 ft (1524 m), and they were hampered by strong winds. The factory was devastated, however, and the raiders flew on to land in North Africa. When they returned, on 23/24 June, 52 Lancasters attacked La Spezia en route.

Firestorm

Even as the Battle of the Ruhr continued, Harris briefly turned his attention to Hamburg, Germany's second city and premier port, attacking it on 24/25, 27/28 and 29/30 July, and on 2/3 August. Three thousand sorties were launched against the city, and 10,000 tons of bombs were dropped on it. Some 1,500 people were killed on the first night, and 40,000 on the second. Most of the latter were killed by carbon monoxide poisoning as the raging firestorm sucked the oxygen from their shelters. In the immediate aftermath of this second raid 1.2 million people fled the city, thereby reducing the casualty figures on the last two nights. Losses were low, thanks in part to the introduction of radar-confusing Window.

Window had actually been developed months before, but its use had been embargoed to prevent the Germans from discovering its secret and using a similar invention against British radar. During the so-called 'Window embargo', however, Bomber Command lost more than 2,200 aircraft, a figure that might have been halved had Window been used, and the decision was taken to use it for the attack on Hamburg. Window consisted of strips of foil-covered paper, 27 cm (10.6 in) long and 2 cm (0.78 in) wide, which provided a blizzard of false returns. These could blind German fighter control radar, and even the Lichtenstein AI radar used by night-fighters.

Peenemünde

On 17/18 August the whole Main Force attempted a precision raid against a single target. Some 324 Lancasters, 218 Halifaxes and 54 Stirlings attacked the German rocket research establishment at Peenemünde. The Pathfinders marked three aiming points in succession as the raid developed – first the scientists' accommodation, then the rocket factory, and finally the experimental facilities. Peenemünde was devastated, 180 senior personnel were killed, and work on the V-2 rocket programme was set back by several months. The cost was high, however, with 40 bombers (including 23 Lancasters) being shot down.

Berlin

Harris believed that Germany could be defeated through bombing alone, and further believed that the best way of achieving this would be by destroying the German capital, which, he said, would "cause the heart of the German nation to stop beating". As early as 1942 he had predicted that he could win the war alone, claiming that a continental land campaign would have no use except for "mopping up".

"We can wreck Berlin from end to end if the USAAF will come in on it. It will cost us between 400 and 500 aircraft. It will cost Germany the war," Harris told Churchill. In a letter to the Air Ministry on 7 December, Harris claimed that he would have destroyed Berlin, and brought the war to an end, by 1 April 1943. When pressed to use a higher proportion of incendiaries, he argued the case for high explosive, saying: "I do not agree with this policy. The moral effect of HE is vast. People can escape from fires, and the casualties on a solely fire-raising raid would be as nothing. What we want to do in addition to the horrors of fire is to bring the masonry crashing down on top of the Boche, to kill Boche and to terrify Boche..."

Berlin was the third largest city in the world; the capital of the Third Reich, it housed the Nazi government machinery and all of the ministries which ran the German war machine. The city was also a vital industrial centre, containing tank, aircraft engine, instrument and machine tool factories, and one-quarter of Germany's electrical engineering industry. It was a crucial communications hub between the Western and Eastern Fronts. Had Bomber Command been able to cause the same level of damage in Berlin as it had done in Hamburg, it might have fulfilled his own prophecy and conceivably have brought the war to an earlier end. But Berlin was not Hamburg, and it was never likely that Bomber Command would be able to hit it anything like as hard, or to such devastating effect.

The vital targets in Berlin were scattered within a massive, sprawling, very well-defended city, and they could easily be missed without accurate marking. Berlin was 150 miles (240 km) farther inside Germany than Hamburg, well beyond the range of Oboe and a poor target for H_2S. The city could be reached only after flying through miles of hostile and heavily-defended airspace, and forced the Lancaster units to make difficult compromises between bombload, performance, range and armour.

By the time Bomber Command began its campaign against the Reich capital, it was winter, and although the long winter nights allowed the bombers to travel to and from the target in darkness, the enemy's defences had been improved, and the winter weather posed its own problems. Many of the German night-fighters were by then equipped with the new SN-2 Lichtenstein radar, and this was much better able to differentiate between real targets and Window. These fighters were able to operate with much greater autonomy, allowing a move away from the previous rigid system of boxes which a bomber stream could quite easily overwhelm. Using *zähme Sau* (Tame Boar) tactics, the radar-equipped night-fighters were vectored to the bomber stream where they then found and engaged their own targets. Berlin's defenders made greater use of searchlights to illuminate the attacking bombers for guns and non-radar equipped *wilde Sau* fighters, and even used high-flying Ju 88s to drop parachute flares above the bomber stream.

Worse still, the Luftwaffe had discovered the secrets of Bomber Command's primitive Monica tail warning radar, and of H_2S, and had developed new equipment which effectively turned both into homing beacons for suitably-equipped enemy fighters. Increasing numbers of enemy fighters also carried new upward firing 20- or 30-mm (0.78- or 1.18-in) cannon (known as *schräge Musik* – literally 'sloping music', or jazz). These allowed a night-fighter to sit

Left: The use of chaff (called Window by the RAF and Düppel by the Germans) was first employed on 24/25 July during the Hamburg fire raids. On the first night a total of 46,000 bundles was released – 92 million individual strips – which effectively blinded the German radar system. While waiting for improved radar with greater target discrimination, the Luftwaffe returned to visual night-fighter operations, which were guided by radar to the areas of concentrated Window returns.

An 8,000-lb (3629-kg) 'block-buster' and smaller bombs tumble from a Lancaster of No. 101 Squadron during a raid on Duisburg. No. 101's jamming specialists routinely flew with full bombloads to add their weight to the attack, while at the same time providing jamming cover thanks to the use of Airborne Cigar which disrupted FuG 16ZY fighter control radar. ABC was used from October 1943.

Airborne Cigar

B.Mk I, No. 101 Squadron
Named *Oor Wullie*, this aircraft was equipped with Airborne Cigar transmitters which were used to jam German fighter control frequencies. The aircraft was also one of the first to be fitted with a Rose tail turret with twin 0.5-in (12.7-mm) guns.

© Osprey Publishing Ltd

In the months either side of D-day the Lancaster force was mainly employed on sorties in aid of the invasion, striking a wide range of tactical targets. Many of the short-range sorties were conducted in daylight with full fighter cover. This pair is seen unloading over northern France in August 1944.

B.Mk Xs of No. 419 (Moose) Squadron line up at Middleton St. George. Canadian squadrons were gathered into No. 6 Group, which had nearly completed its Lancaster conversion at the war's end.

underneath a bomber (in its undefended blind spot) and pour a concentrated stream of fire into its unprotected belly, bomb bay and fuel tanks.

All the ingredients were in place for a bloodbath when Harris finally launched his assault on Berlin. Bomber Command initially mounted a series of three attacks against Berlin, spread over a 12-night period. These raids, undertaken on 23/24 August, 31 August/1 September and 3/4 September, may have been intended as the start of a full-scale onslaught against Berlin, but heavy losses and poor accuracy led to the postponement of further attacks.

The Battle of Berlin began in earnest on 18/19 November 1943, and continued well into 1944. During the winter of 1943-1944 Bomber Command attacked a number of different targets, making it impossible for the Luftwaffe to concentrate solely on the defence of Berlin and giving the RAF bombers some element of surprise when they did attack the city. Had Harris attempted to attack Berlin whenever the weather permitted, losses would have been enormous and might have reduced Bomber Command to no more than a cadre within days.

Casualties in Berlin mounted steadily, and after the second major attack on 22/23 November, 50,000 troops (equivalent to three full divisions) were sent to the city to help with the clean-up operation. After only four major raids, Bomber Command had killed 4,330 Berliners, destroyed 104,613 homes and displaced 417,665 people. There were still nights when Bomber Command losses outweighed those on the ground: 37 Lancasters, two Halifaxes and a Mosquito failed to return on 2/3 December, and the scattered bombing resulted in the death of only 36 on the ground. On 20/21 January, 495 Lancasters and 264 Halifaxes managed to miss Berlin altogether, and 35 bombers were lost. By the end of January, the Battle of Berlin had cost Bomber Command 384 aircraft in 14 raids. The last major attack against Berlin was made on 24/25 March 1944, by 811 aircraft including 577 Lancasters. Some 44 Lancasters and 28 Halifaxes failed to return from the mission, which killed 150 German civilians.

Failure over Berlin

The Battle of Berlin had failed, and the collapse predicted by Harris had not occurred. Despite the bombing, Berlin's war production actually increased during the period, and although fully one-fifth of the city's residents were made homeless there was little crisis of civilian morale, and the city's utilities and services continued to function. Heavy losses had resulted in the withdrawal from operations of the Stirling and the Merlin-engined Halifax. Plummeting morale led to a growing problem of creep-back, as aircraft dropped their bombs early and the bombing extended back into the countryside away from the aim point. Some crews even jettisoned some of their bombs over the North Sea in order to be able to fly higher. On some nights, when No. 1 Group (in which there was the greatest enthusiasm for loading the aircraft with ever-heavier bombloads) was flying, the course of the bomber stream was marked by a series of explosions far below.

One of No. 617 Squadron's 'big-belly' Tallboy-carriers sets out for an operation, possibly one of the Tirpitz *raids. The 'Dambusters' squadron had special SABS (Stabilised Automatic Bomb Sight) equipment fitted for the accurate delivery of the oversize weapons. In the background is a Mosquito, a type which was also used by the squadron (along with a Mustang Mk III) for target-marking.*

G-H Master Bomber

B.Mk 1, No. 57 Squadron
Coloured tailfins were introduced to allow rapid identification of Master Bomber aircraft and those carrying G-H during daylight raids. This aircraft is seen as it appeared during early 1945, before its loss in March.

© Osprey Publishing Ltd

Grand Slam – B.Mk I (Special)

B.Mk I (Special), No. 617 Squadron
PD119 was one of 32 'Specials' produced to carry the 22,000-lb 9979-kg) Grand Slam, 24 being delivered to No. 617 Squadron in early 1945. They lacked all guns except for those in the tail, which had reduced ammunition to save further weight.

© Osprey Publishing Ltd

The Lancaster crews did have some 'help'. The aircraft of No. 101 Squadron carried a high-powered jammer, known as ABC or Airborne Cigar, and used this to jam the 38- to 42-MHz FuG 16ZY radio used by the Germans for transmitting a running commentary of the bomber stream's position, course and altitude, and to disrupt the Benito navigation aid used by enemy night-fighters. The No. 101 Squadron aircraft usually operated as part of the Main Force, with a full bombload and a Special Operator as an eighth crew-member. A German linguist, the Special Operator found enemy transmissions, listened to them to ascertain whether they were fighter control frequencies, and then tuned his jammers to prevent them from being used. RAF bombers used Mandrel to jam German Freya early warning radar, and Tinsel to broadcast engine noise over fighter control frequencies.

The bombers were also supported by Serrate-equipped Mosquitoes and Beaufighters, which used their equipment to home onto the AI radar carried by enemy night-fighters. Other Mosquitoes flew Flower missions (attacks against enemy night-fighter airfields as the enemy aircraft took off, while the bomber stream crossed the coast). The RAF night-fighters were not able to make a significant dent in the enemy night-fighters' sortie rate, nor to make a real difference to Nachtjagdwaffe morale.

While attacks against the Ruhr had become more and more successful (though still costly), the failure of the offensive against Berlin seemed to show that missions against farther-flung, largely symbolic targets represented a wasteful use of resources. Harris, however, pressed on with attacking such symbolic targets, next turning his attention to Nuremberg which he and others viewed as the cradle of Nazism, and as its shrine. On 30/31 March Bomber Command unleashed 569 Lancasters and 212 Halifaxes against the lightly-defended city. The raid was badly planned, and diversionary attacks were recognised as such by the Germans, who refused to be drawn. The route chosen was poor, taking the bomber stream directly between two major fighter assembly points, and there was full moonlight and no cloud cover. Sixty bombers had fallen even before the stream turned south near Frankfurt. Nuremberg itself was covered by cloud, and marking was inaccurate, resulting in quite heavy attacks on Lauf and Schweinfurt. Bomber losses ran at 12.13 per cent, including 55 Lancasters.

Tactical 'Lancs'

Bomber Command's attacks on Peenemünde, the dams, Augsburg, Le Creusot and Friedrichshafen showed that it was capable of exemplary precision, and it was transferred to Eisenhower's control on 14 April 1944. The Command was tasked with attacking a range of tactical targets in France in support of Operation Overlord, the impending Allied invasion, and traditional area attacks against German cities were suspended from 27/28 April (when Bomber Command attacked Friedrichshafen) until 23/24 July (when Kiel was attacked). Bomber Command aircrew were enthusiastic in embracing an effort which was largely directed against military targets and which promised directly to

Above: An 8,000-lb (3629-kg) 'block-buster' is manoeuvred towards a waiting Lancaster. The weapon was essentially two 'Cookies' joined together. A triple-section 12,000-lb (5443-kg) weapon was used by No. 617 Squadron but was found to be unsuitable.

Grand Slam – the 'earthquake' bomb

Developed in parallel with the 12,000-lb (5443-kg) Tallboy, the 22,000-lb (9979-kg) Grand Slam was the largest bomb ever dropped in wartime. The bomb had a similar shape to the Tallboy, and was finned so that it struck the ground cleanly, with maximum kinetic energy so that it penetrated deep into the earth before exploding. The resulting tremors caused a local earthquake, which would shake the foundations of targets such as bridges. The first targets were strategic communications links. The Bielefeld viaduct was attacked on 14 March 1945, and the Arnsberg viaduct on 15 and 19 March. Bridges were hit at Nienburg (22 March), Bremen (23 March) and Arbergen (25 March, illustrated right). The final wartime target were the U-boat pens at Farge (27 March).

Aircraftsmen work on the port inner Merlin of LM577/ 'XH-Q'* Edith*, of No. 218 Squadron's 'C' Flight at Chedburgh. The impressive mission tally numbers 84, including a swastika mark – presumably to signify a claimed kill. The aircraft flew a further 14 missions to drop food to the Dutch during Operation Manna, and to bring back released PoWs in Operation Exodus.

The Lincoln-style nacelles identify this aircraft as one of the few high-altitude Mk VIs converted from Mk IIIs. JB675 was converted in November 1943 and flew on service trials with Nos 7, 405 and 635 Squadrons, before being turned over to Rolls-Royce at Hucknall for use as an engine testbed. Following the initial nine aircraft, no further B.Mk VIs were produced as it was deemed more important to concentrate on commonality across the Lancaster squadrons rather than to introduce new variants.

contribute to the invasion of occupied Europe. In the run-up to invasion Bomber Command's Lancasters attacked coastal gun positions, German radio listening stations, radar installations and radar jamming facilities, and dropped arms and equipment to the resistance.

On the night before the invasion, Bomber Command flew a record 1,211 sorties, attacking enemy troops, gun positions, ammunition and oil dumps, and road and rail communications. Enemy E-boats and other vessels were bombed in their French ports. After the invasion, the tactical bombing effort continued. Sometimes the Lancasters even flew close air support missions, attacking enemy positions that were within a few hundred yards of Allied troops.

Actions in support of Operation Overlord included the operational debut of the 12,000-lb (5443-kg) Tallboy bomb. Cast and then machined from solid steel, Tallboy was fitted with a carefully designed tail unit which imparted a gentle spin as the bomb fell, preventing it from toppling as it reached compressibility. This was routine, since the weapon could attain a terminal velocity of 2,500 mph (4023 km/h). Tallboy was easier to aim with much greater precision, and could penetrate deeply before exploding, allowing it to shake large structures to pieces like a small earthquake. Tallboy was first used in anger on 9 June 1944, when No. 617 Squadron dropped 25 of the new bombs on the railway tunnel at Saumur. This was buried in a hill and had been previously thought to be impregnable, but it collapsed quickly, preventing a Panzer division from southern France from getting to the front. The weapon was subsequently used with similar success against E-boat pens and V-1 rocket flying bomb sites. On 15 September, 28 Lancasters attacked the *Tirpitz* with Tallboys and 500-lb (225-kg) Johnny Walker mines, damaging the ship so badly that it was decided to use it as a coastal defence battery, until it was hit again and capsized on 12 November.

More panacea targets

Although control of Bomber Command's Lancasters reverted to the Air Staff on 14 September, it continued to attack tactical targets, supporting the airborne operation Market Garden (Arnhem) and the assault against Calais. Harris desperately wanted to return to the general bombing of German cities in the belief that this would cause a complete collapse, saving the Allied armies the losses which would inevitably accompany a bloody assault on the rest of the Reich.

Before the D-Day interlude, Harris had been left to fight the bomber war pretty well as he saw fit, but afterwards came under increasing pressure to pursue a more focused campaign. Sir Charles Portal and most of the Air Ministry preferred to concentrate on the German oil and synthetic oil industry, which, they felt, would rob the Wehrmacht of its mobility and would effectively ground the Luftwaffe, while Eisenhower's air commander (and second in command), Air Chief Marshal Sir Arthur Tedder, wanted the Lancasters to continue to pound the German communications and transport facilities.

Harris viewed both of these as panacea targets and obeyed directives with little enthusiasm, finding every possible excuse to continue with his own strategy. He camouflaged area attacks on cities by describing his aiming points as oil or transport targets within those cities.

Bomber Command's task became progressively easier during the final year of the war, and Lancaster losses plummeted. One year after the Battle of Berlin, in October 1944, Bomber Command's loss rate had fallen to less than 1 per cent. During the last six months of the war, Bomber Command lost only 649 heavies – representing significantly less than 1 per cent of the total despatched. Morale soared, as it became clear that, statistically, two of three bomber crews could now expect to complete a 30-mission tour of operations. Occasional raids would still suffer heavy losses, but many more missions were 'milk runs'. The last of the Stirlings and Merlin-engined Halifaxes had been withdrawn from Main Force squadrons, which were now dominated by the Lancaster. Bomber Command had 51 Lancaster squadrons by the end of 1944, and 57 by VE Day (8 May 1945). The remaining Halifaxes with Nos 4 and 6 Groups (the latter already re-equipping with 'Lancs') were radial-engined Mk IIIs or, increasingly, the improved Mk VIs.

Despite the shrinking perimeter and area of the Reich, and despite the fact that many targets had been bombed until they were in ruins, the intensity of Bomber Command operations did not falter. In fact, the Command dropped

B.Mk VI – high-performance Lancaster

In an attempt to increase the performance of the Lancaster while development of the Mks IV and V (which subsequently became the post-war Lincoln Mks I and II) continued, Avro fitted Merlin 85/87 engines – with their characteristic nacelles with annular chin intakes – into the Mk III. Two development airframes, DV170 and DV199, were followed by seven aircraft (JB675, JB713, ND418, ND479, ND558, ND673 and ND784) converted for service trials, the new variant being designated B.Mk VI. They were used at various times by a number of front-line squadrons (Nos 7, 83, 405, 583 and 635) for high-altitude pathfinding duties. The high-time aircraft was ND673, which completed 23 operational missions.

B.Mk VI, No. 635 Squadron
ND673 was one of the five B.Mk VIs which saw operational service with No. 635 Squadron from July 1944. They were used as high-altitude Master Bomber aircraft and for jamming duties, for which they carried extra Window and jamming equipment.

© Osprey Publishing Ltd

almost as great a tonnage of bombs during the last nine months of the war as it had done during the whole of the rest of the conflict, representing an astonishing 46 per cent of the total. Bomber Command mounted some of its largest and heaviest attacks – many raids involved more than 700 Lancasters – and destroyed a number of cities which had previously defied its efforts, and then turned its attentions to smaller, less industrialised towns.

As the Allies advanced, Germany lost many of the early warning sites on which it had relied, while the RAF was able to extend the range of systems such as Gee, G-H and Oboe by using navigation beacons in liberated mainland Europe. Few airfields in liberated France would have been capable of hosting a squadron of fully-laden Lancasters, so the bombers themselves remained at their bases in eastern England. Significant numbers of twin-engined fighters which might otherwise have been pressed into use against Bomber Command were destroyed by the USAAF's long-range escort fighters, and more were lost when they were thrown into ground attack missions against the advancing Allied armies. Marauding Allied fighter-bombers wreaked havoc at German night-fighter aerodromes, while Mosquito night-fighters took a steady toll of the defenders. The rest found themselves perilously short of fuel.

Battle of the airwaves

If German night-fighters did get airborne, and evaded Allied intruders, their task was more difficult than ever before. No. 100 Group was increasingly effective in jamming the night-fighters' SN-2 airborne radar using Piperack, while Jostle was used to jam the entire Luftwaffe VHF radio band. German-speaking operators in Britain broadcast confusing and contradictory instructions to night-fighter crews using Corona. In a desperate attempt to give some direction to the night-fighters, German controllers resorted to broadcasting music, using different musical genres to indicate the likely target (jazz meant Berlin, a waltz meant Munich, etc.). The British responded by blanketing the airwaves with the 'Old Comrades' march – the Luftwaffe's musical code which indicated that the bombers had departed.

Finally, if a night-fighter pilot managed to find and engage a Lancaster, he would find that his target was now better defended. Some Lancasters were fitted with a rear-ward-facing radar (known as AGLT Automatic Gun Laying Turret, or Village Inn) which could be used automatically to 'lay' the guns, while others had a new Rose Brothers tail turret with two 0.50-in (1.27-cm) guns and a more open, high-visibility design.

The Hercules-powered B.Mk II flew its last operation, against Neuss, on 23/24 September. The aircraft was in many respects a better aircraft for the job at this late stage of the war, when it was less critical to fly at the highest possible altitude. The type's air-cooled radial engines were less vulnerable and improved performance at slightly lower altitudes. The withdrawal of the B.Mk II left Bomber Command operating what was effectively a common fleet of Lancasters. Some had US-built Packard Merlin engines and were known as B.Mk IIIs, while others had been built in Canada and were designated B.Mk X.

The Lancaster B.Mk IV and B.Mk V were improved, longer-range versions intended to replace the B.Mks I, III and X, which became the Lincoln B.Mk I and B.Mk II, though they were too late to see wartime service. The Lincoln's Merlin 85/87 engine was fitted to a handful of B.Mk IIIs to produce the interim Lancaster B.Mk VI, five of which gave brief wartime service.

No. 3 Group's Lancasters began operating independently on 18 October 1944, when they attacked Bonn. Previously unbombed, Bonn was targeted so that post-strike reconnaissance would give an accurate impression of the damage caused by the raid. About one-third of No. 3

Above: Lancaster B.Mk Is which differed little from those of the first production batch were still being delivered at the war's end. Most had H_2S radar fitted, and they were powered by slightly uprated Merlin 24 engines. NG347* Piccadilly Princess *was assigned to the RCAF's No. 424 Squadron at Skipton-on-Swale in June 1945. Note the lack of guns in the nose turret.

Above left: This Lancaster B.Mk III, TX273, was the last to be built at Avro's Yeadon plant near Leeds, and certainly had the highest serial number of any completed aircraft. It was delivered on 31 October 1945. Aircraft were still being delivered from Armstrong Whitworth until early 1946, the last of which (and probably the last of the 7,377 Lancasters to fly) was TW911.

Lancaster B.Mk VII – low-drag turrets

The Lancaster B.Mk VII was built exclusively by Austin Motors Ltd. This version introduced a Martin dorsal turret, mounted further forward than the standard turret. It was a low-drag design and was armed with two 0.5-in (12.7-mm) guns. The tail turret was the Frazer-Nash FN 82, also with two 0.5-in machine-guns. Due to a delay in turrets arriving from the US, the first 50 were built with the standard FN 50 – albeit in the forward position – and were sometimes known as B.Mk VII (Interim). This batch was delivered from February 1945, but the definitive Mk VII production did not reach the RAF until April, and continued until September. Many of the Mk VIIs were tropicalised for Tiger Force duty, like the aircraft below, seen post-war, which wears No. 1689 Flight/20 MU's '9X' codes.

Tiger Force – Lancs for the Far East

With Germany defeated, Bomber Command immediately began to prepare Lancasters for participation in the war against Japan, assembling the Tiger Force. This consisted of a strategic bombing force with 36 (later 24) Lancaster squadrons at its heart. The squadrons were equipped with white-painted tropicalised aircraft – either Mk I(FE)/III(FE) aircraft with the dorsal turret removed, or the new Mk VII(FE) with Martin turret. The war ended before the Tiger Force could be brought to bear, although several of the units were deployed to the Far East in the aftermath.

Right: B.Mk VII(FE)s of No. 617 Squadron fly over Benares in India during their brief deployment to Salbani in India, which lasted from January to May 1946.

Below: No. 7 Squadron received Lancaster B.Mk I(FE)s for Tiger Force duties, but remained in the UK. In January to March 1947 it was deployed to Kallang in Singapore during Operation Red Lion.

Below: RF310 was the first Lancaster to be fitted to carry an air-droppable lifeboat. It served as 'RL-A' with No. 279 Squadron at Beccles in December 1945. Cunliffe Owen built the boat (Airborne Lifeboat IIA) and performed the conversion, which changed the Lancaster's designation to ASR.Mk III (for Air-Sea Rescue).

Group's Lancasters were equipped with G-H, and these were decorated with prominent coloured fins. The group used G-H to find its targets when the ground was obscured by cloud, and when cloud tops did not exceed 18,000 ft (5486 m). Non G-H-equipped aircraft were expected to find and fall into formation with a G-H aircraft in the target area and to bomb when the G-H aircraft did. The technique proved devastatingly effective, tearing the heart out of the old city and burning its historic buildings and university.

The 22,000-lb (9979-kg) Grand Slam had been developed in parallel with the Tallboy, but could be carried only by specially-modified Lancasters with more powerful Merlin engines, a strengthened undercarriage and a cut-away bomb bay. The first of these were delivered to No. 617 Squadron on 1 February 1945, and the aircraft immediately mounted raids against the Bielefeld and Arnsberg viaducts, and later against bridges and U-boat pens.

Dresden

More controversially, Bomber Command extended its operations to eastern cities which lay in the path of the Soviet advance or through which German reinforcements might flow to the Eastern Front, thereby gaining a slough of new targets in the eastern part of Germany, including the historic city of Dresden. This vital port on the upper Elbe was an important communications hub for supplies and reinforcements for the Eastern Front, then a scant 150 miles (240 km) away.

The operation was to have opened with a raid by USAAF day bombers, but bad weather prevented this, and the first attack was made by night by the RAF. On 13/14 February some 796 Lancasters and nine Mosquitoes attacked Dresden in two waves. They dropped 2,600 tons of bombs, creating an immense firestorm and killing more than 50,000 people. The USAAF joined the operation the next day, dropping another 700 tons of bombs.

The tremendous destruction and epic human cost of Dresden hid the fact that accuracy had been first rate, with a heavy concentration on the aim point. It shook Churchill's support for Harris's strategy: on 28 March, the Prime Minister observed, "It seems to me that the moment has come when the question of bombing German cities simply for the sake of increasing terror (though under other pretexts) should be reviewed, otherwise we shall come into control of an utterly ruined land. I feel the need for more precise concentration on military objectives, rather than mere acts of terror and wanton destruction, however impressive."

But similar attacks against smaller towns continued, interspersed with more tactical missions in support of the advancing Allied armies. The last day of the war for the Lancaster force was 25 April 1945, when it attacked coastal gun batteries on the island of Wangerooge (which covered the approaches to Bremen and Wilhelmshaven) and Hitler's mountain-top retreat at Berchtesgarden. That night (25/26

Right: No. 38 Squadron was based at Ein Shemer in Palestine from September 1946 to April 1948. It operated a mix of ASR./GR.Mk IIIs (centre aircraft) and B.Mk VIIs (flanking aircraft). Its ultimately futile task was to identify vessels carrying Jewish immigrants attempting to enter Palestine illegally.

Engine tests and other post-war trials

The size, load-carrying ability and ready availability of the Lancaster made it a natural choice for many trials programmes in the late-war and post-war periods, especially those concerned with turbine powerplants. The first was the prototype Lancaster (BT308), which mounted a Metrovick F.2/1 jet in the tail. Its tasks were taken over by a Mk II (LL735) in 1944. Another UK jet testbed was Mk VI ND784, which flew with the Armstrong Whitworth ASX engine in the bomb bay and, much later, tested the Adder and Viper jets in a tail installation. In Canada the Avro Canada Orenda jet was test-flown in the outer wing nacelles of Lancaster Mk 10O FM209 from July 1950 (above right), while the Swedes used RA805/80001 to test the STAL Dovern engine (and later the de Havilland Ghost) in an underfuselage nacelle. A number of turboprops were also fitted to Lancasters, either in a nose installation (Armstrong Siddeley Mamba in ND784, Rolls-Royce Dart in NG465) or in the outer wing nacelles (TW911, illustrated right, which first flew with two Python turboprops in January 1949).

As well as the engine testbeds Lancasters also served on other important programmes. Two aircraft (PP755 and RE131) were used to test elements of the control system for the Bristol Brabazon airliner, while ME540 was fitted by Boulton Paul with a long nose probe as part of a gust alleviation system (below right). Other tests were aimed at improved combat systems, such as the remotely controlled ventral 20-mm cannon barbette and rear turret fire control radar tested by LL780 in 1946.

Above: One of the best-known post-war testbeds was PA474. Having served as a PR.Mk I with No. 82 Squadron (from whom it inherited its 'M' code), the aircraft was loaned to Flight Refuelling Ltd in 1952, before being transferred to the Cranfield Institute of Technology in 1954. It was used to test wing sections (this is a Handley Page laminar-flow section) on a dorsal mount. It was retired in 1964 and was presented to the RAF to maintain as a flying tribute. It now flies with the Battle of Britain Memorial Flight.

Right: Following the conversion of the prototype to carry a Metrovick F.2/1 jet engine in the tail (and later the W.2B/33), a Mk II (LL735) received a similar conversion, mounting a F.2/4 Beryl engine. It was subsequently fitted with an F.2/4 in the bomb bay.

April 1945) Lancasters bombed an oil refinery at Vallo (Tonsberg), while others mined Oslo Fjord. These were destined to be the Lancaster's last offensive missions.

The reckoning

Once the RAF decided to follow a strategy of despatching great numbers of large heavy bombers against area targets and to bomb with relatively little precision from great height, the Lancaster was undeniably the best aircraft for the job. Early Lancaster operations had often been hopelessly inaccurate, and frequently inflicted no more than a handful of civilian casualties, perhaps killing a cow or two, sometimes destroying a couple of houses, often at the cost of several aircraft and their precious crews. Again and again, Bomber Command's losses far exceeded the German civilian death toll.

The Lancaster's accuracy did improve exponentially, until what Bomber Command regarded as area attacks were often more accurate than what the USAAF described as pinpoint "precision attacks". This was especially true when poor weather intervened to prevent purely visual bomb-aiming, and when the USAAF's bombardiers had to forsake their much-vaunted Norden bombsights for H2X bombing radar, in the use of which they were more poorly trained than their RAF counterparts. By 1944, even in poor weather, RAF Lancasters still tended to saturate their planned aim point, while USAAF B-17s would scatter their bombloads over huge areas of the German countryside. This was confirmed by a surprising source, Germany's Armaments Minister, Albert Speer, who commented, "The RAF night attacks are considerably more effective than the US daylight attacks, since heavier bombs are used, and an

Air Photographic Development Unit Lancaster PR.Mk I TW661 beats up its base at Benson in May 1948. The Oxfordshire airfield was the RAF's main centre for recce and survey, parenting No. 82 Squadron which flew Lancaster PR.Mk Is on survey duties in Africa.

Avro Lancaster B.Mk I Special

No. IX Squadron
No. 5 Group, Bomber Command
RAF Bardney, 1944

This Lancaster B.Mk I (LM220) was one of those was fitted with bulged bomb bay doors to allow it to carry a 12,000-lb (5443-kg) Tallboy bomb, and it used this formidable weapon in anger against the *Tirpitz*. Moored in a succession of Norwegian fjords, *Tirpitz* appeared to be poised to attack the Arctic Convoys, or to sally forth into the North Atlantic, and was felt to represent a grave threat. Bomber Command made a series of unsuccessful attempts to sink the ship, and then in mid-1944 decided that the new Tallboy might represent the best means of attacking the vessel. LM220 was one of the 28 aircraft from Nos IX and 617 Squadrons which participated in Operation Paravane, attacking *Tirpitz* on 15 September 1944 as it lay at anchor in the Altenfjord. The aircraft was flown by Flight Lieutenant W.D. Tweddle and his crew during the operation. The Lancaster bombers, supported by a Lancaster camera ship from No. 463 Squadron and a meteorological Mosquito, actually operated from Yagodnik in Russia, returning to the Russian base after the attack, before flying home some days later. The aircraft attacked *Tirpitz* with Tallboys and 'Johnny Walker' mines, and although a dense and effective smokescreen saved the German ship from destruction, one hit on the bows caused so much damage that the German Navy was forced to conclude that its career was over, and began converting it to serve as a coastal defence battery, before two more attacks by Nos IX and 617 Squadron finally sank it in Tromsø Fjord.

Hitting the target

Most Lancasters were fitted with primitive Mk IXA course-setting bomb sights which could not take account of even gently banked turns, or (from 1942) the much improved and gyro-stabilised Mk XIV computing bomb sights, which were capable of great accuracy under ideal conditions. They were housed in the extreme nose, behind the bomb-aimer's transparency. This incorporated an optically flat panel for aiming, and was a different shape (longer) on the B.Mk III and on some retrofitted B.Mk Is. No. 617 Squadron's Tallboy aircraft were fitted with a new SABS (Stabilised Automatic Bomb Sight) sight, a Farnborough-developed equivalent to the more famous US Norden sight. The SABS sight was a telescopic sight in which the bomb-aimer acquired the target, having first fed in all necessary bomb ballistics, wind and altitude information. It was mounted on a gyro-stabilised platform, and automatically generated aiming corrections as long as the bomb-aimer held the sight onto the target. The equipment required a 20-second straight and level bombing run, but proved capable of extraordinary accuracy, such that 75 per cent of weapons dropped would fall within a 150-yd (137-m) diameter circle on the ground. But to hit the target, the Lancaster first had to find it, and in European weather inexperienced aircrew were often hard-pressed to find their targets using dead reckoning and celestial navigation alone. Fortunately, new navigation aids were soon introduced. Gee used pulsed phased transmissions from British ground stations in conjunction with a special set of charts carried by the aircraft's navigator. Gee was not accurate enough to allow blind bombing, but was an excellent aid for target-finding, and was invaluable in helping bombers find their own bases, and it was difficult to jam. The Pathfinders used Oboe to find targets, though the equipment required a high degree of expertise to use, and was never widely available. Oboe ground stations transmitted pulses which could be detected by a suitably equipped aircraft, which could measure its range from the ground station with great accuracy. Each aircraft used two ground stations, one transmitting a series of dots or dashes on each side of a curved track to allow the pilot to ascertain which side of the track he was flying, with the second station transmitting an intersecting beam which bisected the track at the planned release point. Only the Mosquito could fly high enough to receive Oboe over enemy territory, but these marked targets for the Lancasters. From 1942, Lancasters received H_2S, a centimetric wavelength radar with a real ground-mapping capability, able to pick up distinctive features (and especially water features) on the ground. H_2S Mk III was similar, but was actually the US H_2X radar, while the K-band H_2S Mk VI gave better discrimination. Gee-H was introduced in October 1944, using a transmitter receiver in the aircraft to determine distance from ground stations. This gave a very high degree of accuracy, such that Gee-H equipped aircraft gained special markings, and other aircraft were encouraged to 'bomb on their command'. The Pathfinders further refined accuracy by broadcasting corrections using VHF radio.

Propellers
Those Lancasters powered by British-built Merlin engines used Rotol constant-speed airscrews, while the radial-engined B.Mk II used Rotol RE.6 propellers. Packard-engined aircraft used Hamilton Standard propellers, or increased-area 'paddle-bladed' Nash-Kelvinator propellers. The latter improved the rate of climb, and increased performance at high altitude, adding about 8 mph (13 km/h) to cruising speed and a useful 1,500 ft (457 m) to the operational ceiling, but gave poorer performance below 16,000 ft (4877 m). Although the later Lincoln used four-bladed airscrews, these were never fitted to the production Lancaster.

Accommodation and emergency egress
The ceiling of the bomb bay also formed the main cabin floor, on which were positions for the pilot, flight engineer, navigator and wireless operator. Aft of these stations, in the centre section, was a rest bunk (replaced in No. 101 Squadron aircraft equipped with ABC by a station for an eighth crew member – the Special Operator). In the rear fuselage was an Elsan chemical toilet, just forward of the tailplane carry-through. When the Lancaster was used in the passenger-carrying role, around 25 people could be carried in some discomfort in the centre-section and rear fuselage. Access to all crew positions was made through the side door in the rear fuselage. In an emergency the 'front end' crew would aim to bale out through a hatch in the floor of the bomb-aimer's position in the nose, while the mid-upper gunner would head for the crew access door. The rear gunner had to swivel the turret so that the entry door was facing out into the airstream. In the event of a ditching, the crew used hatches above the flight deck, the centre section and rear fuselage forward of the mid-upper turret. A dinghy was housed in the starboard wing root.

Tallboy
The original 12,000-lb (5443-kg) HC bomb was a primitive weapon, a crude three-section cylinder to which armourers attached a simple and rudimentary tail unit. It was inaccurate and thin-skinned, like the 4,000-lb (1814-kg) and 8,000-lb (3629-kg) bombs, so while it had tremendous 'blast' and was ideal for knocking down buildings in heavily built-up areas, it was poorly suited for use against pinpoint targets or against targets which required a weapon to penetrate before exploding. There was clearly a need for a more accurate heavy bomb, which would be easier to aim with greater precision, and which might also penetrate deeply before exploding, allowing it to 'shake' large structures to pieces like a small earthquake. The 12,000-lb (5443-kg) Tallboy and the 22,000-lb (9979-kg) Grand Slam were developed in parallel to meet this requirement. The weapons were the brainchild of Dr Barnes Wallis, who had already produced the Upkeep bouncing bomb used by the Dambusters, and who had earlier pioneered the geodetic construction used by the Vickers Wellesley and Wellington. The new bomb casings were cast and then machined from solid steel, and were fitted with a carefully designed and precision-engineered tail unit which imparted a gentle spin to the weapons as they fell, preventing them from toppling as they reached compressibility. Both weapons could attain a terminal velocity of 2,500 mph (4023 km/h), so this precaution was vital. Manufacturing the weapons was difficult, expensive and time consuming, and work on the ten-ton Grand Slam was slowed down to allow the smaller Tallboy to be rushed into service. The first combat drop of Tallboy was on 9 June 1944, when No. 617 Squadron dropped 25 on the railway tunnel at Saumur, previously thought to be impregnable, and the weapon was then used to great effect against E-boat and U-boat pens, V-1 sites, *Tirpitz*, canals and viaducts, and finally against Hitler's 'Berghof' mountain retreat at Berchtesgarden.

'Tail-end Charlie'
Early Lancasters were fitted with a Frazer-Nash FN 20 tail turret, with four 0.303-in (7.7-mm) Browning machine-guns. Because most attacks would be made from the rear, the tail turret had 1,900 rounds per gun in the 'boxes' located just behind the bomb bay and 600 more in the long 'feed' mechanisms. From May 1943, the FN 20 gave way to the similar FN 120, but it was not until the introduction of the FN 121 that the aircraft gained Village Inn Automatic Gun Laying (AGLT) radar. Some aircraft were retrofitted with a new turret manufactured by Rose Brothers of Gainsborough. This was much more roomy than the Frazer-Nash turret, and gave a much better view. The gunner could wear his parachute in the turret and was even able to abandon the turret over the guns, instead of having to go back into the fuselage to get his parachute and then wind the turret round to escape through what were usually its 'back doors'. The Rose turret also gave improved firepower, with a pair of heavyweight 0.50-in (12.7-mm) machine-guns.

Defensive armament
Most Lancasters had the same defensive armament, though there was some variation, especially by the end of the war. The Frazer-Nash FN 5 nose turret contained a pair of 0.303-in (7.7-mm) Browning machine-guns. Later in the war this was sometimes removed from heavyweight B.Mk I Specials to save weight. The mid-upper turret was usually a Frazer-Nash FN 50, again with two 0.303-in Brownings. From March 1942, a fairing was added, incorporating a 'cam track' which prevented the guns from hitting parts of the airframe. It had been hoped to introduce a Bristol B.17 mid-upper turret, with two Hispano 20-mm cannon, on Austin-built Lancasters during late 1944. The weight of these weapons, and their ammunition, made it necessary to relocate the turret further forward, and though the aircraft eventually received standard Frazer-Nash turrets with 0.303-in guns, these were fitted in the new forward position. The B.Mk VII introduced a Glenn Martin mid-upper turret that had once been planned for the Canadian-built B.MK X. Though this contained a pair of longer range, harder-hitting 0.50-in (12.7-mm) Brownings, it also caused centre of gravity problems and was therefore moved further forward, making the aircraft more cramped. The Dambusters Lancasters (and those sent out against Tromsø) had their mid-upper turrets removed altogether. Small numbers of Lancasters (with a higher proportion of B.Mk IIs) were fitted with ventral turrets, usually a Frazer-Nash FN 64 faired into the rear of the bulged bomb bay doors. The ventral turret was of limited traverse, capable of moving through only 100° on each side of the centreline, and could be aimed only via a periscope which gave the gunner a very restricted view of the outside world. Alternatively, some aircraft (especially within No. 6 Group) had a single 0.303-in Vickers K gun (drum fed) on a crude mounting to protect the bomber's vulnerable belly. The ventral turrets were not fitted to aircraft with H_2S, but were initially retained by those aircraft with bulged bomb doors, though many squadrons, especially in Nos 1 and 5 Groups, quickly dispensed with them. The Lancaster's gunners could do little to stop the slaughter, however many turrets were fitted. There were some who proposed stripping the Lancaster of all of its gun turrets, claiming that the resulting 50-mph (80-km/h) increase in speed would actually reduce the loss rate. But gunners were a deterrent, and provided an invaluable lookout, even if they shot down relatively few enemy aircraft.

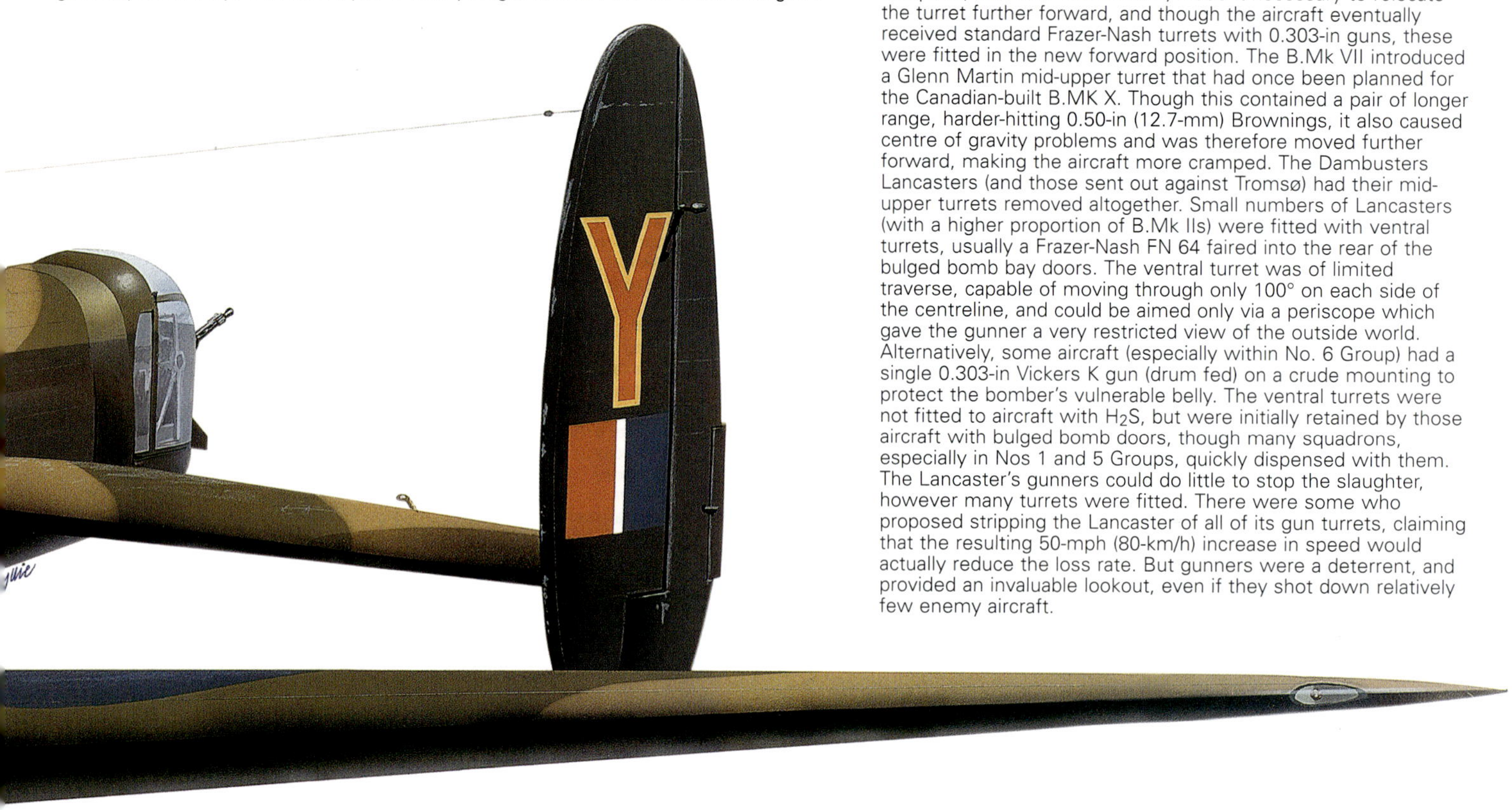

with Nash gned this r easier

s the h had no b nb that nclosed east 20

e four which small sing

Wing
The secret of the Lancaster's success lay largely in its wing. Other RAF heavy bombers were handicapped by artificial constraints on their wingspan, imposed to ensure that they could fit inside peacetime hangars. The useless Manchester's wingspan was increased from 80 ft 2 in (24.43 m) to 90 ft 1 in (27.46 m) when plans to catapult-launch the aircraft were abandoned, but even this was inadequate, and it was not until the wing was further lengthened to 102 ft/31.09 m (on the Manchester Mk III/ Lancaster) that the aircraft gained the high-altitude performance which would come to characterise it. After three extensions, the final wing was unusually slender and thin, producing very little drag. The Lancaster's wing also housed the aircraft's capacious fuel tanks, which contained 2,154 Imp gal (9792 litres) of fuel in six cells. The very earliest aircraft had only four cells, with 1,710 Imp gal (7774 litres).

Construction
The Lancaster's all metal monocoque fuselage was built in five sections (nose, front fuselage, intermediate centre section, rear centre section and rear fuselage), with transport joints between all but the nose and forward fuselage, which were permanently bolted together. Primarily constructed from light alloy, the fuselage included pressed mild steel formers supporting the bomb bay, with tie rods to support heavy bombs of up to 2,000 lb (907 kg). Some steel plates were also incorporated to provide armour protection for the pilot and navigator. The cantilever wing was built around front and rear mainspars, with the wing centre section constructed integrally with the intermediate fuselage centre section, and with separate outer wing sections.

Powerplant
Except in its Mk II form, the Lancaster was powered by four Rolls-Royce Merlin liquid-cooled, inline V-12 piston engines, each driving a three-bladed propeller, with right-hand rotation. The Lancaster Mk II used Bristol Hercules radial engines, and its propellers turned in the opposite direction. The Lancaster prototype used Merlin X engines, each rated at only 1,145 hp (854 kW), while the B.Mk I used Merlin XXs or 22s, each rated at 1,280 hp (955 kW), though the Merlin 22 could deliver more power at sea level. A handful of B.Mk Is had the more highly boosted Merlin 24, rated at 1,610 hp (1201 kW). The Lancaster B.Mk III and Canadian-built B.Mk X had Merlins produced by the US Packard motor company, in Merlin 28 (1,300 hp/970 kW), 38 (1,390 hp/1037 kW) or higher boosted 224 (1,640 hp/1223 kW) forms, with Bendix-Stromberg pressure injected carburettors instead of SUs. The handful of high-altitude Lancaster B.Mk VIs were fitted with Merlin 85 engines, rated at 1,750 hp (1305 kW), while the B.Mk VIIs for Tiger Force had the 1,610-hp (1201-kW) Merlin 24.

Markings
Apart from a handful of aircraft at the end of the war, all Bomber Command Lancasters wore the same basic colour scheme, with dark green and dark earth camouflaged topsides, and night undersides and fuselage sides. National insignia were toned down, with only a thin ring of white separating red and blue in the roundels and fin flashes, and with dull dark shades of red and blue being used. Though the interesting exceptions have been often illustrated, most Bomber Command Lancasters were exceptionally anonymous, apart from their three-letter codes and occasional bomb logs recording the number of missions flown. Nose art and names were unusual in most units, though from late 1944, aircraft equipped with G-H often had distinctive tail markings, and from August 1944 many No. 5 Group Lancasters had codes thinly outlined in yellow. The same group also occasionally applied the full codes above the tailplanes, with the two-letter unit prefix on the top surface of the port tailplane, and the individual letter identifying an aircraft within the squadron to starboard. More usual was the practice of painting the individual code letter on the outer face of each tailfin, and of repeating the code letter in a smaller form on the nose. Other groups had other unusual markings, No. 1 often applying a gas detection patch as a disc on the nose, and No. 3 Group sometimes using stripes across the fins and coloured airscrew spinners, though such markings were still always the exception rather than the rule. LM220 was therefore an unusually colourful aircraft – though it was less unusual within No. IX Squadron, which was something of an elite unit, many of whose aircraft wore colourful and individualistic markings.

During the war No. IX Squadron flew 3,495 Lancaster sorties, losing 177 aircraft in the process. One of the squadron's members, Flight Sergeant G. Thompson, was awarded a posthumous VC. The aircraft's 'bomb log' included 15 red bomb symbols representing night operations and 27 yellow bombs indicating daylight operations. The 39th and 40th bombs had the letter R (for Russia) superimposed, denoting a shuttle-bombing mission that landed in Russia. It also included two silhouettes of the *Tirpitz*, representing attacks against the German battleship. Nose art and individual aircraft names were often based on an aircraft's individual code letter, while many No. IX Squadron aircraft had names reflecting an alcoholic theme – celebrating various brewers and distillers. The cartoon of an elderly gentleman with a pint of beer applied to LM220's nose, and the slogan 'Getting Younger every day' echoed a contemporary advert for the products of William Younger's brewery. One of this aircraft's sisters was W4964/WS-J, which had the figure of Johnny Walker and a contemporary whisky slogan 'Still Going Strong' on its nose. (The latter aircraft was commemorated by the RAF's Battle of Britain Memorial Flight, which painted its aircraft, PA474, in the same markings between 1993 and 1999). Though damaged in March 1945, LM220 survived the war, but the aircraft was then scrapped at No. 5 MU in November 1946.

Avro Lancaster B.Mk I

No. 467 Squadron, RAAF RAF Waddington, 1944

Herman Goering's famous quote, "No enemy plane will fly over Reich territory", was the legend applied to Lancaster R5868 'PO-S' of No. 467 Squadron – right below the mission tally which had recorded 137 successful bombing sorties by the war's end. It is depicted here with 98 mission marks as it approached the magic 'century'. The aircraft was built by Metropolitan-Vickers in Manchester and assembled at Woodford before delivery in June 1942 to No. 83 Squadron, becoming 'Q-Queenie'. It then transferred to the Australian squadron, becoming 'S-Sugar'. Its preservation status was assured in February 1956, and today it is the centrepiece of the Bomber Command display in the RAF Museum at Hendon.

At least 34 other Lancasters achieved 'century' status, notable aircraft being No. 61 Squadron's ED860/'QR-N' (130 operations), No. 100 Squadron's ND458/'HW-A'/'A2' (127 operations), No. 15 Squadron's LL806/'LS-J' (134 operations) and ED888, which flew at least 135 and possibly 140 operations with Nos 103 and 576 Squadrons with a variety of codes. All Lancaster 'centurions' were Mk I or Mk III aircraft, and included aircraft delivered as late as June 1944, such as No. 550 Squadron's PA995/'BQ-V'/'K' and No. 625 Squadron's PB150/'CV-V'/'V2'.

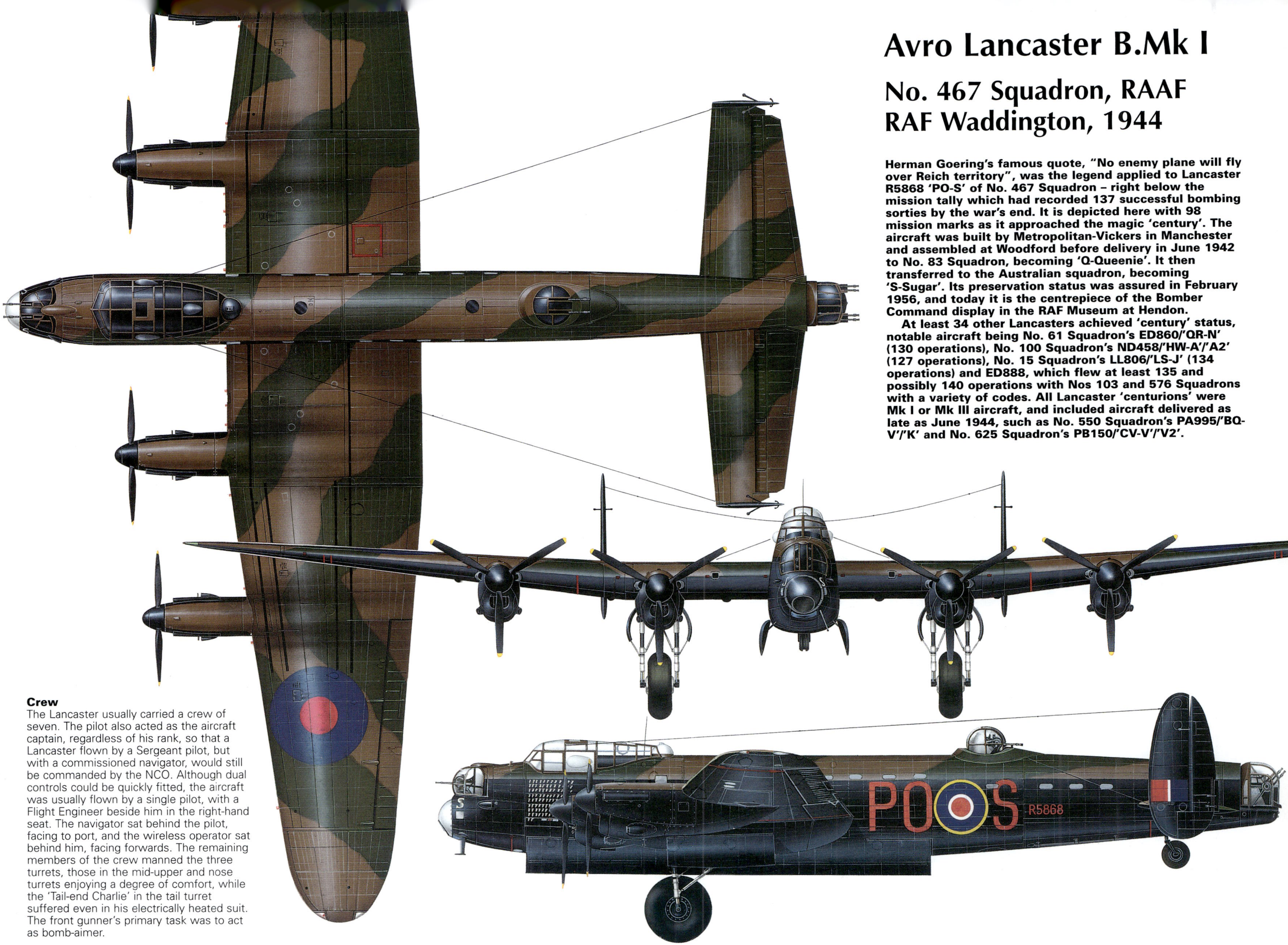

Crew
The Lancaster usually carried a crew of seven. The pilot also acted as the aircraft captain, regardless of his rank, so that a Lancaster flown by a Sergeant pilot, but with a commissioned navigator, would still be commanded by the NCO. Although dual controls could be quickly fitted, the aircraft was usually flown by a single pilot, with a Flight Engineer beside him in the right-hand seat. The navigator sat behind the pilot, facing to port, and the wireless operator sat behind him, facing forwards. The remaining members of the crew manned the three turrets, those in the mid-upper and nose turrets enjoying a degree of comfort, while the 'Tail-end Charlie' in the tail turret suffered even in his electrically heated suit. The front gunner's primary task was to act as bomb-aimer.

Flight Refuelling Ltd

One of the most important experiments for which the Lancaster was used was the development of inflight refuelling by FRL's fleet of converted aircraft. Lancasters were first assigned to Staverton for trials in May 1944, leading to the development of a Lancaster tanker/receiver system. The tanker deployed a hose/drogue unit through a ventral hatch, which was snagged by the receiver and winched in to mate with a receptacle in the tail. On 24 April 1949 a Meteor Mk III equipped with a nose probe refuelled from a Lancaster, the start of what has become the RAF's standard refuelling method to this day.

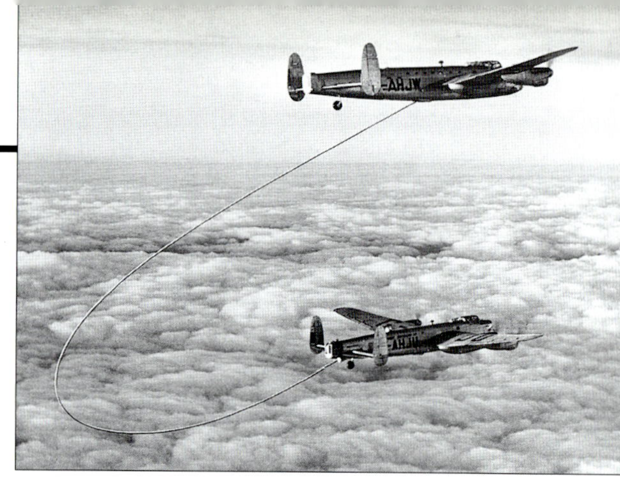

Above and top right: Between May and August 1947 two of Flight Refuelling's Lancasters (G-AHJW, tanker and G-AHJV, receiver) undertook an inflight-refuelled London-Bermuda service in trials with British South American Airways, which was a major civil Lancaster operator. FRL's Lancasters were heavily involved in the Berlin Airlift, ferrying in fuel for the beleaguered city.

extraordinary accuracy in attacking the target is reported."

Finally, the 'exchange rate' of crews lost compared to damage inflicted was transformed, so that bomber and bomber crew losses were negligible compared to the horrific destruction and carnage which they caused.

Nonetheless, many analysts now believe that the way in which Harris used his heavy bombers was crude, profligate and extravagant, and that many of Bomber Command's 55,000 dead were squandered in a strategic bombing campaign that was both wasteful and largely ineffective. Few believe that the pinpoint bombing of individual factories and economic targets would ever "destroy the foundations of the German war machine", or undermine the morale of the German people, and although the bombing offensive did cause enormous disruption and diverted resources from the front, it was an extremely expensive way of achieving a relatively limited effect.

Had the four-engined heavies been used more in tactical operations like those which led up to and accompanied D-Day, and in the war against the U-boat menace, losses would have been much lighter and damage to the enemy war machine may have been much more effective.

Many observers believe that the strategic bombing campaign should have been left until the Luftwaffe had been defeated. The bomber apologists' argument was that heavy bombers were the best (or even the only) way of carrying the war to the enemy during the dark days of 1940-1943, but this does not bear close scrutiny. Everything achieved by the heavy bombers up to the end of 1943 could have been achieved at less cost by fighter-bombers and faster, more survivable light bombers. Such aircraft were always more likely to hit their targets than the four-engined heavies, rendering their relatively small bombloads more effective because they actually hit their intended targets more often.

The Luftwaffe's chief of staff, General Jeschonnek, was moved to remark, "Every four-engined bomber the Western Allies build makes me happy, for we will bring these down just as we brought down the two-engined ones, and the destruction of a four-engined bomber constitutes a much greater loss to the enemy." It is hard to imagine that Jeschonnek would have been so calmly accepting had the industrial capacity and manpower spent on the casually-expended Lancasters instead been used to build and deploy much larger fleets of aircraft like the Westland Whirlwind and de Havilland Mosquito. Had the RAF been launching nightly raids by hundreds (or even thousands) of Mosquitoes and Whirlwind fighter-bombers, Germany's defences would have been overwhelmed. Far greater tonnages of bombs would almost certainly have hit their targets, and larger numbers of aircrew and aircraft would have returned safely home to hit the enemy again.

With hindsight it can be seen that the strategic air campaign against the Third Reich could have been fought more effectively, causing greater damage and costing fewer casualties. It remains uncertain, though, whether any commander of Harris's background and age could have embraced the kind of innovative thinking that would have been required for such a different bomber war to have been fought. For the bomber war that the RAF actually waged, the Lancaster was probably the ideal tool: relatively slow, but with a substantial defensive armament and a heavy bombload, the Lancaster was a supremely effective bludgeon, while de Havilland's unarmed, high-speed 'Wooden Wonder' was a futuristic rapier.

Peaceful and peacetime missions

Even before the shooting stopped, the Lancaster began flying intensive food-dropping missions to the starving Dutch under Operation Manna. Some 17 Lancaster squadrons eventually flew 3,156 sorties between 28 April and 10 May, dropping 6,684 tons of food. Although large areas of Holland were still under German control, a truce was arranged with the local Wehrmacht commander and the Lancasters operated unmolested. For food dropping, the Lancasters made low-altitude deliveries (from 200 to 500 ft/61 to 152 m), dropping food from five panniers in the bomb bay at a speed of 110 to 120 kt (203 to 222 km/h), using half flap.

From 2 May, Bomber Command Lancasters were stripped and fitted with rudimentary seating for 25 passengers and began flying Operation Exodus, ferrying home PoWs from Europe. In a single 24-day period the Lancasters flew 2,900 round trips, bringing home 74,000 former PoWs, suffering only a single loss in the process. Lancasters were

In British, French and Canadian hands the Lancaster had a long post-war career, with the accent on maritime and survey operations. The RCAF developed a number of special variants, including the Lancaster 10SAR for search and rescue duties (below). The last was retired in 1964, the same year the French ended Lancaster operations.

France's last Lancasters served with Escadrille 9S in New Caledonia. Late on in their careers they received an all-white scheme, as displayed by this aircraft visiting Australia.

***Left:* Wearing British civil registration G-ASXX, Escadrille 9S's WU15 (ex-NX611) was put up for preservation on its retirement and flew home to Britain in May 1965. During its departure from Butterworth it was joined by an RAAF Canberra and an RAF Victor. Today it resides at East Kirkby, maintained in taxiing condition.**

later used for ferrying home soldiers and other personnel from Italy and the Mediterranean.

Before the war in the Far East was brought to a close by the atom bombs dropped on Hiroshima and Nagasaki, it was expected that Britain's Tiger Force of heavy bombers would be deployed to support a long invasion campaign. Tiger Force consisted of three 22-squadron groups (one RAF, one RCAF and one composite RAF/RAAF/RNZAF/SAAF) each including 12 bomber squadrons (later reduced to eight). They were to have been equipped with the Lancaster B.Mk I (FE) and the B.Mk III (FE), which traded the mid-upper turrets for increased fuel capacity, and the B.Mk VII which had a Rose tail turret and a new Martin mid-upper turret, each with twin 0.50-in (1.27-cm) machine-guns. To maintain the correct centre of gravity, the new Martin mid-upper turret was located farther forward than the old turret.

Bomber Command shrank dramatically in the immediate aftermath of the war, but the Lancaster remained its mainstay. Nos 5 and 8 Groups simply disbanded, surplus to the RAF's requirements, No. 4 Group transferred to Transport Command, and No. 6 Group's squadrons returned to Canada. The Lancaster bomber soon gave way to the Avro Lincoln. The last front-line bomber unit, No. 49 Squadron, converted to the new type in March 1950.

The numerical dominance of the Lancaster, and its close relationship to the new Lincoln, had ensured that it was retained post-war rather than the Halifax, which would arguably have been better suited to the roles which the Lancaster fulfilled after its withdrawal from the bomber role.

A small number of aircraft were converted as PR.Mk Is, with defensive armament removed and a camera crate in the bomb bay, and these served in the survey role with Nos 82, 541 and 683 Squadrons until December 1953. Larger numbers of aircraft were converted to ASR.Mk III standards, carrying an Airborne Lifeboat Mk II or Mk IIA below the fuselage, and equipped Nos 179, 203, 210, 279, and 621 Squadrons. Most were converted to GR.Mk 3 standard, with ASV radar in place of H_2S, to serve in the broader maritime reconnaissance role (still with a search and rescue commitment, and still capable of carrying an airborne lifeboat). These aircraft served in the air-sea rescue and maritime reconnaissance roles until October 1956, when they finally gave way to the Avro Shackleton.

On 15 October 1956 the RAF bid a final farewell to the Lancaster when GR.Mk 3 RF325 was flown by Wing Commander Brooks from St Mawgan to Wroughton, where it went into storage. The final user of the Lancaster was the School of Maritime Reconnaissance, which had converted to the Shackleton.

Elsewhere, Lancasters remained even longer. The Royal Canadian Air Force retained a handful of aircraft in the SAR and Arctic surveillance roles until 1964, and the French Aéronavale kept a small number of aircraft active in the air-sea rescue and maritime reconnaissance roles in the South Pacific until the same year.

Avro and its industrial partners eventually completed 7,377 Lancasters, a smaller total than was achieved by the less well known Wellington but more than any of its rivals. Of this number, about 24 survive today, two of which are presently airworthy, one in Canada and one flying with the Battle of Britain Memorial Flight at RAF Coningsby in Lincolnshire. Another (privately owned) Lancaster is maintained in taxiing condition at the wartime aerodrome of East Kirkby, owned by a pair of local farmers, the Panton brothers.

Jon Lake

***Below:* C-GVRA/FM213 is one of only two airworthy Lancasters. Based in Canada, it wears the markings of KB726 of No. 419 Squadron.**

Preserved 'Lancs'

Including the two airworthy examples (the RAF's PA474, left, and the Canadian Warplane Heritage's FM213), there are 16 complete Lancasters on display. Of these, eight are in Canada, four are in the UK, two are in Australia and one each in the US and New Zealand. Remains of a further eight aircraft are either on display or undergoing restoration, including one of the ex-Aéronavale aircraft which is being restored to full display condition at the Musée de l'Air at le Bourget. The 24 aircraft above comprise six Mk Is, four Mk VIIs (all ex- Escadrille 9S in New Caledonia) and 14 Canadian-built Mk Xs. The most historically significant aircraft is the RAF Museum's R5868 which, with Nos 83 and 467 Squadrons, flew what was probably the highest number of wartime missions (137) by any Lancaster.

Lancaster operators

No. 207 Squadron was the RAF's third Lancaster squadron and, like most units to convert to the type, operated it for the duration of the war. In 1949 it was one of the last units to convert to the Lincoln, albeit briefly before gaining Boeing Washingtons.

Royal Air Force – Wartime Units

The Lancaster won the unswerving support of Bomber Command's AOC, and as a result, all production of the type went to meeting Bomber Command's needs. None was diverted to other Commands, nor to overseas-based bomber units. Expansion of the Lancaster force was thus extremely rapid, despite heavy losses. By April 1943, Bomber Command's front-line strength was already dominated by the Lancaster, with one group (No. 5) fully equipped, and another (No. 1) converting. The front line thus included 17 Lancaster squadrons, 11 with Halifaxes and seven with Stirlings, giving a total of 35 heavy bomber units with 682 aircraft. In mid-1943 Bomber Command also had 20 squadrons of light and medium bombers with a further 454 aircraft. Most of these 'Medium' bombers were Wellingtons, which could carry a 4,000-lb (1814-kg) bomb as far as Berlin – a heavier load carried further than was routinely carried by the USAAF's contemporary 'heavies'. The supposedly 'light' Mosquito gave the RAF a light bomber with the same payload/range capability as the Wellington. The Command's Wellingtons and Stirlings gave way to more Lancasters during 1943 and 1944.

By April 1945, Bomber Command's front-line strength was almost entirely dominated by the Lancaster, with three groups (Nos 1, 3 and 5) fully equipped, and another (No. 6) converting, and with the Lancaster serving alongside the Mosquito within No. 8 PFF Group.

The front line thus included 13 Lancaster squadrons with No. 1 Group, 12 with No. 3 Group, 17 with No. 5 Group, 11 with No. 6 Group (two of which were still converting) and six with the Pathfinders, making a total of 59 squadrons. Bomber Command's only other 'heavies' were the Halifaxes of No. 4 Group, totalling 11 squadrons, and two remaining Halifax units within No. 6 Group.

The table lists the wartime Lancaster squadrons and Heavy Conversion Units – those which continued after the end of 1945 are included in the post-war table.

Squadron	Base	Codes
No. 1 Group		
No. 12 Sqn	Wickenby	PH, 'C' Flt GZ
No. 100 Sqn	Waltham, Elsham Wolds, Scampton	HW
No. 101 Sqn	Holme on Spalding Moor, Ludford Magna	SR, 'C' Flt MW
No. 103 Sqn	Elsham Wolds	PM
No. 153 Sqn	Kirmington, Scampton	P4
No. 166 Sqn	Kirmington	AS
No. 170 Sqn	Kelstern, Dunholme Lodge	TC
No. 300 'Masovian Polish' Sqn	Faldingworth	BH
No. 460 Sqn, RAAF	Breighton, Binbrook	AR, UV?
No. 550 Sqn	Waltham, North Killingholme	BQ
No. 576 Sqn	Elsham Wolds, Fiskerton	UL
No. 625 Sqn	Kelstern, Scampton	CF
No. 626 Sqn	Wickenby	UM
No. 1 Lancaster Finishing School	Lindholme, Blyton, Faldingworth, Hemswell	3C
No. 3 Group		
No. XV Sqn	Mildenhall	LS, 'C' Flt DJ
No. 75 (NZ) Sqn	Mepal	AA, 'C' Flt JN
No. 90 Sqn	Tuddenham	WP, 'C' Flt XY
No. 115 Sqn	East Wrething, Little Snoring, Witchford	KO 'C' Flt A4 and IL
No. 138 Sqn	Tuddenham	AC, NF
No. 149 'East India' Sqn	Methwold	OJ, 'C' Flt TK
No. 186 Sqn	Tuddenham, Stradishall	XY, 'C' Flt AP
No. 195 Sqn	Witchford, Wratting Common	A4, 'C' Flt JE
No. 218 'Gold Coast' Sqn	Methwold, Chedburgh	HA, 'C' Flt XH
No. 514 Sqn	Foulsham, Waterbeach	JI, 'C' Flt A2
No. 622 Sqn	Mildenhall	GI
No. 3 Lancaster Finishing School	Feltwell	A5
No. 5 Group		
No. IX Sqn	Waddington, Bardney	WS
No. 44 'Rhodesia' Sqn	Waddington, Dunholme Lodge, Spilsby	KM
No. 49 Sqn	Scampton, Fiskerton, Fulbeck, Syerston	EA
No. 50 Sqn	Skellingthorpe, Swinderby, Skellingthorpe	VN
No. 57 Sqn	Scampton, East Kirkby	DX, 'C' Flt QT
No. 61 Sqn	Woolfox Lodge, Syerston, Skellingthorpe	QR
No. 83 Sqn	Scampton	OL (to No. 8 Gp)
No. 97 'Straits Settlements' Sqn	Coningsby	OF (to No. 8 Gp)
No. 106 Sqn	Coningsby, Syerston, Metheringham	ZN
No. 150 Sqn	Fiskerton, Hemswell	JN, IQ
No. 189 Sqn	Bardney, Fulbeck	CA
No. 207 Sqn	Bottesford, Langar, Spilsby	EM
No. 227 Sqn	Bardney, Balderton, Strubby	9J
No. 463 Sqn, RAAF	Waddington	PO, JO
No. 467 Sqn, RAAF	Scampton, Bottesford, Waddington	PO
No. 617 Sqn	Scampton, Coningsby, Woodhall Spa	AJ, KC, YJ
No. 619 Sqn	Woodhall Spa, Coningsby, Dunholme Lodge, Strubby	PG
No. 630 Sqn	East Kirkby	LE
No. 5 Lancaster Finishing School	Syerston	CE, RC
No. 6 Group (Canadian)		
No. 408 'Goose' Sqn, RCAF	Linton on Ouse	EQ
No. 419 'Moose' Sqn, RCAF	Middleton St George	VR
No. 420 'Snowy Owl' Sqn, RCAF	Tholthorpe	PT (non-operational)
No. 424 'Tiger' Sqn, RCAF	Skipton on Swale	QB
No. 425 'Alouette' Sqn, RCAF	Tholthorpe	KW (non-operational)
No. 426 'Thunderbird' Sqn, RCAF	Linton-on-Ouse	OW (to Halifax 5/44)
No. 427 'Lion' Sqn, RCAF	Leeming	ZL
No. 428 'Ghost' Sqn, RCAF	Middleton St George	NA
No. 429 Sqn, RCAF	Leeming	AL
No. 431 'Iroquois' Sqn, RCAF	RCAF Croft	SE
No. 432 'Leaside' Sqn, RCAF	East Moor	QO (to Halifax 2/44)
No. 433 'Porcupine' Sqn, RCAF	Skipton on Swale	BM
No. 434 'Bluenose' Sqn, RCAF	Croft	WL
No. 6 Lancaster Finishing School, RCAF	Ossington	
No. 8 Group (Pathfinder Force)		
No. 7 Sqn	Oakington	MG, 'C' Flt XU
No. 35 Sqn	Graveley	TL
No. 83 Sqn	Wyton, Coningsby	OL (back to 5 Gp 4/44)
No. 97 'Straits Settlements' Sqn	Bourn	OF (back to 5 Gp 4/44)
No. 156 Sqn	Warboys, Upwood	GT
No. 405 'Vancouver' Sqn, RCAF	Gransden Lodge	LQ
No. 582 Sqn	Little Staughton	6O
No. 635 Sqn	Downham Market	F2
No. 7 Group (heavy conversion)		
No. 1651 HCU	Woolfox Lodge	BS, QQ
No. 1653 HCU	Lindholme, North Luffenham	A3, H4, M9?
No. 1654 HCU	Swinderby, Wigsley	UG, JF
No. 1656 HCU	Lindholme	EK, BL
No. 1659 HCU	Topcliffe	FD, RV
No. 1660 HCU	Swinderby	TV, YW
No. 1661 HCU	Winthorpe	GP, KB
No. 1662 HCU	Blyton	KF, PE
No. 1664 HCU	Dishforth	DH/ZU?
No. 1666 HCU	Dalton, Wombleton	ND, QY
No. 1667 HCU	Lindholme, Faldingworth, Sandtoft	GG, LR
No. 1668 HCU	Balderton, Syerston, Bottesford, Cottesmore	J9, K2
No. 1678 HCF	East Wretham	SW
No. 1679 HCF	Foulsham, Waterbeach	
No. 1323 AGLT Flight	Warboys	QF
G-H Training Flight	Feltwell	

Royal Air Force – post-war

Many of the wartime Lancaster units remained in service briefly after the war, but those that disbanded before the end of 1945 are not included in the following list. In the home-based Bomber Command squadrons of Nos 1 and 3 Groups (No. 5 having disbanded), Lancasters gave way to Lincolns by the end of 1949, but the type served on in the maritime and survey roles. The RAF's last Lancaster retired from the School of Maritime Reconnaissance at St Mawgan in October 1956, when it was finally replaced by an Avro Shackleton.

Above: No. 82 Squadron was formed post-war to conduct aerial survey work. Its Lancaster PR.Mk Is were used from 1947 to 1952 to conduct a thorough survey of 'British' Africa from Eastleigh in Kenya, and Takoradi on the Gold Coast (now Ghana).

Left: At the end of the war Coastal Command's main type was the Liberator, but these were swiftly returned to the US. Lancasters took up the strain after being converted to GR.Mk 3 standard. This aircraft served with No. 210 Squadron.

Bomber Command

Squadron	Base	Dates/codes
No. 1 Group		
No. IX Sqn	Waddington, Binbrook	to Lincoln 7/46
No. 12 Sqn	Binbrook, Waddington	to Lincoln 8/46
No. 50 Sqn	Sturgate, Waddington	to Lincoln 7/46
No. 57 Sqn	East Kirkby, Elsham Wolds, Scampton	to Lincoln 4/46
No. 61 Sqn	Sturgate, Waddington	to Lincoln 5/46
No. 83 Sqn	Coningsby	to Lincoln 7/46
No. 97 Sqn	Coningsby	to Lincoln 7/46
No. 100 Sqn	Elsham Wolds, Scampton	to Lincoln 5/46
No. 101 Sqn	Ludford Magna, Binbrook	to Lincoln 8/46
No. 106 Sqn	Metheringham	disbanded 18/2/46
No. 617 Sqn	Waddington, Binbrook	to Lincoln 9/46
No. 3 Group		
No. 7 Sqn	Mepal, Upwood	to Lincoln 8/49
No. 15 Sqn	Mildenhall	to Lincoln 2/47
No. 35 Sqn	Stradishall	to Lincoln 2/47
No. 49 Sqn	Mepal, Upwood	to Lincoln 10/49
No. 90 Sqn	Tuddenham/Wyton	to Lincoln 4/47
No. 115 Sqn	Graveley/Stradishall	to Lincoln 9/49
No. 138 Sqn	Tuddenham/Wyton	to Lincoln 9/47
No. 148 Sqn	Upwood	from 11/46-1/50
No. 149 Sqn	Methwold, Tuddenham, Stradishall, Mildenhall	to Lincoln 10/49
No. 207 Sqn	Methwold, Tuddenham, Stradishall, Mildenhall	to Lincoln 3/49
No. 214 Sqn	Upwood	to Lincoln 2/50
No. 230 OCU	Lindholme	A3 and SN

Survey and reconnaissance

No. 82 Sqn	Benson, Leuchars, Eastleigh, Takoradi, Wyton	from 10/46, disbanded 12/53
No. 541 Sqn	Benson	from 6/46, re-equipped 9/46
No. 683 Sqn	Fayid, Kabrit, Eastleigh, Khormaksar, Habbaniya	from 11/50, disbanded 11/53

Coastal Command

No. 120 Sqn	Leuchars, Kinloss	11/46 to 4/51 BS
No. 160 Sqn	Leuchars	8/46 to 9/46
No. 179 Sqn	St Eval	2/46 to 9/46 OZ
No. 203 Sqn	Leuchars, St Eval, St Mawgan, Topcliffe	8/46 to 3/53 CJ
No. 210 Sqn	St Eval, St Mawgan, Topcliffe	6/46 to 10/52 OZ
No. 224 Sqn	St Eval	10/46 to 11/47 XB
No. 279 Sqn	Beccles	9/45 to 3/46
No. 1348 ASR Flight	Pegu, Burma	2/46 to 5/46
No. 236 OCU	Kinloss	from 7/47 K7
School of Maritime Reconnaissance	St Mawgan	1/51 to 9/56

Egypt/Palestine-based bomber and GR/ASR*

No. 18 Sqn*	Ein Shemer	from 9/46, disbanded 9/46
No. 37 Sqn	Fayid, Kabrit, Shallufa	from 4/46, disbanded 1/4/47
No. 38 Sqn*	Luqa, Ein Shemer, Shallufa, Ramat David	from 4/46, re-equipped 12/53
No. 40 Sqn	Abu Sueir, Shallufa	from 1/46, disbanded 1/4/47
No. 70 Sqn	Fayid, Kabrit, Shallufa	from 4/46, disbanded 1/4/47
No. 104 Sqn	Abu Sueir	from 11/45, disbanded 1/4/47
No. 178 Sqn	Ein Shemer, Fayid	from 11/45, became 70 Sqn
No. 621 Sqn*	Aqir, Ein Shemer	from 4/46, disbanded 9/46

Fuerza Aérea Argentina

Argentina obtained 15 surplus Lancaster B.Mk Is (B-031 to B-045) between May 1948 and January 1949 to augment 30 Lincolns, replacing ageing Martin 139s. The aircraft served until 16 April 1951, when they were withdrawn for storage. They returned to service in April 1952. A handful of aircraft were converted for the transport role (as TC-034, T-038 and T-040) but the remainder served on in diminishing numbers until 1965.

II Grupo Bombardeo, III Brigada Aérea – Reconquista
Paratroop Training School – Cordoba

Lancasters were supplied to Argentina alongside Lincolns, and had a brief career in the bombing role. This aircraft (B-034) was one of three which were later converted for transport duties.

Royal Australian Air Force

Aircrew from the RAAF played a major role in the wartime RAF Bomber Command offensive, including the full manning of three Lancaster squadrons. The aircraft of the RAAF squadrons remained under Bomber Command control, and wore RAF serials. However, two aircraft were sent to Australia. The first was ED930, which flew to Australia in May 1943 and later became A66-1 (illustrated). The second aircraft was W4783, which had flown combat with No. 460 Squadron before being sent to Australia on a War Bonds drive in October 1944. It was re-registered as A66-2, and is now on display in the Australian War Museum in Canberra. Other Lancaster airframes were delivered to Australia to act as targets for weapons tests.

Starfighter in Vietnam

Lockheed's Mach 2 Starfighter deployed twice to Southeast Asia. In terms of recordable achievements – such as aerial victories or major bombing strikes – its combat career was unspectacular, and history has largely passed over the type's contribution to the war in Vietnam. Yet, in the mid-1960s, the mere appearance of the Starfighter in the war zone was enough to keep North Vietnamese MiGs firmly on the ground, from where they could do no harm to the USAF's fighter-bombers and high-value command posts.

Two F-104Cs from the 476th TFS rest in their revetments at Da Nang some time during the Starfighter's first deployment to Southeast Asia in 1965. The F-104 retained a polished natural metal finish throughout the first phase of its Vietnam combat career.

Conceived in the famous Lockheed Skunk Works during the early 1950s, the XF-104 first flew in February 1954, only seven months after the Korean War had ended. After many modifications brought about by extensive testing, the first F-104As went into operational service with the USAF in early 1958. As the assembly line was just getting started, more changes were made to the aircraft, and a few months later the definitive F-104C began to appear. It was delivered to Tactical Air Command (TAC) in the fall of 1958. It was this model that would be called upon to carry out a variety of missions in the Vietnam War.

A pure air superiority fighter, the F-104A was built in only limited quantities (153). The 'C' model enabled the type to carry bombs, external fuel tanks or missiles on its fuselage or wing stations. Although only 77 of these models were delivered to TAC, it was this versatility that allowed it to remain in Vietnam as long as it did. In the hands of the 479th Tactical Fighter Wing's superbly trained Starfighter pilots, it carved out a credible reputation in the air-to-ground role.

There was a clear-cut reason for the US Air Force to rush the F-104 into Southeast Asia, and it was not politically motivated. Republic F-105s were shouldering a heavy responsibility, tasked with taking the war all the way up into the Hanoi/Haiphong area. They were constantly forced to defend themselves against MiG-17s, which the North Vietnamese flew in relatively large numbers. F-105 strikes were co-ordinated by airborne control aircraft (EC-121s) known as AWACS. They usually set up station over the Gulf of Tonkin and were considered to be amongst the USAF's most highly valued assets committed to the war.

In an effort to counter the MiG threat, the weapon of choice was to be the Lockheed F-104C. The communists' knowledge of this aircraft, beforehand, would prove to be deterrent enough once the first squadron was deployed from George AFB, California, to Da Nang AB in Vietnam. The deployment of F-104s

An F-104C of the 476th TFS arrives at Da Nang in April 1965 after flying in from Kung Kuan AB in Taiwan. The latter base was to be the main support location for the Vietnam deployment, performing major servicing and holding spare aircraft.

F-104 Vietnam deployments

476th TFS	19 April 1965 to 11 July 1965 (Da Nang)
436th TFS	12 July 1965 to 14 October 1965 (Da Nang)
435th TFS	14 October 1965 to 21 November 1965 (Da Nang)
435th TFS	6 June 1966 to 19 July 1967 (Udorn)*

** During the final deployment, the 435th TFS was under the control of the 8th TFW at Udorn RTAB. Its parent 479th TFW was in the process of converting to the F-4 Phantom and was no longer in the F-104 business.*

Above and right: An F-104D two-seater leads four F-104Cs during the initial trans-Pacific deployment in 1965. The aircraft were flying the Hawaii-Guam sector.

Below: Captain Mike Korte prepares to start the engine before a high-value asset escort mission from Da Nang. Such missions typically lasted from three to five hours.

was triggered by the Rolling Thunder strike campaign, which had started in March 1965 and which had escalated the war far beyond what some planners had predicted. The number of bombing sorties by F-105s increased and, at the same time, so did the resistance thrown up by North Vietnam, in the form of the MiG-17 and the soon-to-appear MiG-21 fighters. These two aircraft had a distinct tactical advantage because they were being controlled by ground radar controllers. This allowed them to be directed into a most advantageous position to attack the 'Thuds' before engaging.

In 1965, US military strategy in Vietnam was to bomb the North into giving up its quest for South Vietnam. In order to make this work, the bombers had to be protected. In its F-8 Crusaders, the Navy had what was considered to be the most effective dogfighter, while the USAF initially used F-100s as escorts. It also brought in some F-102 Delta Dagger interceptors from the Philippines to bolster the air defences. These two fighters could generally hold their own against the MiG-17s, but the appearance of the MiG-21 had put complete air supremacy at risk.

Request for Starfighters

On 4 April 1965 a decisive event occurred when a group of 'Thuds', escorted by F-100s, was attacking the Thanh Hoa Bridge. Suddenly, an element of two MiG-17s appeared out of nowhere and attacked the bombers, shooting down two F-105s and causing major disruption to the entire mission. An earlier request for F-104s to relieve the overworked F-100 escorts had been rejected, so it took an incident of this magnitude to get the Starfighter deployment rolling. Another factor involved was the increased use of the EC-121D College Eye airborne command centres. These were crucial to directing the bombers once they got into the heavily defended areas. The Air Force wanted a high-Mach interceptor to provide protection for the EC-121Ds, and the Starfighter would provide this service.

The initial deployment of 28 (24 primary aircraft and four spares) F-104Cs took off from George AFB, California, on 7 April. Two of the fighters had to turn back to George because one had a problem with its tip tanks. The first leg took the squadron to Hickam AFB, Hawaii. Two more aircraft aborted after take-off from there, so a total of 24 made the long flight over to Kung Kuan AB on Taiwan, which would be their main operational base for deployments into Da Nang AB. This original deployment by

Above: Neither the F-104 or the F-102 Delta Dagger received much press coverage in Vietnam. While the F-104 was mainly used as an escort and attacker, the F-102 provided air defence, standing alert at a number of bases in Thailand and Vietnam. These aircraft are at Da Nang in May 1965.

Left: This F-104C, seen during the 1965 deployment to Da Nang, is completely 'clean'. The wingtip stations were used to mount either AIM-9B Sidewinder missiles or supersonic fuel tanks.

the 476th Tactical Fighter Squadron was part of a large jet build-up in South Vietnam known as Operation Two Buck. The parent wing (479th TFW) retained two F-104 squadrons (435th and 436th TFS) back at George AFB.

Once the squadron had settled in on Taiwan, the decision was made to deploy 14 Starfighters to Da Nang for combat operations, while the remaining 10 would remain at Kung Kuan AB. From there they could be rotated into the combat zone, and the high-timers could periodically return to Taiwan where major maintenance would be carried out. Needless to say, the unusually high demand put on the 14 deployed airframes kept the maintenance personnel at Da Nang extremely busy. It got to the point that some four to six of the F-104s had to be rotated back to Taiwan every 10 days for more serious maintenance and, in turn, fresh aircraft would be flown into Da Nang to fill the gap. Lockheed Technical Representative Benton McAvoy stated that "a seven-day work schedule had enabled the squadron to fly 12 aircraft (three flights of four aircraft each) as many as 90 hours a day in support of the mission. Squadron crew chiefs, maintenance and munitions personnel were working in shifts around the clock."

There was never any question as to why the Lockheed Mach 2 fighter was brought over to Vietnam – to maintain air superiority over anything thrown up by the North Vietnamese. This was accomplished by numerous escort missions and MiGCAPs (MiG combat air patrols).

Protecting the College Eye

Statistics recorded by the 476th TFS state that slightly over 50 per cent of the missions it flew on its deployment were to protect the EC-121s. These were not easy missions for the pilots because they usually lasted up to five hours in a cramped cockpit. Even though the F-104s were carrying external fuel tanks, they still had short legs on missions of this length. They usually remained close to the KC-135 tankers, and on some of the missions two tankers accompanied them. This allowed three flights, each of four fighters, to have the stamina to remain on station to counter any MiG opposition.

For obvious reasons, the MiGs were absent during these missions. NVAF intelligence knew that the Starfighters were in-theatre and their ground control intercept (GCI) operators also knew when they were airborne. For good reason, the MiGs did not come in close and,

Close encounters with Chinese MiGs

Captain Thomas 'Sharkbait' Delashaw, 476th Tactical Fighter Squadron

The F-104's high speed and simplicity of systems allowed it to reach targets 250 nautical miles [463 km] from Da Nang within 40 minutes of alert – and this included the 10 minutes required for the pilot to travel the half-mile to his aircraft. We were over there strictly to counter any efforts by the MiGs to intercept our bombers or the EC-121s. The first close encounter involved a pair of our Starfighters which were vectored after a MiG-21 that had just departed Hainan Island. Directed by Red Crown, the two '104s engaged in a supersonic chase over North Vietnam before the MiG ran across the Chinese border. One of our pilots, Captain George Wells, related to us how his flight was rapidly closing on the MiG at Mach 1.4 when they entered the buffer zone and were forced to turn back before crossing the Chinese border.

The second close encounter occurred during the return from a MiGCAP mission approximately 30-35 miles [48-56 km] south of Hanoi. Four F-104s were proceeding back to Da Nang, low on fuel, when a People's Republic of China J-6 popped out of the undercast only about a mile in front of the flight, facing away. Before any of the four pilots could react, the J-6 lit both of its afterburners (confirming it as a 'Farmer') and dove into the clouds. In my opinion, the J-6 had been under GCI control and had been warned of the F-104s approach by ground radar.

The reputation of the 479th Tactical Fighter Wing was well known even before the introduction of the F-104C. Somewhat akin to the US Navy's F-8 Crusader community, the 479th continued to regularly develop ACM skills even when the practice became politically incorrect in the late 1950s and early 1960s. Furthermore, the 479th was the first USAF wing to adopt Riccioni's 'Double-Attack' tactics, which proved ideally suited to the F-104 and were instrumental in the performance of the 479th's F-104s against other aircraft in ACM practice.

By 1962, the Starfighter had established a reputation as almost unbeatable in ACM. This reputation was later justified during the USAF's Project Featherstone evaluations, the US Navy's F-4/F-104 manoeuvring target testing and numerous F-104 wildcard appearances at the USAF-FWS. Notably, the Zipper's ACM capability was assessed by the USAF as "superior to all other aircraft evaluated at altitudes below 20,000 ft [6096 m]."

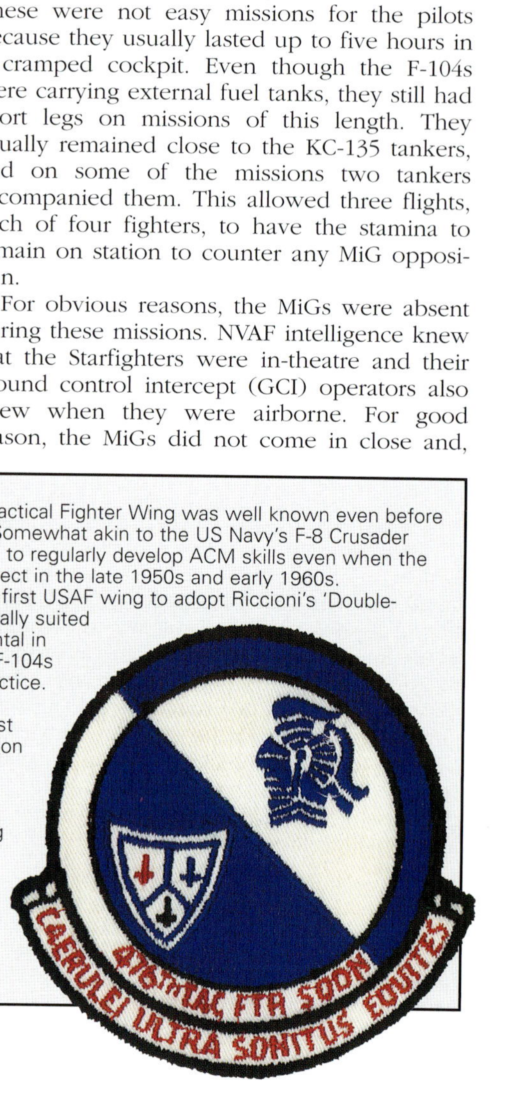

therefore, there were never any recorded dogfights between the F-104 and MiG-21. This was not due to the fact that the hostile defenders were too far away to be effective, because the barrier patrols were always situated about 225 to 275 miles (362 to 442 km) NNW of Da Nang, and their on-station times varied from about 40 minutes all the way up to an hour and a half.

Official AF records state that when the F-104s were in place and covering for the strikers, the MiGs avoided any contact or interference with the formations. This also included the People's Republic of China MiGs that would have had easy access to the EC-121s due to their close proximity to Hainan Island. Up until the time of the Starfighter's first appearances over the Gulf of Tonkin, the Chinese MiGs had frequently launched numerous harassment sorties in their direction. This situation became very frustrating to the 476th pilots that had hoped to test their skills against the MiG.

Based on the publicity that the F-104 had received since going operational, and the fact that most of the attention was focused on its blinding speed, any potential adversary would naturally be wary of a close-in encounter with the fighter. But this trait was not the only factor that made it a deadly opponent. It was very difficult for an enemy pilot to detect, especially when it was rapidly closing in from the six o'clock position. To gunners on the ground, it did not present much of a target and its speed made it very difficult to track. However, it was the weapons that the aircraft carried that made it such a threat to the MiG-17, and eventually to the MiG-21. AIM-9B Sidewinders gave it killing power from a distance and the M61 cannon gave it credibility in close. The 20-mm weapon was highly accurate and this was insured by its radar-controlled gyro-computing gunsight. These two weapons covered the engagement range from several miles down to approximately 600 ft (183 m).

Squadron records state that during the 476th's entire deployment, there were only two occasions when the proximity of the two opposing fighters came close, but a fight never developed. The absence of MiG activity gave reason for PACAF to re-evaluate the Vietnam mission for the F-104, because its primary goal of MiG suppression had already been achieved without firing a shot. The options for the 476th were limited as they were approaching the end of their deployment. However, rather than be sent back to the US early, they were tasked with flying weather reconnaissance and air-to-ground sorties. They quickly became proficient at both, even though they had been trained in air superiority.

Above: 435th TFS F-104Cs take a break at Clark AFB in the Philippines on their way back to George. When the F-104s were withdrawn, MiG activity increased.

Right: Aerial refuelling was vital in the deployment of F-104s – and on most missions. Flown by Morgan Lilly, this F-104 suffered an unusual accident when the drogue broke off, although thankfully the aircraft had enough fuel on board to reach its destination safely.

Weather reconnaissance

Due to their speed, the F-104Cs proved to be very effective in the weather recce role. They were sent up into North Vietnam on these missions and they covered the general area that housed the true targets for that day's bombing missions. They were able to relay accurate weather data back to the planners without really giving away the actual target, thus keeping the enemy from bringing in more Triple-A around the suspected target. Normally, these missions required two Starfighters.

Ground attack was the other assignment taken on by the 476th. Squadron pilots quickly became proficient in delivering various types of ordnance in the close support role and airborne forward air controllers gave them high marks. In some tight situations, the F-104s were called in because of their quick reaction time. However, these missions were not flown deep into North Vietnam due to several reasons. Only 21 close air support (CAS) missions were executed against targets over the north, while the majority of these were flown in-country.

On 29 June 1965, one of the squadron's F-104s went down while on a CAS sortie. The squadron report for the loss of F-104C (56-0937) ran as follows: *On 29 June at 1505 hours, one of the squadron's aircraft was shot down by enemy ground fire. The location was approximately 100 km [62 miles] SSW of Da Nang AB in South Vietnam. The aircraft, piloted by Captain Richard Cole, was on a dive bomb and strafe mission with two 750-lb [340-kg] GP bombs and 700 rounds of 20-mm.*

435th TFS F-104Cs share the Da Nang flightline in December 1965 with an F-105 Thunderchief. The 'Thud' was the USAF's primary vehicle for fighter-bomber missions 'up north', and was protected by F-104s during strikes in the early part of the Rolling Thunder campaign against North Vietnam.

Above: F-104Cs from the 435th TFS are seen on the ramp at Kung Kuan AB in January 1966, shortly after the return of the combat detachment from Da Nang. During their deployment MiG activity had dropped to virtually nothing, but would increase again after their departure.

The main criticism voiced over the F-104 was its short range. During overwater escort missions and deployments the Starfighter pilots rarely strayed far from the tanker. The KC-135 employed a short hose/drogue attachment to the boom for refuelling probe-equipped fighters like the F-104.

After identification pass on the targeted area, the aircraft gained altitude and began to roll in on its bombing dive. At this time, both hydraulic systems indicated zero pressure and the pilot had no control of the aircraft. The pilot immediately ejected and was picked up by a chopper seven minutes after he hit the ground.

Squadron rotation

The 476th TFS's deployment to Vietnam ended on 11 July 1965 when the unit was relieved by its sister squadron, the 436th TFS. Statistics compiled by the 476th showed that it had completed 96 days of combat duty and flown 1,182 combat sorties. Over one half (52 per cent) of these were in escorting the EC-121 aircraft, 24 per cent were MiG Screen missions that protected the bombers from being attacked by enemy interceptors. Some of the most effective missions were in the ground attack mode and these totalled 18 per cent, while the remaining 6 per cent was logged in the weather reconnaissance role. Perhaps, the most impressive figure to emerge was due to the tireless efforts of the maintenance personnel. The 476th had maintained an outstanding 94.7 per cent in-commission rate.

The 436th TFS relieved its sister squadron on 11 July 1965 and began flying missions the next day. As the MiG situation had changed, so had the types of missions that the F-104 was tasked with. The presence of the original cadre of Starfighters had reduced enemy interceptor activity down to a very low level. This meant that the number of EC-121 escort missions and MiGCAPs would decrease, but other types would move up in priority. Close air support became the main focus of the F-104's attention and they had already built up a reputation here, so there would be no dramatic transition. The new pilots quickly adapted to the change.

Two weeks after assuming Starfighter operations over Vietnam, the 436th was ordered to keep four of its aircraft on a 15-minute alert. The alert birds were fully loaded with ordnance and, if called upon by forward air controllers, they could be over a target in a very short time. Their quick response to support ground troops was already well known among the FACs. This

Captain Mike Korte of the 436th TFS poses in his freshly-camouflaged F-104C at George AFB. At this base the 479th TFW's primary role was nuclear strike with a single centreline weapon. This aircraft carries a practice bomb dispenser, which allowed pilots to rehearse flying the nuclear delivery profile.

By the time of the second phase of Starfighter deployments to the war zone the aircraft had adopted T.O. 1-1-4 tactical camouflage. This group is from the 435th TFS, seen at George AFB in October 1966 prior to a deployment to reinforce the Udorn detachment. The parent 479th TFW was converting to the F-4 at the time, and ended F-104 training in early 1967.

stance would be required all the way through the month of September and, at that time, the escort missions also began increasing in tempo. MiG activity was, once again, on the move. This did not eliminate the other assignments, but when the 90-day deployment was finished, the number of ground attack missions flown by the squadron accounted for 56 per cent of its total. This was a drastic change from the earlier tour by the 476th TFS.

Ground attack loss

Sleek and quick, the Starfighter had a good chance of survival in the ground attack mission, as it provided a much smaller target for the Triple-A gunners. Even with this advantage, they still ran a high risk, and on 22 July 1965 Captain Roy Blakely's aircraft was hit while at the bottom of a strafing pass. He attempted to land, gear up, at Chu Lai but his luck ran out. He was listed as KIA.

The squadron reported the loss of F-104C (56-0908) thus: *The aircraft had been in-country on a bombing/strafing mission when, after completing two dive bombing passes, the aircraft took hits on its first strafing pass and the left leading-edge flap separated from the plane. Captain Blakely's wingman confirmed this and also stated that there was a huge gash in the fuselage above the left wing. The pilot reported that his oil pressure was zero. The nozzle closure handle was pulled to lock the nozzle closed and the aircraft proceeded toward Chu Lai. Just prior to touchdown, the left wing dropped about 10°, causing it to touch down on the left side of the runway with the gear up. It skidded off the runway, hit a sand dune and exploded. The pilot was killed instantly. It was determined that the loss of the left leading-edge flap was caused by enemy ground fire and that, due to control difficulties, the pilot was unable to lower the landing gear prior to touchdown.*

It was during the month of September, on the 20th, that the squadron hit bottom on the bad luck chart. Captain Philip Smith was flying an escort mission out over the Gulf of Tonkin and, in an effort to rendezvous with his tanker, he became lost (a combination of erroneous steering commands from Da Nang and the tanker). He accidentally flew over Hainan Island and was shot down by a communist Chinese fighter. The day was not over yet – later that evening, two F-104s were returning from an aerial search for Captain Smith and, while penetrating the bad weather on approach to Da Nang, there was a collision between the two and both pilots had to eject. The only good news for the day was that both of the pilots that ejected were recovered in good shape. In a report turned in by Lockheed Tech Rep Ben McAvoy, he stated that the cause of the collision was due to failure of the navigation lights on both aircraft. At the time, the only lights operating on the aircraft were the ones on the sides of the intake ducts.

In mid-October 1965 the 436th TFS completed its tour, and although it had suffered some losses, the unit still made a good showing during the deployment. Records state that the squadron flew a total of 1,382 combat sorties, which translated into 3,116 flying hours. The wear and tear on the airframes was beginning to show, but due to the efforts of the ground crews, they still maintained an 88 per cent in-commission rate. Their air abort rate stood at 4.6 per cent and the ground abort at 0.45 per cent.

The final squadron

The third and final squadron that would deploy from George AFB and fly the 'Zipper' in Vietnam was the 435th TFS. It arrived on 14 October 1965 to replace the 436th. Once again, the new unit would have a change of assignments, and the first to be all but eliminated was close air support. As in the early days of the F-104's Vietnam experience when

8th TFW – the 'Wolfpack'

Above: Robin Olds's deputy was Colonel Daniel 'Chappie' James. He is seen here during a visit to Udorn with 435th TFS pilots in the back of a C-130.

Above: For its Udorn deployment the 435th TFS was assigned to the 8th TFW (itself headquartered at Ubon). Under the guidance of the legendary Colonel Robin Olds, the 'Wolfpack' rapidly became the premier air combat unit in the theatre, although its success was built almost exclusively on the Phantom. Nevertheless, the F-104 squadron provided useful back-up. Here Olds (left) converses with Major Hugh Spencer.

Right: Compared with some in SEA, the quarters of the F-104 troops at Udorn were sturdy and comfortable, although seasonal rains brought their own problems.

Escort and attack

Major Marvin D. Roupe, 435th Tactical Fighter Squadron Udorn RTAFB, July 1966

One of our tasks was to protect Command and Control aircraft, along with the Electronic Warfare types that were positioned over the Gulf of Tonkin. On those missions, we customarily patrolled as a flight of two along the coast of North Vietnam on an arc between the mainland and the Gulf. Usually, we refuelled a couple of times from a KC-135 and then, after being relieved by a fresh pair of fighters, we returned to Udorn. On one occasion, however, our relieving flight was just one aircraft rather than two, because the leader had been forced to abort due to mechanical problems. I sent my wingman home and picked up the reliever to make a flight of two.

After another series of refuelling, I was ready to go home, but the same thing happened again. I picked up the new wingman and stayed for another tour, only to find at the end that I must stay for yet another stint because of more mechanical problems with one of the F-104s. Finally, when a complete flight of two arrived for relief, I refuelled a final time and streaked for home (I lost track of the total number of times that I had to hit the tanker). When I landed, it was eight hours and 15 minutes after my initial take off. This was a record for a single flight in an F-104 up to that time, but a few weeks later, a squadron mate broke the record with a flight of 8 hours and 25 minutes!

Our little fighters could carry two 750-lb [340-kg] conventional bombs and many times we were assigned bombing missions against ground targets in the north. One day, just after my wingman and I established ourselves on a vector to our assigned target, the ground controller came on the radio to tell us we were being diverted to another, more critical target. It seemed the enemy had moved artillery and triple-A guns into position overnight and they had a unit of our ground troops pinned down in a valley. We immediately picked up our new vector and headed in to see if we could help. When we arrived on the scene, I could see right away that it would not be easy! The enemy was dug in on a hillside and our only access meant we would have to lay our bombs in on a downhill run which increased our chances for error. The Starfighter was a really stable bombing platform, however, and I elected to drop our bombs singly rather than both at once. This gave us two chances to get it right. I rolled in on the first run and put a '750' dead-on a bunker. Then my wingman came in and did the same thing. The return fire from ground positions indicated we still had some work to do. We lined up and came in for our final drops and we repeated the same precision as we had done on the initial run. Later, we found out that we had nearly wiped out all of the emplacements. In any case, our guys were able to come out and mop them all up for good. It was the single-most rewarding and satisfying mission of all the 100 that I flew over there.

the 476th initially brought its fighters over, the primary mission would revert to escorting the command and control EC-121Ds and C-130Es over the Tonkin Gulf. Both of these sophisticated aircraft were escorted separately, with the EC-121s requiring a four-ship escort and the Silver Dawn C-130Es requiring a pair. Squadron records show that this duty required the aircraft to be airborne for at least 10 hours each day. This was worked in shifts with ample support from the tankers.

The presence of F-4 Phantoms was ever increasing and it had become evident that it was a matter of time before they would replace the F-104s, not only in Vietnam but back home with the 479th TFW. On 21 November, all of the Starfighters were brought back to their operational base on Taiwan for re-deployment back to CONUS. All of the equipment and aircraft were finally moved back to George AFB by Christmas Day 1965. Sure enough, the F-4Cs from the 390th TFS took over all responsibility of escorting the EC-121s and C-130Es from their base at Da Nang. The F-104s had done their job all too well, because the MiG threat over the Gulf of Tonkin had all but disappeared and the concept of maintaining a single-mission aircraft in Southeast Asia had become too expensive.

Just a few months later (early 1966), MiG-17 appearances began to increase and, in addition, the MiG-21 began to show up in ever-increasing numbers, with the largest number being seen on 23 April. Three days later, two MiG-21s attacked a couple of F-4Cs that were escorting an EB-66C over North Vietnam. One of the MiG-21s was shot down with a missile and the other escaped without any damage to American aircraft. However, the writing was on the wall that the missiles-only Phantom might have trouble with the speedy little MiG-21.

Return to combat

A request was immediately sent through channels to once again bring in the F-104. It was approved, and on 6 June 1966 eight Starfighters from the 435th TFS landed at Udorn RTAB in Thailand. This time there was an upper management problem; the F-104 was in the process of being phased out of the active inventory from the sole fighter wing that operated it (479th), so the 435th had to become attached to PACAF's famous 8th TFW while it was in Southeast Asia.

This new deployment would see the 'Zippers' escorting F-105 Thunderchiefs over North Vietnam. These missions were shared with F-4s already assigned to the 8th Wing. This protective umbrella was, perhaps, too effective because no MiGs were encountered. The F-104s stayed close to the 'Thuds' because they, in turn, depended on them to provide warnings when SAM sites had locked on. The Starfighters were not equipped with RHAW gear as were the '105s.

Major Marvin D. Roupe, one of the F-104 pilots that brought their aircraft into Udorn, recalls that deployment, "When word came down that we were headed back to Vietnam, I was in the first group to leave. We ferried our fighters all the way to Thailand, using KC-135 tankers and making three layover stops along the way. Our maintenance crews had deployed ahead of us, so we quickly set up shop and began flying missions over the north. Our first ones were in support of fighter-bombers whose primary role was to attack ground targets. The MiGs had been a bother to them and it was our job to engage any hostile aircraft that got in the way. However, the MiG threat quickly waned when we arrived and we were soon assigned other roles."

Covering the Weasels

Early in the F-104's deployment to Udorn, the new F-105F Wild Weasel aircraft were brought over from the US. These SAM hunters were at home in a high-risk environment, but they needed protection from MiGs, especially when they were working up in the higher

Pussycat *flies over Udorn in 1967. The projection under the nose indicates that the aircraft had been fitted with APR-25/26 RHAWS equipment to provide the pilot with SAM warning. At the start of the year the 435th TFS participated in Operation Bolo, four flights of F-104Cs being assigned a back-up role to the Phantom force which shot down seven MiG-21s in a single fight.*

Above: Ben McAvoy, in the cockpit, was the Tech Rep for Lockheed assigned to the 435th, and was a key figure in maintaining a commendably high in-commission rate for the squadron.

Right: A 435th TFS F-104C refuels from a KC-135A in February 1967. With ECM equipment fitted the F-104 made a brief re-appearance over North Vietnam, although it was withdrawn after a few months.

Route Packages. At first, the F-4 was assigned the escort role, but it had difficulty keeping up with the Wild Weasels on their high-speed runs. Consequently, the F-104s were given the mission. This worked well because of the excellent speed offered by the Lockheed fighters. The use of F-104s gave the Weasel aircrews breathing room to complete their missions without the worry of being ambushed by GCI-guided MiG-21s. The demand was so great for this particular mission that another dozen Starfighters were sent to Udorn on 22 July to bolster the eight that were already there.

Life in the fast lane can be dangerous and two F-104s were lost to SAMs on 1 August 1966. Both pilots were killed and this triggered a major re-assessment of the Weasel escort mission. These losses amounted to 10 per cent of the deployed Starfighter combat force in Southeast Asia. PACAF's leadership determined that, with the speed and low altitudes at which the F-105F Weasels operated, the MiG threat was minimal. It was also reasoned that it was not wise to expose the F-104s to the kind of SAM threat that the Weasels were used to handling. Losing such a significant percentage of a single aircraft-type asset was not worth the risk. The decision was made to withdraw the 'Zipper' from escort service over North Vietnam

First to 100 combat missions

Captain Harold R. Alston, 435th Tactical Fighter Squadron
Udorn AFB, 30 September 1966

My final mission was on 30 September 1966, flown from our base at Udorn RTAFB. I led a flight of four F-104s on an armed reconnaissance mission that gave us the flexibility to attack targets of opportunity. This included targets that we might spot, or one being worked by a forward air controller. We had two A-1Es, which found and pinpointed a target for us. It was an ammunition dump/storage site with an opening in the side of a hill.

I was armed with finned napalm which allowed me to deliver the ordnance from a 30° dive. I rolled into my dive and kept the opening locked as I released my napalm. As I pulled off the target and climbed for altitude, the A-1 pilots reported that my nape scored a direct hit right into the opening, setting off numerous secondary explosions. Everyone was impressed, including me, as it was a great way to finish up my final combat mission.

I brought the flight back to Udorn and we pitched out at three-second intervals. I made a missed approach and came back to do a victory roll at 100 ft [30 m] after the rest of the flight was on the ground. Since I was the first pilot to log the magic '100', I got away with the roll without any jeopardy or threat from anyone. You could not survive something like that nowadays or even after the Vietnam War ended. I just did not ask or tell anyone what I was going to do and got away with it.

The reception I got after taxiing in was a surprise and very welcome. Colonel Ed Gaines poured the champagne. I am a non-drinker, so I posed for the pictures with my glass full, and then gave it to my crew chief. He later brought the glass to me and had me autograph it with a grease pencil – '100' and '30 September 1966'. Captain Mike Korte was also honoured at the party because he had recently flown the 435th Squadron's 1,000th combat mission. Since they had just built a new Officers Club at Udorn, I threw the first '100-mission' party in the new facility. I was so happy to survive and go home to my wife and kids that I did not mind drinking Coke while others got into the hard stuff. Drinks were cheap and it was a worthwhile $100 spent!

Later, when most of us were at Luke AFB training German F-104 pilots, Lockheed brought us to their facility in Burbank for a big banquet to honour all of the pilots that had flown their Starfighter in actual combat. Tony LeVier presented me with an engraved bronze tray acknowledging my being the first pilot in the USAF to fly 100 combat missions over North Vietnam. I still treasure it to this day.

'Hal' Alston's F-104 comes to a rest at Udorn at the end of his 100th combat mission over the north. A few days previously the 435th TFS had notched up its 1,000th combat mission.

This view shows the 435th TFS's parking area at Udorn in 1966, with 16 of the squadron's 18 F-104Cs being visible (including one with its tail removed in the far left revetment). At the time the base was very busy, and there were insufficient revetments for the Starfighters. Behind them are the RF-101C Voodoos of the 20th TRS, assigned to the main operating unit at Udorn (432nd Tactical Reconnaissance Wing).

until it could be fitted with ECM gear.

By late August, the role of the fast Lockheed fighters had been shifted to mainly ground attack over the lower Route Packages which covered targets in South Vietnam and Laos. This did not mean the end of tragic losses. On 1 September, Major Norman Schmidt was shot down (in 57-0913) by Triple-A while conducting a road recce mission over northern Laos, and he subsequently died while being held as a POW. Schmidt had flown 100 combat missions 14 years earlier over North Korea. A month later (on 2 October) Captain Norm Lockard was shot down by a SAM (in 56-0904) and was rescued. It was the loss of Captain Charles Toffieri on 20 October 1966 that spelled the death knell for the close air support mission for the F-104. He was flying the mission in 56-0918, and was shot down over northern Laos and listed as KIA. Captain Toffieri had won the William Tell competition in 1962 flying in a Starfighter. In early December, the remaining F-104s at Udorn RTAB were reduced back to escort missions only. From this date until the summer of 1967, most of the missions flown were of the escort type over the Gulf of Tonkin.

During the latter months of 1966, under Project Pronto, all of the 435th aircraft were fitted with the APR-25/26 RHAW gear. This upgrade enabled them to commence flying missions over North Vietnam again. Operation Bolo was one of the most publicised efforts by the F-4 force to lure out and destroy MiGs in the air. Being attached to the 8th TFW gave the F-104s a chance to participate, but it was not to live up to their initial expectations. They were never given a chance to go face-to-face with the MiGs; instead, they were assigned the

After experiencing undercarriage problems, Major Joe Nevers veered off the runway and on to the grass at Udorn, but thankfully with no damage to pilot or aircraft.

responsibility of protecting the F-4s as they egressed out of the hot areas.

The saga of the F-104 in a combat environment came to a close on 19 July 1967. At this time, all of the force was withdrawn from Thailand and it was replaced by the F-4D Phantoms of the 4th TFS. The Pentagon gave three reasons for the withdrawal: there was an urgent need to protect the dwindling F-104 assets in case the MiG threat re-appeared in Southeast Asia or elsewhere; the F-104s were rapidly being phased out of active USAF service; and their inability to carry heavy loads of ordnance hampered their effectiveness in the close air support mission.

Deployment assessment

Starfighter pilot Captain Tom 'Sharkbait' Delashaw and Lockheed Tech. Rep. Benton McAvoy compiled accurate records of the 435th TFS's deployments into Vietnam. Quoting records furnished by both individuals, "during the second deployment to Southeast Asia, the Starfighters from the 435th TFS had flown a total of 5,306 combat sorties, for a total of 14,393 combat flight hours. Due to increasing parts shortages and unrelenting sortie rates, aircraft in-commission rate dropped from a high of 85+ percent to a low of 62 percent. Nevertheless, despite its tired aircraft, the 435th maintained the reputation of the F-104 among the warriors in Southeast Asia. If the Starfighter is judged against other US aircraft for its ability to sustain battle damage, to deliver heavy bomb loads or to conduct operations in bad weather, then the F-104 rates as an 'also-ran'. If, however, the F-104C is judged for its ability to deter MiGs, to ensure the safety of the aircraft entrusted to its escort or to out-perform any aircraft in existence at the time, the 'Zip-4' is unrivalled!"

The final assignment for the 435th pilots in Southeast Asia was to fly their remaining F-104s to Puerto Rico by way of Guam. The long flight was made with stopovers at Clark AFB, Philippines, Andersen AFB, Guam, Hickam AFB Hawaii, and on to their destination. The ANG unit in Puerto Rico (198th TFS) flew the F-104s for many years after they had been removed from operational duties with the USAF. Records show that the PRANG operated the F-104s from mid-July 1967 until sometime in 1975. The 198th was the only ANG squadron to fly the tactical fighter version of the Starfighter.

Warren Thompson

Farewell, Starfighter – a KC-135 leads six 435th TFS aircraft as they leave George for the trip to a new career with the Puerto Rico ANG at San Juan. By the time the Vietnam F-104 deployment ended, the 479th TFW at George had already ceased F-104 operations.

Starfighter names

Right: *For much of their career the USAF's F-104Cs were relatively anonymous, carrying a serial, national insignia, a TAC badge on the fin and not much else. However, during the final deployment to Udorn in 1966/67 a fair number of the aircraft received nicknames. This aircraft, 56-0892, was named* **Miss Pevie J. VI** *and was flown by Major Floyd Trotten. The aircraft is photographed here from the Silver Dawn C-130E command post that it, and its un-named squadronmate, were escorting.*

Below: **Pussycat**, *56-0910, was the mount of Lieutenant Jim Trice. The cat was based on the 'Pink Panther' cartoon and was painted in pink. Posing here on the aircraft is one of the ever-popular USO troops, complete with flightsuit, squadron badge and non-standard footwear!*

Right: **Lil Poo II** *was one of the 435th TFS's F-104Cs. Note the RHAWS antenna under the nose and fixed refuelling probe.*

Above: **Debbie Sue** *was the personal aircraft of the 435th TFS's Captain Rod Trimble.*

Above: **Snoopy Sniper** *was one of the more elaborate artworks – it was assigned to Major Herb Drisko.*

Left: **Sex Machine** *was the Starfighter flown by Major Hugh Spencer. Names were often painted on the starboard side (rather than the more traditional port) as that was the side from which the pilot accessed the cockpit.*

Right: **Hellooo Dolly** *was one of the 435th aircraft. Here a group of crew chiefs poses with Lt Colonel Robert Preciado, the squadron commander (seated, centre). F-104 maintenance troops worked hard in Southeast Asia, at times keeping 12 aircraft flying up to 90 hours every day in hot and humid conditions.*

Left: *Lieutenant Addison 'Ace' Rawlins flew F-104C* **Miss Judy.**

Right: *The 435th TFS's Major Bobby Bedsworth was from Missouri – the 'Show Me' state which features a mule in its emblem. His aircraft was adorned with this artwork.*

Convair F2Y Sea Dart

Supersonic Seaplane

Few aircraft marked such a radical departure from the norm as the Sea Dart. Not only did it embrace the latest in aerodynamic fashion – in the form of the delta wing – but it also introduced an entirely new concept for operating from water.

Like a vision from the future, or the inspired doodlings of an imaginative schoolboy, the first Sea Dart stands proudly on its skis on the ramp at San Diego. The view highlights the simple and aerodynamically clean boat hull, and the hydro-ski installation with retracting front struts and retracting/extending main struts.

Throughout aviation history designers of seaplanes have walked a tightrope, caught between the conflicting requirements of airworthiness and seaworthiness. In the early years of heavier-than-air flight, a clever blending of aerodynamics and naval architecture had resulted in seaplanes that could rival and occasionally eclipse land-based aircraft, culminating in the Schneider Trophy racers of the late 1920s/early 1930s. However, in the 1930s, as aircraft steadily grew heavier and faster, a significant performance gap appeared between the seaplane and the landplane – one which would magnify over the coming years. Among the many factors which allowed the landplane to outstrip the seaplane was the development of much better runway facilities.

Waterborne aviation played a valuable part in World War II – large flying-boats ranged out over the oceans to monitor enemy shipping and to search for submarines. Smaller flying-boats and floatplanes flew coastal patrols, and provided limited fighter cover and spotting for warships. Many airmen and sailors were plucked from the seas by seaplanes. In regions where land bases were scarce or non-existent, or unsuitable for high-performance aircraft, the seaplane ruled. However, as the global conflict developed, so the seaplane was increasingly sidelined.

One of many facts harshly underscored by World War II was the growing disparity in performance between seaplanes and their land-based counterparts. Seaplanes could only operate truly effectively in areas where land-based opposition would not be encountered. That seaplanes achieved so much in such areas, especially in the early part of the war, was testament to the skill and bravery of their crews, although losses were heavy.

However, the inescapable fact remained that, given suitably equipped airfields, land-based aircraft were generally superior for any task that did not actually require the aircraft to land on water (such as SAR). For example, the float-equipped Nakajima A6M2-N fighter was around 70 mph (112 km/h) slower than the landplane

Mitsubishi A6M2 upon which it was based. On the same power of four 1,200-hp (895-kW) engines, a Consolidated PB4Y Liberator could carry a broadly similar load across a similar range as the PB2Y Coronado or Short Sunderland, but did so around 30-40 mph (48-64 km/h) faster. Indeed, versions of the Liberator came to dominate the long-range patrol task and did more than any type to close the crucial 'Atlantic gap' exploited by U-boats.

The reasons for such performance disparities were obvious to see: the floats and supporting struts of the floatplane were clearly heavier and draggier than the neat retractable undercarriage of land-based aircraft, while the flying-boat was just as its name suggested – a boat with wings. Water is around 800 times denser than air, and the traditional flying-boat hull had to be strong enough to cope with the battering it received when operating across waves at high speed, and this extra strength meant extra weight. Furthermore, the shape of the hull was not aerodynamically efficient, and by adding the weight penalties of the necessary marine accoutrements such as anchors, the resulting flying-boat was invariably heavy and cumbersome compared to a similarly powered landplane. Of all the wartime 'boats, only the Kawanishi H8K could rival land-based aircraft like the PB4Y in terms of performance, but it paid its own penalties by being over-large for the limited load it carried, and it needed very powerful engines.

Closing the gap

By the end of World War II, three new developments had emerged which, combined, might just allow a seaplane to be created with performance that equalled that of a land-based counterpart. These were the jet engine, a new generation of hull designs, and the hydro-ski, all of which would be brought together in the Convair F2Y Sea Dart.

In piston-powered seaplanes the propeller had to be mounted high above the water, well out of any spray effects. The jet engine was marginally less affected, and its adoption would allow the design of much more compact aircraft. The ability to place the engine much closer to the centre of the aircraft would also improve handling, both in the air and on the water. Britain took the lead in designing a jet seaplane, in the form of the Saunders-Roe SR.A/1. This aircraft nevertheless retained the classic flying-boat hull, and was consequently too heavy to ever compete with contemporary landplanes.

Advanced hull design studies were undertaken in the closing stages of World War II, and pointed the way for hulls that were much cleaner aerodynamically than the traditional stepped and chined shapes, yet would also reduce the amount of stress encountered during take-off and alighting, enabling the hulls to be made lighter. Formed in 1943, the Consolidated Vultee Aircraft Corporation, or Convair, was at the heart of this work.

Through the Consolidated arm of the company, Convair had a long pedigree as a flying-boat manufacturer and had, in the shape of the PBY Catalina, produced arguably the most effective flying-boat of the war (if one of the slowest). Before the war ended Convair's Hydrodynamics Research Laboratory, under the direction of Ernest G. Stout, had begun a series of trials under a US Navy Bureau of Aeronautics (BuAer) contract to develop new methods of recording the water characteristics of flying-boat hull designs. Such methods included towed hull models and radio-controlled flying models.

This impression shows what Convair's Skate fighter design might have looked like if it had been preceded with. The aircraft had swept wings and a shallow, blended hull. Like the Sea Dart, it would have floated very low in the water

In furtherance to this work, BuAer issued a follow-on contract in 1946 under which Convair would undertake to develop a high-performance hull that would allow the seaplane to bridge the performance gap with land-based aircraft. Various advances in hull design had made the shapes more aerodynamically efficient, including an increased length/beam ratio, planing tails and retractable steps. Ironically, increasing aircraft perfor-

The prototype XF2Y-1 Sea Dart was hoisted on to a trolley in Convair's experimental workshops before being towed across to the seaplane ramp for final completion and painting. Note the box protecting the sensitive air data vanes on the nose probe.

Hydro-ski history

Development of the snow ski is generally attributed to the Sámi people of northern Scandinavia, well over 1,000 years ago. A millennium of snow ski development has resulted in an object that has been considerably refined, and also one which has spawned a number of derivatives. Of these the most obvious is the water ski, which was developed for sport in the early 1920s.

Whatever the application, the basic concept remains the same: by using an upturned nose, or by being driven at an upturned angle, the ski provides sufficient dynamic lift at speed to rise up to the surface of whatever medium it is travelling through – be it snow, water, sand, mud or similar. Once on the surface, the ski can be propelled across it with reduced drag or friction, providing a means of transporting people or objects with great energy efficiency. It is the all-important dynamic lift function which differentiates the ski from other concepts – like the snow shoe – which merely distribute weight over a wider area.

From the early days skis were attached to aircraft to allow them to operate from snow and ice, but the application of water skis for aircraft was not seriously studied until after World War II. That is not to say that the principle of what became known as the hydro-ski had not already been employed: the ever-resourceful bush pilots operating in Canada and Alaska had often used the concept in their daily operations, especially in the spring and autumn when the land was still snow-bound (and therefore requiring the fitment of snow skis) yet the rivers and lakes had begun to thaw. Ski landings on stretches of water and taxiing at speed to a suitable beaching point was routinely practised and, in cases where there was nowhere to beach, pilots would even let their aircraft sink to a rest in shallow water. Maximum power in the shallows would lift the aircraft back on to the surface, and they could take off again.

Immediately after the war formal studies began into the use of the hydro-ski for combat aircraft, under the guidance of NACA – the National Advisory Committee for Aeronautics – and the Stevens Institute of Technology. The first tests were performed in a model basin, using a model of the Douglas D-558-2 high-speed research aircraft fitted with two hydro-skis. The tests showed that the model aircraft could take off and land from water, with gentle, stable splashdowns even when the aircraft was banked. These results led to further work to explore the possibilities of creating an operational aircraft. There were two avenues to be explored.

The first, handled primarily by the Edo Corporation (noted for its manufacture of aircraft floats), studied the use of hydro-skis on aircraft that were already inherently buoyant. In this concept the aircraft floated, like a conventional flying boat, but as speed built up the aircraft rose under hydrodynamic lift up on to its hydro-skis, accelerating to take-off speed as it planed across the surface. Edo used a Grumman JRF Goose to investigate hydro-skis, demonstrating a significant improvement in take-off and landing performance.

An altogether more ambitious avenue was researched by All American Airways – the use of hydro-skis on aircraft that were not inherently buoyant. In this concept the aircraft would not sink all the time it maintained sufficient forward speed, and therefore hydrodynamic lift. The aircraft could operate from ramps, or possibly ships, which would allow it to build up sufficient speed before it entered the water, and which provided a means of beaching from a water taxi. Such hydro-skis were tested on a Piper Cub and a North American SNJ and, while they proved workable, the non-floating ski aircraft was deemed to be less practical in operational terms, and the Navy had no real plans to develop a combat aircraft under this concept. Nevertheless, All American's hydro-ski work – along with that of NACA and Edo – contributed to the Sea Dart's development.

The Sámis of a 1,000 years ago could not have begun to imagine that, one day, their simple invention might possibly provide the key to bridging the performance gap between seaplanes and landplanes.

This series of photographs depicts a take-off by Edo's ski-equipped Grumman JRF-5 Goose. Here the aircraft is fitted with hydrofoil-type skis, which generated lift under the water by virtue of their airfoil-like shape, as well as by their angle of attack.

mance was also forecast to help close the gap between land- and water-based aircraft: as land-based aircraft flew faster they required stronger fuselages to cope with growing aerodynamic loads, to the point where strength levels approached those needed for waterborne operations. This would consequently reduce the weight disparity between the two branches of aviation.

Convair's work resulted in a series of radical new designs which incorporated what became known as blended hulls. Under this concept the aircraft rode very low in the water – so low that its wings would aid in the craft's buoyancy. As power was applied, the aircraft rose out of the water to skim across the surface on the lower part of the hull like a hydroplane (an invention of the Englishman Reverend C.D. Ramus in 1872). A number of potential designs were drawn up, including a swept-wing fighter which was christened the Skate.

Convair's Skate approach was to develop a good hydrodynamic shape and then attempt to turn it into a high-performance aircraft. Meanwhile, the National Advisory Committee for Aeronautics (NACA), was exploring the reverse option – to begin with a high-performance aircraft and then adapt it for operations from water. The key to this approach was the hydro-ski. Initially, tests were conducted in a model basin using a model of the D-558-2 Skyrocket (which in November 1953 became the first aircraft to exceed Mach 2) fitted with a pair of skis underneath.

Advantages of the hydro-ski

A test report issued in late 1947 indicated that hydro-skis would allow the aircraft to rise clear of the water at relatively low speeds, restricting surface contact to the skis themselves and sparing the main fuselage from the punishment of high-speed water operations. Once the aircraft was airborne, the slender skis could be retracted into the fuselage, restoring it to a clean aerodynamic shape. The attractions of the hydro-ski concept were obvious, but it suffered from the law of diminishing returns as aircraft size and weight grew, and could only be effectively applied to small aircraft such as fighters.

By 1948 both the blended hull and hydro-ski concepts were showing considerable promise, and were deemed suitable for further experimentation, leading to the development of an operational aircraft. In May the Chief of Naval Operations recommended that a requirement be drawn up to provide a more concrete objective for the ongoing experimental work, and on 1 October 1948 BuAer announced a design competition for a Mach 0.95 seaplane night-fighter able to operate from 5-ft (1.5-m) waves.

In charge of the Sea Dart project was F. Herbert Sharp. He had joined Consolidated in 1935, and worked on the PBY, B-24, XB-46 and XP5Y before taking on the F2Y.

'Sam' Shannon flew the XF2Y-1 on its first flight. Before becoming a test pilot he had served in the USAAC (2nd BG) in the early 1930s, and advised the Chinese air force.

Two competing designs were tendered, one a blended hull aircraft proposed by Convair, and another with hydro-skis. On paper the hydro-ski aircraft lost out in terms of rate of climb and ceiling, and there was yet no tangible proof that the skis themselves would work on a full-size aircraft. Consequently, the Skate was chosen as the basis for preliminary development of a research aircraft.

While the US Navy wished to remain firmly at the forefront of seaplane design, the ongoing development of carrier-based aircraft took priority, and the seaplane fighter programme proceeded at a leisurely pace. The project was formally brought into the Navy's R&D planning system through the issue of Operational Requirement CA 05501A on 30 November 1949, which called for an advanced seaplane fighter capable of operating from forward bases in all weather conditions. This required Convair to not only continue preliminary development of the Skate, but also to conduct further comparative tests of the Skate and hydro-ski aircraft.

Such was the early promise of hydro-skis that research had continued, and by 1950 the programme was producing results that could no longer be ignored. Based largely on the latest hydro-ski results, the US Navy realigned the seaplane fighter programme, instructing Convair to examine various ski configurations for the Skate. The Navy also revised its performance requirements upwards, while the aircraft's size decreased. Both of these revisions would be better served by a ski-equipped aircraft.

The Sea Dart takes shape

A series of detailed hydrodynamic tests was performed by Convair, along with wind-tunnel tests at the David Taylor Model Basin (low speed) and Ordnance Aerophysics Laboratory (high speed). From this work Convair created a design known as the Y2-2. Aerodynamically, the Y2-2 was closely based on the delta wing YF-102 interceptor then under development for the US Air Force. It promised a top speed of Mach 1.5 and a rate of climb at sea level of 30,000 ft (9144 m) per minute, figures which impressed the Navy greatly. At rest the aircraft would float so low that the trailing edge of the wing floated on the surface, and the hull was based on the Skate designs. However, it was to be fitted with two retractable hydro-skis, on to which it would rise as forward speed increased hydrodynamic lift.

On 19 January 1951 Letter of Intent for Contract 51-527 was issued, covering the procurement of two Y2-2 airframes to 'provide a research and development program to develop the hydrodynamic features necessary for VF class seaplanes'. Power was to be provided by two Westinghouse J46-WE-2 turbojets but, as these would not be developed in time, Westinghouse J34-WE-32 engines were to be substituted for initial trials. In February F. Herbert Sharp was appointed as project manager.

The 'VF' in the contract referred to the US Navy fighter squadron, and alluded to the aircraft's primary intended role. At this early stage a number of operational scenarios for the seaplane fighter were envisaged. The most obvious was the aircraft's use in providing air defence over a beachhead, freeing a Marine amphibious force from the need to operate under the protective umbrella of carrier-based aircraft. Sea Darts (as the aircraft became known) could operate from coastal waters, beaching on ramps that could be swiftly prepared as soon as the Marines 'hit the beach'. Other options included the operation of Sea Darts from small escort carriers. The aircraft could be catapulted from the ship and then land alongside, taxiing back aboard up a water ramp into the hangar deck. This concept was tested in model form at an early stage. A more ambitious plan called for the basing of Sea Darts in submarines.

With North Island in the background, the XF2Y-1 flies over San Diego during an early test flight. The extension below and behind the jetpipes was a split brake which could be used both in the air and in the water.

Such schemes had been hatched long before the first aircraft was built, but served to illustrate that what was taking shape was more than just a research machine. In August 1951 the Y2-2 became the XF2Y-1, a clear indication that this was the prototype of an operational aircraft. On 2 February 1952, while construction of the first of two XF2Y-1 prototypes was under way, the BuAer issued Letter of Intent to Contract 52-835. This covered the development of the F2Y-1 operational aircraft and the construction of a mock-up for review, which was undertaken between 18 and 20 August.

Production contract

As a result of the mock-up review, Contract 52-835 was amended to include 12 F2Y-1 production aircraft for delivery from mid-1954. It was quickly realised that the date was impossible to meet, so a contract amendment resulted in the first four production F2Y-1s being switched to a pre-production status, allowing them to be built in the experimental workshops and thus expediting their delivery for test work. These aircraft were therefore redesignated as YF2Y-1s. In December 1952 another production F2Y-1 was removed from the contract to be completed for static test purposes. On 25 March 1953 yet another amendment added five more F2Y-1s, restoring the initial production batch to 12. Thus, shortly before the first official flight,

During the take-off run the Sea Dart rose very quickly on to its skis. It also decelerated rapidly on landing or when power was cut. The skis were carefully aligned to displace water outwards, away from the engine intakes.

Shannon brings the XF2Y-1 in to land in San Diego Bay during an early (possibly the first) test flight. The Convair seaplane flight test facility lay across the road from the main factory at Lindbergh Field, San Diego's main airport. Ryan also shared the airfield, its plant being visible in the background.

the order book stood at two XF2Y-1s (BuNos 137634 and 137635), four YF2Y-1s (BuNos 135762/135765), 12 F2Y-1s and a static test article (BuNos 135766/135773, 138530/138534).

Sea Dart anatomy

When the XF2Y-1 was completed in late 1952, fighting was still raging in Korea, and the swept-wing Sabre and MiG-15 were the yardsticks by which fighters were measured. The Sea Dart, by comparison, looked like a vision from a time way in the future, and in its technologies represented a host of novel features. Not least of these were the hydro-ski arrangement and the delta wing.

As demonstrated by the XF2Y-1, the Sea Dart floated with its wing trailing edges resting on the water surface. Here the no. 1 aircraft taxis at a few knots across San Diego harbour – the skis are already beginning to lift the nose out of the water. Great care had to be taken on the water in what was the US Navy's busiest Pacific coast port, while in the air flight tests had to co-ordinate with operations from both Lindbergh Field and nearby NAS North Island.

Convair had no difficulty in choosing the 60° delta layout for the seaplane fighter, so spectacular was its speed potential. In the Sea Dart the trailing edges were cropped near the tip, and a cropped delta shape was adopted for the fin. Primary control surfaces consisted of hydraulically activated rudder and elevons on the trailing edges. The two Westinghouse J34-WE-32 turbojets were housed in straight trunks on the upper rear fuselage. The D-shaped intakes, with boundary layer splitter plates and auxiliary inlet doors, were located on top of the fuselage, where they would be shielded from the worst of the spray thrown up by the skis.

The pilot sat on a rudimentary ejection seat under a canopy incorporating two flat-pane windows, which came together to form a V-shape at the front. This arrangement, similar to that fitted to the X-15, was far from ideal as the seat had to eject forwards as well as upwards, and would have presented unacceptable difficulties in an operational aircraft. If the Sea Dart had been accepted for production, a change to the F-102/F-106 layout – with a V-shaped windscreen and separate conventional canopy – would have been adopted to facilitate emergency egress.

Underneath, the XF2Y-1 drew on Convair's blended hull designs, and featured a shallow V-shaped keel with a small ventral fin. Narrow anti-spray strakes were added to the sides of the forward fuselage as further protection for the engine intakes. The hull ended in a streamlined projection below and beyond the jetpipes, which was split to form braking surfaces. They could be used symmetrically in the air or water for braking, and differentially in the water to act as a rudder. At rest the Sea Dart floated with around 2° nose-up incidence to prevent any tendency of the wing leading edge to submarine, while the trailing edge rested on the water surface. The elevons could be used to provide a measure of lateral control while on the water.

Extending skis

Most radical of all were the hydro-skis, each of which was attached to the aircraft on two extending oleo struts. The forward struts were attached below the cockpit area, while the main struts were attached level with the engine intakes. Once extended, the main struts could telescope further to alter the incidence of the skis relative to the fuselage. Each ski was fitted with a small wheel, as was the ventral fin, to allow the aircraft to taxi in and out of the water under its own power via a ramp. In flight, the skis were designed to be fully retractable, lying flush with the hull. As initially completed, the XF2Y-1 was fitted with instrumentation gear which prevented the skis from being fully retracted.

Two views show the XF2Y-1 emerging from the waves, like a mythical sea monster, on to the San Diego ramp. The skis are in the beaching position, which was also the minimum extension position used for the middle part of the take-off run. The ability to self-beach, and to move (albeit very slowly) on land was one of the keys to the operational versatility of the Sea Dart concept. In the scene above the prototype XP5Y-1 Tradewind can be seen in the background, plus a redesigned engine nacelle for the R3Y-1. Sea Dart chief test pilot 'Sam' Shannon also performed the first flight of the XP5Y-1.

As exhibited by the mock-up, the 'production' F2Y-1 fighter would have featured a primary armament of 44 2.75-in (7-cm) folding-fin aircraft rockets. The FFARs would have been held in 22-tube packs housed in the sides of the forward fuselage. The packs would have swung out for firing and then retracted again. Additional armament comprised four of the then-standard Mk 12 20-mm cannon. Fire control equipment was to have consisted of the Aero 13E system and APQ-50 radar in the nose.

Ease of maintenance at sea was a key consideration in designing the F2Y-1, and key components were to be easily reached from panels on top of the aircraft. The radar and other nose-mounted avionics could be accessed through the cockpit, while the engines were to be installed on flexible mounts which raised them from their normal positions by a few inches. In this way components underneath the engines and on the inner sides could be reached. The engines themselves could be completely removed upwards by a portable crane.

Into the water

Taxi tests got under way in San Diego Bay on 14 December 1952. At the controls was Ellis Dent 'Sam' Shannon, the chief of experimental flight test at Convair's San Diego plant. Shannon was a highly experienced test pilot who had worked for Martin until moving to Convair in 1943. As well as the wartime types, Shannon had flown the B-36 and XB-46 bombers, XP5Y-1 flying-boat and the XP-81 turboprop fighter. His most illustrious achievement was the first flight on 18 September 1948 of the Convair XF-92A – the world's first delta-winged aircraft.

On 14 January 1953, during a high-speed run, the XF2Y-1 lifted off for about 1,000 ft (305 m), but the official first flight was not to occur until some while later. The delay was caused by a highly uncomfortable and potentially dangerous phenomenon which was dubbed 'ski-pounding'. At low speed the Sea Dart performed well, and it rapidly rose up out of the water to plane on its skis. However, when about 60 mph (96 km/h) was reached, a thundering vibration set in which got worse the greater the wave height. The vibration stemmed from the blunt aft end of the skis as they skipped across the waves. The buffeting resonated and amplified as the skis flexed, and was transmitted to the airframe through the struts. The forward struts were located under the cockpit, and the hammering experienced by the pilot got so bad that it was impossible to read the instruments. There was also a very real danger of the vibration causing life-threatening damage to key systems.

Convair tried several minor changes in ski shape to lessen the vibration, and also improved the damping to reduce the transmission of vibration to the airframe, but the 'ski-pounding' phenomenon remained, although diminished to the point where an official first flight could be made on 9 April 1953. This preceded by some months the maiden flight of the YF-102. Just as with the land-based interceptor prototype, which stubbornly refused to punch through the sound barrier, there was to be disappointment in the Sea Dart's projected flight performance.

Firstly, the intended powerplant – the J46-WE-2 – was failing to provide the promised thrust levels, a flaw compounded by the less than optimum position of the engine intakes in the Sea Dart installation. Partly because of the poor power output, and also because of the same aerodynamic problems that dogged the YF-102 (and which were subsequently resolved in the YF-102A by Richard Whitcomb's area-ruled 'Coke bottle' fuselage), the Sea Dart's projected speed performance was reduced from an initially promised Mach 1.5 to a disappointing Mach 0.99, well below the Navy's Mach 1.25 requirement.

At this juncture it should be remembered that Convair was sailing in largely uncharted waters – contemporary service aircraft could barely achieve Mach 0.9, and the company was striving to develop two radical delta-winged aircraft with very ambitious top speed figures. It should also be mentioned that the company – with the aid of NACA and Whitcomb – went on to crack the aerodynamic problem in just 117 days to produce the easily supersonic YF-102A. However, at the time, the Navy saw only the poor performance expectations of the Sea Dart and its seemingly insoluble 'ski-pounding' problems.

Beginning of the end

Questions were raised, and when in late 1953 it became necessary to prune BuAer's budgets as a result of the end of the Korean War, the Sea Dart was an obvious choice for cuts. Accordingly, on 14 October the second XF2Y-1 was cancelled, followed by 10 of the F2Y-1s in November. Prophetically, the money was reallocated to procurement of F8U Crusader development aircraft. Further realisation of the vast amount of work still necessary to bring the Sea Dart into operational reality resulted in the March 1954 cancellation of the

The no. 2 aircraft taxis at idle after leaving the launch ramp, exhibiting the natural floating position of the aircraft. The jetpipes of the J46 engines were much longer than those of the J34, giving the rear end a more stepped appearance when viewed in profile.

Development of the Sea Dart was conducted under conditions of some secrecy, but the proximity of Lindbergh Field and the adjacent seaplane base to the heart of San Diego made any attempts to hide the Sea Dart pointless once flight trials began. Here the no. 2 aircraft flies over downtown as it makes its landing approach, with Balboa Park behind.

remaining two F2Y-1s and the static test aircraft. Thus, the order was reduced to just one XF2Y-1 and four YF2Y-1s. In the event, only three of these would take to the air.

Soon after the first flight, 'Sam' Shannon was joined in the test programme by Charles E. Richbourg, and the XF2Y-1 gained its intended powerplant – the afterburning J46-WE-2B. Tests concentrated on reducing the 'ski-pounding'. The aircraft flew in its original twin-ski configuration until mid-1954, when it was grounded for modification.

In early 1954 the second Sea Dart joined the test programme, piloted on its maiden flight by Richbourg. Originally begun as the second XF2Y-1, the aircraft was completed as the first of the YF2Y-1s (BuNo. 135762) and incorporated a number of changes. Most significantly, it was slightly larger, with increases in length, span and wing area. J46-WE-2 engines were fitted from the outset, with an extended tail-cone area, and the aircraft was completed with twin skis. The skis were not fitted with wheels, and the no. 2 aircraft used special beaching gear. Shortly after flight tests began wing fences were fitted to cure a high-speed span-wise migration, although it was the only aircraft to sport this feature.

While the XF2Y-1 continued on its ski-related tests, the second aircraft was used primarily for high-speed aerodynamic handling, stability and control trials. On 3 August 1954 Richbourg exceeded Mach 1 in a shallow dive at 34,000 ft (10363 m) – the YF2Y-1 had become the first, and to date the only, seaplane to fly supersonically. Like the YF-102, the aircraft would not break through the sound barrier while flying straight and level.

In late 1954, at about the same time as Convair was applying Whitcomb's Area Rule to the YF-102 design, the company also proposed an area-ruled, single-ski Sea Dart. Designated F2Y-2, the aircraft represented a considerable redesign, not only incorporating the new aerodynamic principle, but also a change of powerplant to a single Pratt & Whitney J75 or Wright J67 (development of the Bristol Siddeley Olympus). Speed of the F2Y-2 was estimated to be Mach 2. Despite the promise of this design, backed up by the ease with which the redesigned YF-102A achieved supersonic speed (21 December 1954, on its second flight), the Navy remained suspicious of any seaplane fighter proposals until the ski problems had been resolved. Until they were, no serious consideration would be given to follow-on designs.

Marine Sea Darts – the seadrome concept

In 1954 the US Marine Corps Development Center, at Quantico, Virginia, commissioned a study into the potential use of the Sea Dart, examining the benefits and pitfalls of operating seaplane fighters from what it dubbed a 'seadrome' in support of Marine operations, especially the cover of amphibious assaults. The study provided an interesting view of one of the main operational concepts proposed for the F2Y.

In essence, the seadrome concept envisaged a seaplane base that could be established by submarine, surface ship or transport flying-boat. It would be deployed close to the area of operations and installed swiftly to facilitate the rapid deployment of aircraft to the area. In such a way an amphibious operation, and continuing operations ashore, could be launched and supported quickly with immediate air cover. Traditionally, such operations could only be mounted once carrierborne air cover had reached the area. Naturally, the seadrome concept took full advantage of water-basing, but had to take into account the operational limitations imposed by the aircraft themselves.

One of those limitations was sea condition. Sheltered water was essential to ensure that the Sea Darts could meet their operational commitments when required, with minimum loss of sorties to adverse wave conditions. Another limitation was the requirement for a shore base. Although it was envisaged that the fighters would spend much of their time at sea, it was expected that they would have to be beached at least weekly for a complete fresh-water wash-down and maintenance at specially constructed shore facilities.

In normal operations, however, the Sea Darts would be moored at sea. This would allow them to be dispersed widely, so minimising the risk of being damaged or destroyed by air attack. Scramble times would also be minimised, and operational attrition rates would be reduced because damaged or disabled aircraft could land on any suitable water rather than be constrained by the availability of airfields. However, the latter implied that dedicated at-sea recovery techniques and equipment would have to be evolved.

The seadrome would consist of all the equipment needed for daily operations. In order to protect it from air attack, much of it was to be kept below the surface, safe from rockets and guns. Provision of fuel was of primary importance, and it was planned that fuel would be held in underwater tanks with flexible links to individual mooring buoys. Armament, too, would be held in a central underwater dump, with smaller weapon loads being moved to the mooring buoys for short-term sub-surface storage. Communications and landing aids would also be required. Basic servicing could be accomplished at sea, and an internal weapons loading capability in the aircraft was required to keep the armament dry. Obviously, the seadrome would also require a sizeable flotilla of boats to service the needs of at-sea maintenance, personnel and equipment transport, aircraft towing, firefighting and rescue, plus beaching mats, spares packs and servicing hangars for the shore base. And all of the above was required to be air-transportable!

While the seadrome concept had some clear benefits, the Sea Dart could never operate for any length of time completely independent of shore bases (or a large carrier-type vessel), and so it only reduced the dependence on such facilities rather than removed it. Furthermore, the concept was tied closely to inshore 'white' waters and, despite the use of dispersal and underwater storage, nevertheless remained more vulnerable to air attack than a carrier sailing offshore in 'blue' waters. Perhaps the biggest flaw in the scheme was the amount of money needed to develop and procure entirely new support and logistic equipment, as the seaplane fighters could utilise precious little of the existing infrastructure.

This artist's impression shows how the Sea Dart might have been used in support of the Marines' Mobile Base Concept. The aircraft cover the beachhead while equipment is landed ashore by both landing craft and Convair R3Y Tradewinds. Note the matting which has been laid on the shore to provide a ramp for the Sea Darts. A shore base for servicing and washdown was vital to F2Y operations.

The gaudy yellow markings applied to the first two Sea Darts (this is no. 2) allowed Convair engineers to accurately assess the spray patterns when reviewing photographs. The second YF2Y-1 – Sea Dart no. 3 – did not receive these markings, retaining a midnight blue finish with standard Navy titles and national insignia.

Open-sea trials were undertaken by Shannon and Richbourg with the no. 2 aircraft south of Point Loma, involving numerous support vessels, a SAR helicopter and a Douglas AD-5 Skyraider chaseplane. Included in the campaign were operations from the LSD (landing ship dock) USS *Catamount*. The sea trials generally underlined the unsatisfactory nature of the hydro-ski arrangement, particularly when open-sea waves were encountered.

In late 1954 the XF2Y-1 re-emerged from the experimental shop with a new single-ski arrangement. The ski took the form of a large, shallow V-shaped surface, mounted below the centreline on four struts. Although it could be raised and lowered, the single ski was not made fully retractable, as the aircraft was not intended for high-speed tests. It was wide enough, however, to be fitted with twin wheels at the aft end to maintain the Sea Dart's beaching capability.

Throughout the test period Convair toiled relentlessly to cure the 'ski-pounding' problem, but without complete success. Eventually over 100 different ski configurations were tried. When Shannon and Richbourg began taxi trials of the single-ski arrangement they immediately encountered hydrodynamic problems. Although the 'ski-pounding' was appreciably reduced, directional stability on take-off suffered, and a divergent pitch oscillation ('porpoising') was encountered. Partial solutions to these new problems were evolved, including a new variable damping system which sensed the stroke rate of the oleos and altered the size of the hydraulic orifices accordingly. They were sufficiently encouraging for the Navy to convene a preliminary flight evaluation, scheduled for November 1954.

A tragic loss

Before the PFE could take place, tragedy struck the Sea Dart programme on 4 November, when Charles Richbourg was killed in a crash of the no. 2 aircraft. The occasion was a demonstration in front of an invited press tour, which was at San Diego to witness the XFY 'Pogo', R3Y Tradewind and the Sea Dart. During a low-level pass at about 500 kt (926 km/h) the YF2Y-1 disintegrated in mid-air, falling to the water surface in a fireball. Richbourg was barely alive when the rescue boat reached him, but he died soon after.

Examination after the crash revealed that the aircraft had broken up under severe aerodynamic loads caused by the pitch-divergent oscillations which plagued several high-speed programmes of the day. Pitch divergence was encountered at transonic speeds at low level, and the early, immature hydraulic flight control systems were not able to cope with it adequately, which in turn led to pilot-induced oscillations. The Sea Dart had broken up during the second nose-down pitch. As a result, trials of the Sea Dart were suspended immediately and, although ski trials were reinstated, the F2Y never again flew at high speed.

Sea Dart trials were scheduled to begin again on 29 December, but 'Sam' Shannon was suffering from a mild illness. In stepped Billy Jack 'BJ' Long as a substitute. Long, with extensive seaplane experience from World War II, had joined Convair in September 1953 and had flown chase on the Sea Dart and other programmes. He was training to become a test pilot on the XFY 'Pogo', but volunteered to step in to the Sea Dart programme, and went on to become the most experienced of the F2Y pilots, accomplishing 187 flights and 47.6 hours, and flying most of the configuration changes that occurred during the remainder of the programme.

During sea trials the no. 2 Sea Dart is manoeuvred out of the dock of USS* Catamount*, an LSD class vessel. One of the operational concepts envisaged for the F2Y involved the aircraft landing alongside its carrier vessel and then taxiing into a dock similar to this one. The dock would have a ramp, up which the Sea Dart could taxi into the vessel's hangar deck.

As fitted to the XF2Y-1 from late 1954, the single-ski arrangement proved to be superior to the twin-ski installation in hydrodynamic terms, although it was never fully satisfactory.

Faced with yet further delay at a time when interest in the seaplane fighter was waning fast and the 'Forrestal'-class 'super-carriers' were being commissioned, the Chief of Naval Operations recommended in January 1956 that the Operational Requirement be cancelled and that work on the Sea Dart should be terminated. The third and fourth YF2Y-1s, which were essentially complete, never received their engines and played no active part in the trials programme.

Epilogue

Both flying Sea Darts were placed in storage, but at the end of 1956 the no. 1 aircraft was returned to service to test a new hydrofoil-type ski for the Bureau of Aeronautics. The hydrofoil was more than just a planing surface, it also generated hydrodynamic lift in the same way as an aerofoil does in air. The new ski design was much smaller than that previously fitted, and was mounted in a fixed position for taxi trials. The short struts and fixed mounting of the ski precluded the aircraft from reaching the take-off incidence of 17-19°, so flight was impossible.

In April 1957 'BJ' Long, having been reassigned to F-102 development work in the interim, returned to the Sea Dart to undertake three taxi tests with the new ski. As with the very first trials, the aircraft encountered severe 'ski-pounding' at around 60 mph (96 km/h) and this ski design was abandoned. In the autumn the XF2Y-1 returned to the water again, this time to test an even smaller hydrofoil ski that was about 50 percent the size of the previous installation, and about 20 percent of the size of the original Convair single ski. Donald P. Germeraad, who was chief pilot on the P5Y/R3Y programme, performed the tests, which again encountered severe pounding problems. With the small-ski tests abandoned, the Sea Dart programme came to a halt. Over 300 flights and test runs had been undertaken, of which about 250 had been performed by the no. 1 aircraft.

Somewhat surprisingly, all four surviving Sea Darts have escaped the scrapman's torch. The first aircraft (BuNo. 137634) is owned by the National Air and Space Museum. After having spent some time on display at NAS Norfolk, Virginia, the machine was restored for outside display at the Naval Air Test and Evaluation Museum at NAS Patuxent River, Maryland. It was returned to NASM and currently resides in the Paul E. Garber Facility at Suitland, Maryland. NASM received it in poor condition and in pieces, and there are no immediate plans for its restoration. The no. 3 aircraft (BuNo. 135763) was held in storage at NAS North Island, then at Lindbergh Field, San Diego (site of Convair's plant) before taking pride of place on poles outside the San Diego Aerospace Museum in Balboa Park. No. 4 (BuNo. 135764) was displayed at NAS Willow Grove, Pennsylvania, for many years, and is now part of the Wings of Freedom Museum at

'Sam' Shannon clambers into the cockpit of the XF2Y-1 after it had been modified with a single ski. Note the retractable beaching wheels, and small towing cleat fixed to the underside of the nose.

Sea Dart specifications

Dimensions

Wing span: XF2Y-1 – 30 ft 6 in (9.30 m), YF2Y-1 – 33 ft 8 in (10.26 m)
Length: XF2Y-1 – 41 ft 2 in (12.55 m), YF2Y-1 – 52 ft 7 in (16.03 m)
Height (skis up): YF2Y-1 – 16 ft 2 in (4.93 m)
Height (skis down): XF2Y-1 – 21 ft 1 in (6.43 m), YF2Y-1 – 20 ft 9 in (6.32 m)
Wing area: XF2Y-1 – 539 sq ft (50.07 m^2), YF2Y-1 – 563 sq ft (52.3 m^2)

Weights and fuel

Empty: YF2Y-1 – 12,652 lb (5739 kg)
Loaded: XF2Y-1 – 15,000 lb (6804 kg), YF2Y-1 – 16,527 lb (7497 kg)
Maximum take-off: YF2Y-1 – 21,500 lb (9752 kg)
Internal fuel capacity: 630 US gal (2385 litres)

Performance (YF2Y-1 estimated)

Maximum speed: 695 mph (1118 km/h) at 8,000 ft (2438 m) and 825 mph (1328 km/h) at 35,000 ft (10668 m)
Initial climb rate: 17,100 ft (5212 m) per minute
Service ceiling: 54,800 ft (16703 m)
Range: 513 miles (826 km)

Airframe limits (YF2Y-1)

V_{NE} (never-exceed speed): 350 kt IAS below 20,000 ft (6096 m), or higher with skis extended. Mach 1.05 above 20,000 ft with skis retracted
Ski retraction/extension limits (inflight): 180 kt IAS
Ski retraction/extension limits (water): 6 mph (9.6 km/h)

Powerplant

XF2Y-1 (initial): two Westinghouse J34-WE-32 turbojets, each rated at 3,250 lb (14.46 kN) thrust dry and 4,200 lb (18.69 kN) thrust with afterburning
XF2Y-1 (final) and YF2Y-1: two Westinghouse XJ46-WE-2 turbojets, each rated at 4,080 lb (18.15 kN) thrust dry and 6,100 lb (27.15 kN) thrust with afterburning

'BJ' Long poses with the no. 3 aircraft, which he first flew in March 1955. It had what was considered the definitive twin-ski configuration, with a neat wheel installation which involved the rotation of the extreme aft of each ski for deploying the wheels, and was also briefly used for RATO trials. Experience with this aircraft was enough to convince Long, and others, that the single-ski configuration was the only road to follow if the Sea Dart was ever to go to sea in the hands of regular squadron pilots.

the base, overseen by the Delaware Valley Historic Aircraft Association. The last aircraft, no. 5 (BuNo. 135765), spent time at NAS Sand Point, Washington, before moving to nearby Renton. It was then placed on display at the SST Museum at Kissimmee, Florida. Today it has been restored and is pole-mounted outside the Florida Air Museum, part of the EAA's Sun'n Fun complex at Lakeland.

An interesting postscript to the Sea Dart story came in September 1962, when the US air arms rationalised their aircraft designations into a single tri-service system. For some inexplicable reason the Sea Dart was included in the reshuffle, with the YF2Y-1 becoming the YF-7A and the F2Y-1 redesignated as the F-7A. This occurred some six years after the programme's termination, and five years after the final tests had been undertaken.

The supersonic seaplane

When it first appeared, the Sea Dart was undoubtedly ahead of its time, and perhaps if final development of the aircraft had started later than it did, then maybe the Sea Dart would have emerged with a Mach 2 top speed instead of Mach 0.99, and been a real contender for production funds.

From a technical standpoint the Sea Dart was dogged by three main areas of concern: weak powerplant, high transonic drag and unsatisfactory hydrodynamics. The solutions to the first two problems lay in the area-ruled, single-engined F2Y-2 proposal (the potential of which was graphically hinted at by Convair's F-102 and F-106 land-based fighters), and while the hydrodynamic problems were never fully overcome, Convair had reached a point with its late single-ski configuration where the solution was in sight if development had been allowed to proceed further.

As it is, Convair's F2Y Sea Dart – a radical design that encompassed many untried features – has gone down in the history books as a failure, yet it achieved many of its objectives and might have been shaped into a fully operational fighter that could have had a profound effect on naval aviation strategy. Always teetering on the edge of what was technologically possible, the Sea Dart eventually foundered as the US Navy pursued more conservative means of providing airpower at sea. A 'Forrestal'-class 'super-carrier' populated with Vought Crusaders represented a much safer bet than a fleet of smaller vessels carrying Sea Darts. It was not until the advent of STOVL fighters in the 1970s that naval/marine airpower witnessed any kind of major departure from the dominance of the conventional aircraft-carrier.

David Donald

Sea Dart survivors

The third Sea Dart has remained close to its birthplace in the San Diego area. Above it is seen while on display at NAS North Island in 1975. It was later transferred to Lindbergh Field, where it was stored outside pending restoration and display outside the San Diego Aerospace Museum (right). The National Air and Space Museum holds the no. 1 aircraft in its Garber restoration annexe.

The fourth and fifth aircraft remained engineless and never flew. No. 4 (above) is at Willow Grove in Pennsylvania, where it has spent many years. No. 5 spent time awaiting display in Washington state, but then surfaced in Florida. It is now at Lakeland (right), confusingly presented with the BuNo. of the no. 3 aircraft.

Index

Page numbers in **bold** refer to illustration captions.

INDEX

Picture acknowledgments

Front cover: Ted Carlson/Fotodynamics, Jamie Hunter/Aviacom, Marnix Sap/MIAS. **4:** Dassault (three). **5:** Dassault (two), Yaso Niwa (two). **6:** Westland (two), Boeing via Tom Kaminski. **7:** Neil Dunridge, via Tom Kaminski. **8:** via Tom Kaminski, Alexander Mladenov, Shlomo Aloni. **9:** via Tom Kaminski, Nigel Pittaway (two), Michael Stroud. **10:** Dassault, Jamie Hunter/Aviacom. **11:** Shlomo Aloni (four). **12:** Dick Lohuis (four), Riccardo Niccoli. **13:** Dick Lohuis (four), Nate Leong. **14:** Paul Kievit and Carlo Kuit/Bronco Aviation (two). **15:** Neil Dunridge, Jamie Hunter/Aviacom (two), Yaso Niwa. **16:** Dick Lohuis (four). **17:** ERS 1/91 via Gert Kromhout (two), Dave Willis. **18:** Gert Kromhout (three), ERS 1/91 via Gert Kromhout (three). **19:** ERS 1/91 via Gert Kromhout (two), SIRPA Air via Gert Kromhout, Dave Willis, Gert Kromhout. **20:** Gert Kromhout (two), SIRPA Air via Gert Kromhout. **21:** SIRPA Air via Gert Kromhout (two). **22-30:** Marnix Sap/MIAS. **31:** Marnix Sap/MIAS (three), Peter R. March. **32-33:** Marnix Sap/MIAS. **34:** Piotr Butowski. **36-38:** Piotr Butowski. **39:** Piotr Butowski (two), Hugo Mambour/Aviascribe (two). **40:** Hugo Mambour/Aviascribe (two), Piotr Butowski. **41:** Piotr Butowski, Hugo Mambour/Aviascribe. **42:** Hugo Mambour/Aviascribe (three). **43:** Mikhail Kuznetsov (three). **44-45:** Ted Carlson/Fotodynamics. **46:** Ted Carlson/Fotodynamics (two), Robert Hewson. **47:** Robert Hewson, Ted Carlson/Fotodynamics. **48:** Robert Hewson. **49:** Ted Carlson/Fotodynamics (two). **50:** Peter Steinemann, Cees-Jan van der Ende. **51:** Yoshitomo Aoki (two), Yaso Niwa. **52:** via Robert Hewson (two). **53:** Shlomo Aloni (four). **54:** Peter Steinemann (two). **55:** Ted Carlson/Fotodynamics (two). **56:** Ted Carlson/Fotodynamics, US Marine Corps. **57-58:** US Navy. **59:** UK MoD (two). **60:** US Navy, Robert Hewson. **61-62:** US Navy. **63:** Robert Hewson (four). **64:** Ted Carlson/Fotodynamics, David Donald. **66:** General Electric, Ted Carlson/Fotodynamics. **67:** Ted Carlson/Fotodynamics (three), David Donald (two). **68:** Ted Carlson/Fotodynamics (three), David Donald, Robert Hewson (four). **69:** Bell via Robert Hewson (two). **70:** Bell via Robert Hewson (two), Robert Hewson. **71:** Bell via Robert Hewson (two). **72:** Bell, Bell via Robert Hewson, US Marine Corps. **73:** Bell (two), Tom Kaminski. **74:** Jamie Hunter/Aviacom, Robert Hewson, Bell via Robert Hewson. **75:** Peter Steinemann, Bell, Peter R. Foster. **76:** Robert Hewson (three). **77:** Shlomo Aloni (four). **78:** Shlomo Aloni (four). **79:** Yaso Niwa, Peter R. Foster, Yoshitomo Aoki. **80:** Peter R. Foster (three), Peter Steinemann (two). **81:** Peter R. Foster, Cees-Jan van der Ende (three). **82:** Ted Carlson/Fotodynamics (five), Robert Hewson. **83:** Ted Carlson/Fotodynamics (four), Tom Kaminski (two). **84:** Jamie Hunter/Aviacom. **85:** Daniel J. March, Jamie Hunter/Aviacom (two). **86:** Jamie Hunter/Aviacom (two), USAF, Daniel J. March (two). **87:** Jamie Hunter/Aviacom (three), 49th FW, Daniel J. March. **88-89:** Jamie Hunter/Aviacom. **90:** Robert F. Dorr, Stephen Miller, Brian C. Rogers via Tom Kaminski, Ray Leader via Don Spering/AIR. **91:** via Don Spering/AIR, Ted Carlson/Fotodynamics, Ray Rivard, Carey Mavor, Tom Kaminski. **92:** Keith Riddle, Bill Crimmins via Tom Kaminski, Bruce Trombecky, Carey Mavor, Mark Whitehead via Tom Kaminski, Robert F. Dorr via Tom Kaminski. **93:** Carey Mavor (three), Chris Knott/API, Joseph Handleman via Tom Kaminski. **94:** Raytheon via Tom Kaminski (two), via Tom Kaminski. **95:** Tom Kaminski, Carey Mavor, Mart Jan Gerards via Tom Kaminski, Ray Rivard, Ted Carlson/Fotodynamics. **96:** Carey Mavor, TRW Inc. via Tom Kaminski, Bill Crimmins via Tom Kaminski. **97:** Carey Mavor, Don Spering/AIR, Daniel Soulaine via Don Spering/AIR, Tom Ring via Tom Kaminski. **98:** RAAF Museum via Nigel Pittaway (three). **99:** RAAF Museum via Nigel Pittaway, Aerospace (two). **100:** Aerospace, RAAF Museum via Nigel Pittaway (two), RAAF. **101:** RAAF Museum via Nigel Pittaway (three), Dassault. **102:** via Nigel Pittaway, RAAF Museum via Nigel Pittaway (two). **103:** RAAF Museum via Nigel Pittaway, Aerospace. **104:** RAAF Museum via Nigel Pittaway (four). **105:** RAAF Museum via Nigel Pittaway, Aerospace, Terry Panopalis, Dassault. **106:** Terry Panopalis, RAAF Museum via Nigel Pittaway (two). **107:** TRH, RAAF Museum via Nigel Pittaway (three). **108:** RAAF Museum via Nigel Pittaway (three), Nigel Pittaway. **109:** RAAF Museum via Nigel Pittaway, RAAF, TRH. **110:** Peter Steinemann, via Nigel Pittaway, Nigel Pittaway. **111:** RAAF Museum via Nigel Pittaway, Nigel Pittaway, Eddy de Kruijff. **112:** Nigel Pittaway, Terry Panopalis (two). **113:** Nigel Pittaway (five), TRH. **114:** Nigel Pittaway (two). **115:** Aerospace (three), TRH, Nigel Pittaway (two), via Nigel Pittaway, via Terry Panopalis. **116:** IWM, Aerospace (two). **117:** RAF Museum. **118:** Aerospace (three), via Peter R. March. **119:** Aerospace. **120:** Aerospace (two), IWM. **121:** IWM, Aerospace. **122:** IWM (two). **123:** RAF Museum, IWM. **124-127:** Aerospace. **128:** Aerospace, IWM. **129:** Aerospace, IWM. **130:** Aerospace (three). **131:** Aerospace (two), IWM. **132:** Aerospace (two), **133:** IWM, TRH. **134:** IWM (two), Aerospace. **135:** IWM (two). **136:** IWM, Aerospace. **137:** RCAF, Hawker Siddeley, Aerospace. **138:** RAF Museum, MoD (two), Aerospace. **139:** Aerospace (two), Terry Panopalis (two), Boulton Paul, RAF Musuem. **143:** Hawker Siddeley, Aerospace. **144:** Aerospace (three), Hawker Siddeley, MoD. **146:** RAF Museum, Flight Refuelling Ltd, RCAF, TRH. **147:** MoD, Aerospace, Peter R. March, MoD. **148:** Aerospace. **149:** Aerospace (two), MoD, via Santiago Rivas. **150:** RCAF (three). **151:** Aerospace (three), Hawker Siddeley. **152:** via Warren Thompson (WT), Larry Knox via WT (two). **153:** Ray Holt via WT (two), Mike Korte via WT. **154:** Bob Donaldson via WT, Larry Knox via WT, via WT. **155:** Darrell Hatcher via WT (two), Morgan Lilly via WT. **156:** Darrell Hatcher via WT, Morgan Lilly via WT, Marvin Roupe via WT. **157:** Darrell Hatcher via WT, Rod Trimble via WT (three). **158:** via WT, Rod Trimble via WT. **159:** Mike Korte via WT, Bob Craine via WT, Charles Carr via WT, via WT. **160:** Rod Trimble via WT (two), Charles Carr via WT. **161:** Rod Trimble via WT (seven), Rick Drisko via WT, via WT. **162:** Convair. **163:** Aerospace, via Bill Yenne (two). **164:** Aerospace, Convair (two). **165:** via Bill Yenne, Convair. **166:** Convair (two). **167:** via Bill Yenne, Convair. **168:** Aerospace, TRH, via Bill Yenne. **169:** via Bill Yenne, Aerospace. **170:** via Terry Panopalis, Convair. **171:** Convair, via Terry Panopalis, via Bill Yenne. **172:** Convair, via Bill Yenne. **173:** Convair, Terry Panopalis (three), Dave Willis.